MASSIVE SUBURBANIZATION
(RE)BUILDING THE GLOBAL PERIPHERY

Edited by K. Murat Güney, Roger Keil, and Murat Üçoğlu

Providing a systematic overview of large-scale housing projects, *Massive Suburbanization* investigates the building and rebuilding of urban peripheries on a global scale. Conceptual and empirical chapters revisit classic cases of large-scale suburban building in Canada, the United States, France, Germany, and the former Czechoslovakia, and examine new peripheral estates in China, Egypt, Israel, Morocco, South Africa, Turkey, and the Philippines. Offering a universal inter-referencing point for research on the dynamics of "massive suburbia," this book builds a new discussion pertaining to the problems of the urban periphery, urbanization, and the neoliberal production of space.

(Global Suburbanisms)

K. MURAT GÜNEY is a lecturer in the Sociology Department at Acıbadem University in Istanbul.

ROGER KEIL is York Research Chair in Global Sub/Urban Studies in the Faculty of Environmental Studies at York University in Toronto.

MURAT ÜCOĞLU is a PhD candidate and course director in the Faculty of Environmental Studies at York University.

GLOBAL SUBURBANISMS

Series Editor: Roger Keil, York University

Urbanization is at the core of the global economy today. Yet, crucially, suburbanization now dominates 21st-century urban development. This book series is the first to systematically take stock of worldwide developments in suburbanization and suburbanisms today. Drawing on methodological and analytical approaches from political economy, urban political ecology, and social and cultural geography, the series seeks to situate the complex processes of suburbanization as they pose challenges to policymakers, planners, and academics alike.

For a list of the books published in this series see page 381.

Massive Suburbanization (Re)Building the Global Periphery

Edited by
K. Murat Güney, Roger Keil,
and Murat Üçoğlu

UNIVERSITY OF TORONTO PRESS
Toronto Buffalo London

Toronto Buffalo London
utorontopress.com

ISBN 978-1-4875-0526-4 (cloth) ISBN 978-1-4875-2377-0 (paper)

Global Suburbanisms

Library and Archives Canada Cataloguing in Publication

Title: Massive suburbanization : (re)building the global periphery / edited by K. Murat Güney, Roger Keil, and Murat Üçoğlu.
Names: Güney, K. Murat, 1981– editor. | Keil, Roger, 1957– editor. | Üçoğlu, Murat, 1986– editor.
Series: Global suburbanisms.
Description: Series statement: Global suburbanisms | Includes bibliographical references and index.
Identifiers: Canadiana 20189069376 | ISBN 9781487505264 (cloth) | ISBN 9781487523770 (paper)
Subjects: LCSH: Suburbs – Case studies. | LCSH: Suburban homes – Case studies. | LCSH: Housing development – Case studies. | LCSH: Suburban life – Case studies. | LCSH: Urbanization – Case studies. | LCSH: Neoliberalism – Case studies. | LCGFT: Case studies.
Classification: LCC HT351 .M37 2019 | DDC 307.74 – dc23

University of Toronto Press acknowledges the financial assistance to its publishing program of the Canada Council for the Arts and the Ontario Arts Council, an agency of the Government of Ontario.

Canada Council for the Arts
Conseil des Arts du Canada

Funded by the Government of Canada
Financé par le gouvernement du Canada
Canada

Contents

List of Figures and Tables ix

Acknowledgments xi

Part One: Re-thinking the Massive Periphery

Introduction: Massive Suburbanization – Political Economy, Ethnography, Governance 3
ROGER KEIL, K. MURAT GÜNEY, AND MURAT ÜÇOĞLU

1 Peripheries against Peripheries? Against Spatial Reification 35
STEFAN KIPFER AND MUSTAFA DIKEÇ

2 Public Housing, Heroin Addiction, and America's Industrial Suburbs: A Planetary Urbanist Perspective 56
DAVID WILSON, BASMATTEE BOODRAM, AND JASMINE SMITH

Part Two: Legacies

3 Estates under Pressure: Financialization, Shrinkage, and State Restructuring in East Germany 81
MATTHIAS BERNT

4 Learning from the Socialist Suburb 94
STEVEN LOGAN

5 Decline and Renewal in Toronto's High-Rise Suburbs: The Tragedy of Progressive Neoliberalism 111
DOUGLAS YOUNG

6 Redeveloping Montpellier's Suburban High-Rises: National Policy Meets Local Activism in the Debate over Public Space 126
ROZA TCHOUKALEYSKA

7 (De)Constructing Housing Estates: How Much More than a Housing Question? 142
STEFAN KIPFER

Part Three: Spotlight on Istanbul

8 From Kayabaşı to Kayaşehir – A City Grows "Out in the Sticks" 165
ERBATUR ÇAVUŞOĞLU AND JULIA STRUTZ

9 Building Northern Istanbul: Mega-Projects, Speculation, and New Suburbs 181
K. MURAT GÜNEY

10 Massive Housing and Nature's Limits? The Urban Political Ecology of Istanbul's Periphery 201
MURAT ÜÇOĞLU

Part Four: The Suburban Century

11 Morocco's "Pirate Suburbs" from Punishment to Controlled Integration: Neoliberalizing the Regulation of Casablanca's "Chechnya" 223
WAFAE BELARBI AND MAX ROUSSEAU

12 State-Led Housing Provision Twenty-Five Years On: Change, Evolution, and Agency on Johannesburg's Edge 241
MARGOT RUBIN AND SARAH CHARLTON

13 From Informal Settlements to Harmonious Communities: Professional Squatters and the Many Actors of Urbanization in Metro Manila 267
ABIDEMI COKER

14 Suburbanisms of Ethnocracy: Building New Peripheries in Israel/ Palestine 285
ODED HAAS

15 The Making of Cairo's Vast Planned Periphery: Particularities and Parallels Revealed through an Examination of Four Suburban Cultural Assemblages 303
KARL SCHMID

16 Massive Suburbanization, Heterogeneous Suburbs in China 320
TIANKE ZHU AND FULONG WU

Conclusion: Massive Suburbia – From the Legacy of the Habitat to the Financialization of Housing in the Planetary Periphery 344
K. MURAT GÜNEY, ROGER KEIL, AND MURAT ÜÇOĞLU

List of Contributors 355

Index 361

Figures and Tables

Figures

0.1 The book's central concerns 6
2.1 Condition of housing estates in Aurora and Cicero in Chicagoland 58
3.1 Percentage of property owned by municipality, cooperative, and private investors in Am Südpark 85
4.1 A forgotten public space 96
4.2 Jižní Město seen from the east just beyond the artificial lake 102
4.3 People gathering on the pedestrian walkways in spaces between buildings 105
5.1 Typical slab form private sector rental apartment building in Toronto's inner suburbs 113
5.2 Typical slab building in Toronto's inner suburbs, owned and operated as public housing 114
6.1 The outdoor market and the Halles de la Mosson 129
6.2 Le Grand Mail, mid-afternoon in June 2014 134
8.1 TOKİ Kayaşehir massive housing site as seen from a distance 166
8.2 TOKİ high-rise apartment buildings in Kayaşehir, İstanbul 172
8.3 Kayacity Residence, an ongoing luxury housing project, in Kayaşehir, İstanbul 174
9.1 Trucks carrying construction materials to and from İstanbul's third airport construction area 185
9.2 Ongoing construction of a massive TOKİ housing complex in Başakşehir 187
9.3 House price increase in Turkey and Istanbul 192
9.4 Homeownership rate and homeownership rate for low-income households 196

10.1 The production of suburban space with massive housing projects and gated communities in Halkali, Istanbul 204
10.2 "Watergarden" luxury housing project and shopping mall in Ataşehir, Istanbul 211
10.3 Luxurious suburban housing site on the Bahçeşehir-İstanbul Highway 214
11.1 Localization of Lahraouiyine in the metropolitan area of Casablanca 225
11.2 Lahraouiyine 230
12.1 Location map of selected housing sites in or around Johannesburg 242
12.2 Original starter/core house 245
12.3 Evolution of RDP housing post-2001 246
12.4 Current version of state-supplied housing 247
12.5 Formal double-storey backyard rooms in Cosmo City 251
12.6 Unsanctioned backyard dwelling in Bram Fischerville 252
12.7 Main house and backyard units in Bram Fischerville 254
12.8 State-upgraded housing in Orlando 258
13.1 Seaside homes in Ulingan, Manila 269
13.2 Games and charcoal production in Ulingan, Manila 270
13.3 Mixed housing in Ulingan, Manila 279
13.4 New housing community in Pasay, Manila 281
16.1 The location of Jiangning new town in Nanjing 326
16.2 The villas of the Master-Land project in Jiangning, Nanjing 334
16.3 The middle-class gated community of Vanke Paradiso in Jiangning, Nanjing 336
16.4 The estate of Tiandi new town for relocated households in Jiangning, Nanjing 339

Tables

2.1 Public housing projects in Aurora and Cicero 59
3.1 Welfare dependency rates compared: City of Halle 89
8.1 Plans made for Kayabaşı/Kayaşehir between 2007 and 2009 167
10.1 Housing sales in Turkey from 2013 to 2016 209
10.2 Housing sales numbers for Istanbul, 2013 to 2016 209
13.1 The nine stages of the housing provision process 275

Acknowledgments

This book is the result of a workshop called Spotlight on Istanbul: Building and Rebuilding the Periphery that took place in Istanbul from 10 to 12 December 2015. The event was sponsored by the SSHRC funded Major Collaborative Research Initiative (MCRI) on Global Suburbanisms: Governance, Land and Infrastructure in the 21st Century. The workshop was organized by the City Institute at York University in collaboration with the Global Political Trends Center (GPOT) of Istanbul Kültür University, the Department of International Relations at Istanbul Kültür University, and the Department of Urban Planning at Mimar Sinan University. We are grateful to Sara Macdonald who played a major role in facilitating the event. We would like to thank our reviewers for their insightful comments that helped us sharpen the manuscript in a revision and the professional staff at the University of Toronto Press who have been exceptional in their support.

K. Murat Güney wishes to thank all scholars who contributed to this book with their extensive and informative researches on massive suburbanization in diverse parts of the world. This book is possible thanks to the generous help and support of numerous critical academics, activists, and research collaborators. The persistent struggle for the right to the city and suburb by the young residents of old Istanbul, the hometown, has been a powerful source of inspiration that has fuelled the desire to document, contemplate, and write on the ongoing suburban revolution.

Roger Keil wishes to thank all participants of the workshop and the authors that contributed to this volume but especially the people of Istanbul who have been exemplary in hosting the event and inspiring the work presented in this book.

Murat Üçoğlu wishes to thank all contributors that showed great effort and enthusiasm to complete this fascinating book project. This book is an outcome of a long-running academic brainstorming on the politico-economic and socio-economic dynamics of the large-scale housing projects throughout the world. It is, therefore, a great pleasure to thank everyone who has made it real during this long-running inquiry that began in December 2015 in Istanbul. Nevertheless, it is certain that this project will not stop here since the massive suburbanization is an open-ended process.

K. Murat Güney, Roger Keil, and Murat Üçoğlu
November 2018

PART ONE

Re-thinking the Massive Periphery

Introduction

Massive Suburbanization – Political Economy, Ethnography, Governance

ROGER KEIL, K. MURAT GÜNEY, AND MURAT ÜÇOĞLU

This book provides a first systematic overview and globally scaled conceptualization and investigation of the building and rebuilding of urban peripheries with a particular eye on large-scale housing projects. We call these projects (and their infrastructural, industrial, and commercial environments) "massive suburbia." They are by no means flat and horizontalized worlds – as suburbs are often portrayed – but can rather be part of the increasingly stacked and volumetric world in which Stephen Graham (2016) suggests we live. The chapters in this book contain the most comprehensive, global overview of such mostly state- or developer-led housing projects to date. The claim of the book is to change the perspective on suburbanization in the global context, and it aims to offer a universal inter-referencing point for research on the dynamics of new suburbia. This claim could bring about new discussion pertaining to the problems of the urban periphery, planetary sub/urbanization, and the neoliberal production of space. In this sense, this book seeks to contribute to understanding the empirical aspects of massive suburbia, and how this ubiquitous, yet variegated, type of sub/urbanization has given rise to new theorizations of the urban more generally. A selection of conceptual and research-based empirical chapters both revisit the classic cases of large-scale suburban building in Canada, (the former) Czechoslovakia, France, Germany, and the United States and examine the new peripheral estates in China, Egypt, Israel, Morocco, the Philippines, South Africa, and Turkey.

The book presents an anthropology of (re)development as much as a political economy of the global periphery. It engages the built, natural, and social environments of large-scale housing projects (and their adjacent commercial, infrastructural, and industrial estates). Work showcased in this book deals with the politics, financial mechanisms, and technologies of building and rebuilding the periphery in this suburban century (Keil 2013) and discusses traditional and emerging suburbanisms as particular ways of life. The book is concerned with state and corporate policy for building suburban estates as well as with the politics of the inhabitants and "the right to the suburb." In exposing the strategies of economic actors, it deals with capital accumulation through the production of space. It will also touch on the governmentalities and ideologies of homeownership in neoliberal suburbanism that prop up the material processes of building and rebuilding the periphery. Perceptions of suburban life, suburban lifestyles linked to massive suburbanization, and the dialectics of use and exchange value in large peripheral housing projects are subjected to detailed investigation and critique.

The very idea of what the periphery or peripheral may be is itself up for debate in this book. We operate on two registers with regards to this highly contested term. On one register, it refers to the geographical periphery of urban areas. In simple terms, it is the address of what we refer to as suburbanization as a process. This geographical definition comes, of course, with a lot of baggage. Once considered unproblematic and easy to locate, this "periphery" now begs for explanation. Kipfer and Dikeç as well as Wilson in their respective contributions to the opening section of the book discuss the changing notions attached to this term. In this introduction and throughout the chapters that follow various critical uses of the term will also be introduced. On another register, we follow the meaning given to the term by Teresa Caldeira in a recent intervention (2016) where she puts forwards the notion that "peripheral urbanization" is one of the emergent operational terms in urban studies that capture "modes of the production of space that (a) operate with a specific temporality and agency, (b) engage transversally with official logics, (c) generate new modes of politics, and (d) create highly unequal and heterogeneous cities" (Caldeira 2016, 1; see also Hamel and Keil 2015; Keil 2018; McFarlane 2016; Roy 2015). Caldeira explains:

> Many cities around the world have been largely constructed by their residents, who build not only their own houses, but also frequently their neighborhoods. They do not necessarily do so in clandestine ways and certainly not in isolation. Throughout the process, they interact with the state and its institutions, but usually in transversal ways. While they have

> plans and prepare carefully each step, their actions typically escape the framing of official planning. They operate inside capitalist markets of land, credit, and consumption, but usually in special niches bypassed by the dominant logics of formal real estate, finance, and commodity circulation. In the process of house/city building, many make themselves into citizens and political agents, become fluent in rights talk, and claim the cities as their own ... I refer to this mode of making cities as peripheral urbanization. (Caldeira 2016, 1)

We illustrate the book's central concerns in table 0.1. The book engages ongoing conceptual debates on urbanization (see left column) and evolving real empirical cases of massive (post)suburbanization, and works towards a conceptual and comparative centre that will yield new insights into the global phenomenon of massively building and rebuilding the world's peripheries.

As figure 0.1 indicates, the book shifts the debate on the global periphery and offers new entries into the expected typology of the suburban (Harris 2010). It starts from the extant histories of massive peripheries and moves to an exploration of their expanding geographies. It looks at the continued growth of urban regions through the lens of both initial suburban expansion (the explosion of urban society) and the many forms of post-suburban involution (variable implosions) that have been taking place in the erstwhile massive peripheries of the second half of the twentieth century. The book's chapters contain new theoretical thinking and case studies of massive suburbanization in classic (Paris, Toronto, Halle [Saale], etc.) and emerging developments (Cairo, Istanbul, Manila, etc.). While the former cases speak to the centrality of mass housing in the Fordist-Keynesian and socialist periods of the twentieth century, the emerging cases stand for regional or national strategies of building economic power through the privatized production of urban space on the periphery of the burgeoning cities of the Global South.

In Anglo-America and Australia, automobile suburban subdivisions dominate. Europe historically produced its own version€ of the "suburban solution" with a range of suburban development in post-socialist states and even within Western Europe and North America, where we see denser or non-conforming forms of suburbanization (Charmes and Keil 2015; Le Goix 2016, 2017). The large-scale tower estate on the periphery enabled "a consolidated social welfare state to demonstrate what it was able to do" (Harlander

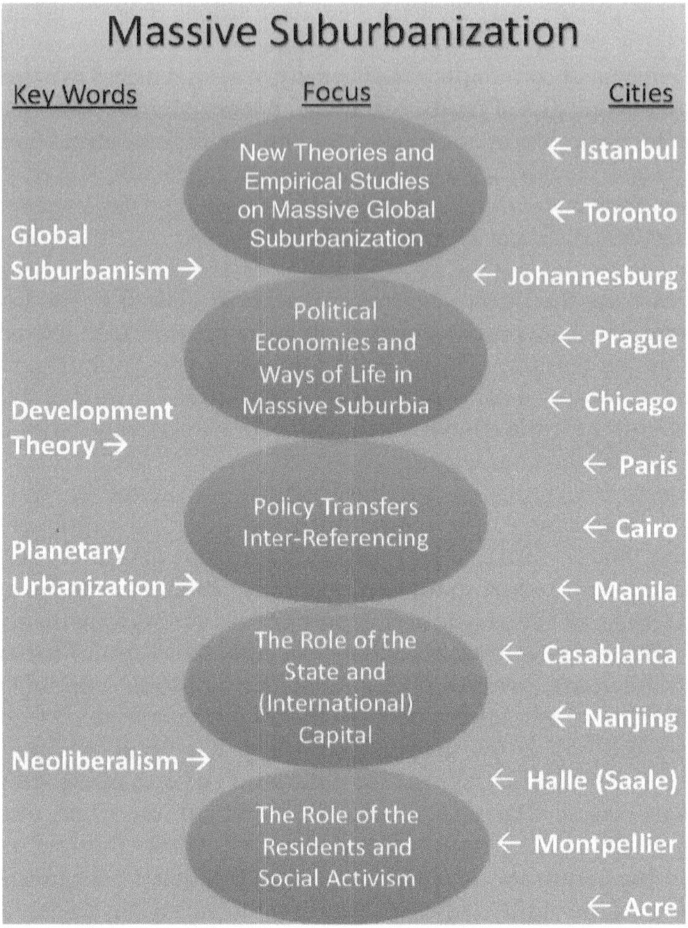

0.1 The book's central concerns

2011, 18; Phelps and Vento 2015). In turn, as Hirt and Kovachev (2015, 181) have observed, "large cities across Eastern Europe were, in the span of about two decades, surrounded by massive pre-made-panel housing blocks with a grandiose but Spartan flavour." These housing estates were the product of long-term planning linked to industrialization and organized by the central state (see the contributions by Bernt and Logan in this book).

At the beginning of the twenty-first century, the world's large cities and small towns, and even some villages (especially in tourist areas), are increasingly ringed by mushrooming settlements with varying morphologies, residential, industrial and commercial spaces, large-scale infrastructures, and compromised open spaces. While vastly diverse in appearance – although there are some recognizable and repeated patterns – these spaces share a certain scale of development. They are expansive in area and often more widespread than the city at their core, they hold varying densities, and they reflect the mode of mass production from where they are born.

When we say that such massive settlements "ring" the centre, we speak only of one type of massive suburbanization that follows the traditional pattern of contiguous settlement away from the core. What we will find more often in this current period, though, is the seeming independence of peripheral settlement from any apparent spatial dependence on the centre and a dissolution of the city into a horizontalized region (Quinby 2011). The massive character of these new settlements therefore is rarely continuous but often appears nested in the soft tissue of urban regions.

The focus of this book, then, is on the built, economic, and social environments of what we call "massive suburbia" in the global urban periphery. We define massive suburbanization as a multi-morphological expansion of urban areas that includes new-built, large-scale suburban building (rental and condominium), the renewal of post–Second World War high-rise estates, developer-built low-rise subdivisions, new town estates, and auto-constructed squatter areas. In that sense, we examine a broad variety of cases that speak to the building or redevelopment/renewal of large-scale peripheral housing estates, tower neighbourhoods, Grands Ensembles, Großwohnsiedlungen, Toplu Konut, and so on around the world. Aspects of massiveness include territorial expansion at the post-metropolitan scale, the demographic dimension (both quantitative and qualitative – segregation by class, ethnicity, and race), morphology (in a spectrum from tower to shack), the technologies of mobility, communication, and metabolism (water, the internet, etc.), and the form of tenure (with rental squeezed out between – although sometimes enabled by – the condominium economy at the top and the squatter economy at the bottom), as well as the rhetoric of expansion as progress and national modernization (especially in China, Egypt, and Turkey).

The book chiefly contributes to an understanding of a particular form of extended urbanization in massive suburban developments and helps grasp the role of the marketization and financialization of land and housing in the building and rebuilding of housing and ancillary suburban landscapes in the peripheries of the cities under investigation. It furthers our understanding of daily life and perceptions of habitation in massive suburbia. The book contributes to the conceptual and theoretical debate on what drives sub/urban development at the beginning of the twenty-first century by engaging literatures from various contexts around the world and using a methodology that subscribes to the principle of "thinking with elsewhere" (Robinson 2016).

The chapters of this book look at historical and current pathways of commodification of housing in particular. This includes the historical process of formalizing peripheral housing in North America – for example from "unplanned suburbs" to "creeping conformity" (Harris 1996, 2004) – and in Turkey (from gecekondu to Toplu Konut [Caldeira 2016; Erder 1996]). While investigating the current financialization and formalization processes of housing, some of the authors examine how housing has become a financial commodity for exchange and how it has created its own financial dynamics: Is rapid and massive sub-urban growth resulting in the tremendous increase in household debt economically and socially sustainable? What do homeowners and residents think? How do they change their behaviour and aspirations as homeowners and citizens?

First, the question of housing (and ancillary commercial, administrative, infrastructural, and employment developments) in massive suburbia is posed from a perspective of urban political economy with a focus on morphological, social, and institutional change in the age of the financialization of land and housing, a process that involves and shapes many aspects of social and everyday life. Chapters in this book investigate the effects of such financial disciplining and increasing rates of indebtedness on the consumption and saving habits and financial choices of home buyers and tenants (for financialization and financial disciplining particularly, see the chapters on Istanbul and the chapter by Matthias Bernt on Germany; see also Aalbers and Christophers 2014; Rolnik 2013; Üçoğlu and Güney forthcoming). Transition is a chief characteristic of this development, as it includes the ongoing replacement of peripheral urban development (Caldeira 2016, 4) and conversion of land that is occurring at a rate greater than that of population growth (Konvitz 2016, 55). This conversion of peripheral land to urban land accelerates the spread of financialization, since more projects will be financed and more housing will have to be sold. Therefore, the housing issue is always related to financialization (see Bernt in this book) and subject to contestation and struggle (Schipper 2013, 2015, 2018).

Second, this book explores the housing question in massive suburbia from the perspective of comparative urban anthropology and in the framework of everyday suburbanisms (Drummond and Labbé 2013). Housing and homeownership have different meanings depending on geographies and social contexts, spanning from being a good citizen and a member of middle classes in the global North (Stout 2016; Weiss 2014) to being seen as an element of "household security" or even an opportunity for upward mobility for the urban poor and rural migrants in the South (Palomera 2014; Zhan 2015; Conley and Gifford 2006; Doling and Ronald 2010). This second area of investigation concerns the different social and cultural motivations of homebuyers in moving in to massive suburbia in the light of financial burdens, and will deal with strategies of survival among the masses that find themselves in the new periphery.

In many countries of the world, housing development is an important part of the economy (Walks 2016a, 23). The proliferation of new suburban housing projects is backed by various levels of government not only with necessary political decisions on land conversion but also with necessary financial facilities provided by the financial institutions and housing development agencies such as TOKI and the Canada Housing Trust, as well as regional policies (Abbruzzese 2016; Addie and Keil 2015; Belina and Lehrer 2016). The sociology, ethnography, and urban studies literature on Canada, for example, deals with a variety of effects of rapid suburbanization on socio-economic structure and community (Hulchanski 2010; Moos and Mendez 2014; Cowen and Parlette 2011; Poppe and Young 2015; Walks and Bourne 2006). The massive suburbanization both in single-family-home subdivisions and densified suburban "places to grow" are subject to intense scrutiny and critique as climate change concerns give suburbanization a new evaluative framework.

While there are robust economic and ethnographic studies on the most vulnerable segments of suburban populations and a generally growing interest in suburbanism as a way of life, especially in relation to the political economy of land development (Drummond and Labbé 2013; Labbé 2014; Walks 2012), the diverging views of homeowners on rapid suburban growth have not been studied in similar detail. Support of certain segments of society for suburban development is mostly disregarded in critical activist accounts and academic studies. Chapters in this book get to the granular level of life in these massive housing developments and begin to analyse this pressing and as yet underdeveloped area, and identify not only disadvantaged groups but also beneficiaries of suburban development in order to understand the reasons underlying public support for suburban development and construction-led economies.

Contemporary suburban growth occurs in vastly diverse forms throughout different parts of the world (Hamel and Keil 2015; Harris and Lehrer 2018). The cities discussed in this book have been chosen as representative examples from the Global North and Global South to elaborate on the diversity in global suburbanization. They all present particular spatial patterns and social characteristics of suburban growth that range from gecekondus (shantytowns) or state-built, high-density public housing projects and gated communities to inner and outer suburbs of variable morphology and density.

Two salient common features invite comparison among many of the places in this book: (1) their status as "arrival cities" (Saunders 2010) and home to "kaleidoscopic suburbia" (Tzaninis 2016), and (2) their experience of rapid construction and housing-sector-led economic growth.

The current book takes note of a growing and increasingly sophisticated literature on comparative urbanism that has now captured the imagination of critical urban researchers across the globe (McFarlane 2010; Tuvikene 2016). We take from Robinson (2015, 187) among others the insight that current "urbanization trends displace the former heartlands of urban theory," from which "urban studies will produce a new generation of scholarship which arises in new centres of authority and expertise." The comparative aspect of this shift is central, as it will be the basis for new theorization from emerging vantage points rather than reducing "new theorizations of the urban to 'particularism'" or "empirical variation' and descriptive color'" (Robinson and Roy 2016). The broad agenda of the new comparative urban theory (summarized in Robinson 2015, 187) – from ethnography to political economy – is inspiration for this book. It is Robinson's notion of "thinking with elsewhere" (188) and with "repeated instances" (Jacobs 2006) in particular that lies at the basis of our work (also Peck 2015, 162–3). Thinking conceptually across conventional divides and looking at case studies around the world, we pinpoint and identify daily life in the rapidly emerging peripheral cities that are part "habitat" and part "uninhabitable" and whose true character may reveal itself not so much in firmly structured built environments but as a city that "hovers across tightly packed nodes dispersed across long distances" (Simone 2016a, 18). The latter admission is central to the work in this book, leading beyond traditional political economic thinking and into the nuances that only anthropological work can deliver (Simone 2011, 2016b). This includes a focus on "investment space, rent space and symbolic space" (Walker 2016, 172). These elements combine into various assemblages of size, density, and diversity in sub/urban form (Tonkiss 2013; Hamel and Keil 2015; McFarlane 2016). Hence, the emerging comparative registers in this book consider "multiple spatialities of density" (McFarlane 2016, 632). Authors in this book investigate the massive suburbanization of today's urban regions not as "a mute artifact of wealth and difference" but as "an active expression of prevailing social hierarchy, class control and state power" (Walker 2016, 173; Ekers, Hamel, and Keil 2012). We do so recognizing vast differences in historical trajectories and path dependencies in urban and suburban systems across the globe that play themselves out on larger-than-city regional scales across the planetary landscape of urbanization (Walker 2016, 175; Brenner and Schmid 2014, 2015). In addition, we are swayed by arguments in theoretical, comparative urban studies that see urban regions as "assemblages" of a variety of contradictory dynamics, of which massive suburbanization is an important instance: it is never just the linear extension, overflow space, and investment area of more central, more powerful interests in surplus value and rent extraction (Addie and Keil 2015; Walker 2016); it is also "peripheral urbanization," a process driven by residents and their politics and economic decisions, which speak back to the structural powerlines of suburbanization (Caldeira 2016; Kohn 2016). The book's chapters examine, deconstruct, and re-establish registers of formality and informality, planning and auto-construction, state and market, ownership and tenancy, and horizontality (low-rise) and verticality (high-rise) in morphology in a comparative perspective. They are guided by engaging "transversal logics" (Caldeira 2016, 5), through which we make visible "multiple formations of inequality, wherein categories such as 'formal' and 'regulated' are always shifting and unstable" (Caldeira 2016, 5; see also Kuyucu 2014).

Suburbanization in the Age of Real Estate Capitalism

Presciently, Henri Lefebvre (2003) portrayed the making of a global urban society as a process of implosion and explosion. The urbanization process is composed of both the concentration and expansion of spatial socio-economic processes (Brenner 2014; Lefebvre 2003). Aware of the inseparable dialectics of these two dynamics, we nonetheless focus in this book predominantly on explosion, the "extended urbanization" (Monte-Mor 2014a, 2014b) that characterizes urbanization presently. For the past half-century, it has been in the periphery, more than in the centre, where the true expression of society's success could be found, whether that was the single-family home subdivision of Anglo-America, the urban extensions of the Soviet Empire, or the tower housing areas of Western Europe and Canada. Garden cities, satellite towns, and subdivisions characterized the automobile metropolis in East and West. While the consecutive waves of twentieth-century suburbanization were still dependent on and related to the urban centre, the latter part of the twentieth century saw an erosion (or perhaps dissemination) of centrality as we knew it. Instead, the urban form becomes polycentric, and the suburbs themselves appear more as free-floating units (Keil 2018).

Accordingly, massive suburbanization in this book is understood as part of the extended urbanization (Monte-Mor 2014a, 2014b), which Lefebvre (2003, 14) considered part of the "explosion" of the city into "large apartment buildings, private homes both large and small, campgrounds, shantytowns," especially in the form of state- and corporate-built housing typical of the late twentieth century, which he called "habitat" (81). Empirically, much of what is called urbanization in the twenty-first century is, in fact, suburbanization, and most of the people who are currently living in or will move into cities will be in urban peripheries (Angel et al. 2010; Moos and Walter-Joseph 2017). Suburbanization is a process of city-building and re-building at the metropolitan edge, including processes of post-suburbanization: "a process that involves densification, complexification and diversification of the suburbanization process" (Charmes and Keil 2015, 581; Charmes and Rousseau 2014). This is happening around the world in a variety of modalities of governance (Hamel and Keil 2015), involving a broad array of land markets (Harris and Lehrer 2018) and infrastructural constellations (Angel et al. 2010; Filion and Keil 2016; Filion and Pulver, forthcoming; Harris and Vorms 2017). The production and retrofitting of these areas have been widely studied in a variety of contexts, especially after the financial crisis of 2008 (for a selection, see O'Callaghan et al. 2015; ARCH+ 2011; Chédiac and Bernié-Boissard 2010; Gururani 2013; Gururani and Kose 2015; Herzog 2015; Jessen and Roost 2015; Hodson and Marvin 2016; Dunham-Jones and Williamson 2011); the planning and policies that led to their existence and redevelopment have been examined (Balducci 2012; Dikeç 2007; Enright 2016; Kipfer, 2016; Logan 2015; Poppe and Young 2015; Touati-Morel 2015; Wacquant 2008); and their significance in a post-suburban landscape has been discussed (Charmes and Keil 2015; Le Goix 2017; Phelps 2015; Sieverts 2003).

Financialization is the dominant feature of the production of suburban space today. Having been cast onto a new trajectory since the 1970s with the emergence of neoliberal policies, financialization includes spreading financial debt mechanisms into all parts of society (Soederberg 2013). Housing had been perceived as a patrimonial investment usually associated with its use value. However, it started to evolve into a commodity traded for its exchange value, subject to market price, especially after the emergence of mortgage-backed securitization (i.e., a pool of mortgage bonds) and the inclusion of the urban poor into the mortgage system (Aalbers 2013, 33; 2009; Haila 2015; Walks 2016a; Weber 2010). Major aspects of the financialization of housing include the transaction between banks and households and the financial profit from land-rent speculation, leading to uneven accumulation of debt and rent. In addition, the creation of new asset streams is followed by speculation (Leyshon and Thrift 2007, 98). As capitalists and rentiers invest in the housing market for profit, and suburbs cement their role as places of capital accumulation around the world (Shen and Wu 2016), working-class homeowners (in the meantime) generally seek financing to acquire "use value" (Lapavitsas 2013, 39). Spatial inequalities and uneven wealth distribution in cities and suburbs follow (Walks 2016b; Foster and Kleit 2015). Public policy choices have shifted the risks of poverty from collective to individual responsibility (Doling and Ronald 2010, 165; Kaika and Luca 2013; García-Lamarca and Kaika 2016; Ronald 2008). Massive suburbanization is a key process both for asset accumulation and the individualization of risk, and part of a redefinition of risk society (Walks 2013; Beck 1999).

In most countries, the suburbanization process has marshalled enormous, often unprecedented, material and intellectual capital and labour. Anthony King (2004, 98) reminds us of "the immense proportion of a society's resources invested in" the suburb, an effort equivalent, as he notes, based on Taylor, to "the great Gothic cathedrals of the high middle ages of feudal Europe." If this was true for America's Levittown and Berlin's Marzahn, it is certainly true for today's expansion of Cairo, Istanbul, and Shanghai. The urban revolution conceptualized by Lefebvre (2003) has gained momentum in the decades since the French writer first put forwards his hypothesis, not only with a spatial extension of so-called urban areas but also with the blurry aspect of urbanization, which eliminates the volatile distinction of urban and suburban. In the imagination, suburbs are contemplated as a place of happy family houses with a backyard and a car. The suburban way of life is known as the place of property ownership, conservatism, and patriarchy (Harris and Keil 2017, 52). Suburbs have been the symbol of auto-dependency but also of the new dynamics of urbanization. We now require a more global vision to explain ongoing urban expansion to the suburban peripheries of the cities. Suburban expansion is a key domain of the production of space (Keil 2013, 2018). However, this expansion is no longer a singular local process; it is a global process of the production of peripheral space. In many countries, the production of peripheral space is now wide open to unprecedented global market dynamics.[1]

The contributions in this book examine the latest forms of suburban development in major metropolitan areas. There are three eminent elements: the role of the state, the political economy, and suburban everyday life (which we also refer to as suburbanism). These three elements are strongly tied to each other. In the past, suburbs were chosen by people due to the image of the suburban way of life. On the one hand, it was assumed that living in the suburbs was healthier than living in dense city centres; on the other hand, suburbanization offered the opportunity of homeownership (Harris and Keil 2017, 57). These dynamics are related to the state (or governance more broadly) (Hamel and Keil 2015), the political economy of suburban land (Harris and Lehrer 2018), and everyday life.

Today, suburbanization is seen as a continued source of housing supply in insecure markets governed by financialization, and simultaneously as a threat to sustainability in an age of climate change. Mainstream neoliberal thinkers view suburbanization as a necessary contribution to balancing housing supply and demand (Perloff and Wingo 1968; Muth 1968; Thompson 1968). Traditionally, suburbs are places of automobile dependency and white middle-class patriarchal social codes (Moos and Mendez 2014). However, there were also many suburbs built by, inhabited, and used by racialized populations and the working class (Harris 1996; Wiese 2004). The suburban profile has been changing in accordance with the new urban policies of neoliberalization, displacement due to inner-city gentrification, and global migration streams: "The suburbs are becoming less middle class and more non-white and immigrant" (Keil 2013, 12). Indeed, with the spread of neoliberal urbanization, suburbs are becoming the realm of financial land-rent speculation, and of household debt. There exist several players in this market, from land owners to informal developers. Richard Harris (2013) states, "

> Huge profits are made from suburban development, especially around rapidly growing cities. They attract big investors in a world of global capital flows. Some money goes into mere speculation, but massive amounts support useful investments in infrastructure. Suburban development, and the millions of people who benefit from it, are vulnerable to anything that stems or redirects such flows, including hiccups in the financial system" (38).

Suburban development is no longer viewed as a stable project to provide necessary housing supply to the growing demand; suburbs are always dynamic and changing. This development is a process in which many actors from the real estate/financial complex (Aalbers, 2012) to mere individuals are taking part. This does not mean that individuals are happy players of this game; rather, many of them are thrown into the pool of mortgage debt (Walks 2013, 2016a).

Naturally, suburbs are an active playing field not only for business actors and the state but also for people who are residing there. Dwellers of the suburbs are also changing themselves in accordance with the hierarchical policies of space. The struggle over the production of space and the hierarchization of space are deeply felt in the suburbs. Dwellers either choose to engage with these new forms of spatial hierarchization or they find new political ways (such as neighbourhood resistance, rent strike, etc.) to cope with them. Urban poverty is increasing in the suburbs with the new changing face of suburban development; the process called the suburbanization of poverty is now finding a significant expression in the newly growing literature on neoliberal suburbanization (see Wilson, Boodram, and Smith's chapter in this book; also see Allard 2004; Cooke and Denton 2015; Kneebone and Berube 2013; Filion 2015).

At the global scale, the recent suburban development is regarded as part of a big picture of new urban politics in which the so-called distinction of urban–suburban has eventually become blurry (Brenner and Schmid 2014). Therefore, urban society could now be portrayed as suburban society and vice versa. This analogy is also relevant with the problems faced by urban society. Poor infrastructure, insufficient transportation links, housing price increases, unaffordability, displacement, and eviction are now common urban problems, and these are widespread both in suburbs and the city centres. Essentially, unaffordability is now the major problem in the urban process. It has been estimated that globally there are now 330 million households that are unable to find an affordable home (Madden and Marcuse 2016, 3). Millions have already been displaced, and the peripheries are dynamic sites of these urban crises. Many of these problems occur due to the strategies of the political economy of the production of space. Madden and Marcuse (2016) explain this neoliberal production of space as the conflict between housing as home and real estate. Housing is not a mere commodity that can be used for profit-making or capitalist accumulation. It is a social issue that captures individual safety, social security, socio-legal relations, and social equity (see Kipfer in this book).

Massive suburban development is now gaining a certain centrality in those power relations and social problems. In other words, in many urban regions around the world suburban development now constitutes the core of the urban process, along with the financialization of major globalizing cities. The ongoing processes of financialization, conurbation, mass housing construction, large-scale investments, displacement, and gentrification are all embedded in global suburbanization. We call it massive suburbia because there is a massive transformation of peripheral space everywhere. Indeed, population growth predictions suggest that this massive process will be a major subject in the Global South even more than in the Global North (Bain and Peake 2017; McGee 2015). The majority of the growing population will inevitably reside in the new peripheral spaces, and the commodification of the housing market will be the inevitable result of this growth and a challenge for these new populations. The housing market is now the major sphere of commodification that overshadows industrial production (Madden and Marcuse 2016, 27). The commodification of housing is a political project, and this commodification works for the well-being of real estate and construction firms that do everything to maintain high prices (Madden and Marcuse 2016, 47). Therefore, as Manuel Aalbers (2012) indicates, there exists a real estate/financial complex that aims at commodifying housing by using several financial tools in order to keep up the demand for housing.

In the meantime, homeownership is promoted by political, social, and psychological discourses. Being a homeowner is seen as a prerequisite of full citizenship or being "free." However, Madden and Marcuse (2016) argue that homeownership causes alienation in society, because dedication to work as a means of owning a home depends on precarization. This leads to what they call residential alienation, since in this new housing market dynamics residents become insecure and afraid of being evicted or failing to pay their mortgage debts. The increase in the insecurity level of residents leads to the spread of precarious forms of life (76). Therefore, in settings as different as Toronto, Istanbul, and Shanghai, homeownership is presented as the antidote to all these problems. It is promoted as a form of empowerment, a market of dignity, and a way to overcome social isolation and estrangement.

Suburban development is triggered by these discourses of homeownership. Commodification of the house goes hand in hand with the promotion of homeownership. However, as Ananya Roy (2015) puts forwards, homeownership in neoliberal times is no longer a guarantee of lifetime stability. It is open to precarization, eviction, and land occupation. Rather than being a safe shelter, housing is now used as an investment that is bolstered by mortgage markets, which once were used as a way to facilitate homeownership (use value) for working and middle-class families. Mortgage markets are now a new tool of profit-making (exchange value). With a new architecture of financialization, housing has become a pure commodity of financial profit-making and rent-maximization. As long as the neoliberal self is perceived as the entrepreneurial agent pursuing her career paths with risk-taking, neoliberal subjects will dedicate themselves to rent-maximization systems. The neoliberal subject is not a passive maximizer, but a constructor of profitable situations that she discovers through her alertness and which she can exploit (Dardot and Laval 2014, 112). The transformation of individuals into entrepreneurial subjects is a new neoliberal strategy, and as Susanne Heeg (2017) indicates, the neoliberal individual is now tied to the individualization of economic responsibility. The entrepreneurial self refers to the spread of responsibilization for further risks. Real estate is a pure realm of investment and everyone in this market is perceived as a potential investor or source of profit-making. Even those who seek housing only for its use value are converted into investors and risk-bearers through debt mechanisms. Massive suburbanization is thereby a giant process that encapsulates all the above-mentioned components of political economy and anthropology. The production of massive suburban areas now designates not only the increasing tendency of profit-making through land-rent speculation but also a new socio-political structure of society.

Massive suburbanization, especially where woven into the notion of national progress, has been part of a "new symbolic economy that had a direct influence on the image and imagination of national territories" (Sevilla-Buitrago 2013, 251). The single-family home was central to this in the twentieth century, as is increasingly the high-rise condominium or apartment in the suburban extension. A mass-produced form of housing and the desires of the development industry to maximize their profits on peripheral greenfields laid the foundation for the role of the bungalow–mansion spectrum of houses at the core of North American suburbia (Ford 1994; Keil 2018). Adds Easterling (2014, 84): "A design idea for suburbia becomes more powerful when it is positioned as a multiplier that affects a population of houses." Massive suburbanization can include the house as the relay station of the Fordist-Keynesian regime of accumulation, a symbol of ostentatious consumption in the age of Metroburbia (Knox 2008, 2017), and the high-rise cell erected by modernist planning and state bureaucratic institutions (De Meyer et al. 1999, 17; Strebel and Jacobs 2013; Keil, 2018).

More recently, massive suburbanization has been seen occurring in line with the spread of a "post-suburban" region that can be highly diverse as, for example, post-socialist countries experience a boom in single-family-home construction (Stanilov 2007, 179–86) resulting in a mixed landscape likened by Hirt and Kovachev (2015, 193) to a "multi-headed deity." In his work on the post-suburbanization of the Global South, Terry McGee (2013, 25) has concluded "that suburbs are best defined as a category of settlement that is one of the many types of the built environment of housing settlement types, commercial and industrial spaces, as well as infrastructures that include high-rise apartments, town houses, condominiums, family homes, and illegal settlements that are now part of the emerging fabric of an urbanized world" (see also Coutard and Rutherford 2015). Such observations have also been recorded for India (Gururani and Kose 2015) and Africa (Bloch, 2015; Mabin et al. 2013). In China, a newly sub/urbanizing country, where Wu (2013, 194) sees a process of "urbanization of suburbs" under way, "suburban development is more about constellations of urbanization and suburbanization, and it is heterogeneous in terms of population composition."

Mixed densities are the rule, although the planning doctrine requires compactness. Yet the utopian massive housing projects that grace the peripheries of emerging cities in most parts of the world won't be "urban" in the classical sense, regardless of their densities. While urban form itself hardly predicts specific ecological damages, the material ecologies and flow metabolisms of the current wave of massive suburbanization are beyond anybody's imagination and far in extension of even the most brutal Corbusian excesses of the twentieth century (Keil 2018). The urban periphery is a major part of the operational landscapes of the extended urban society. The question of extended urbanization is hence both spatial and political (Monte-Mor 2014b, 265). It underlines the domain of political ecologies in massive suburbia.

Massive Suburbanization: What Types? What Modalities?

One of the key points of agreement in the burgeoning literature on suburbs is that diversity is the norm rather than the exception, which means that Levittown is but one form of suburbanization, and it forces us to approach suburbs within urban theories that reach "beyond the west" (Edensor and Jayne, 2012). In terms of the governance of suburbanization, Ananya Roy (2009a, 2009b) pushes us to consider multiple worldly forms of governance, not as derivative of the American experience, but rather as central to the increasing suburbanization of urban-regions in all spaces, including the United States. As the chapters in this book demonstrate clearly, massive suburbanization is governed by a wide range of modalities in which the state, capital accumulation, and private (often authoritarian) forms of governance intersect in different ways (Hamel and Keil 2015). Following Caldeira's lead, we associate peripheral forms of urbanization with auto-construction of living space. This kind of urbanization is rarely associated with large structures and extensive infrastructures. The state is not a builder here, although, as the examples from Morocco, South Africa, and the Philippines in this book demonstrate, the peripheral settlement can receive policy guidance (or even directive) from the government. And the capitalist market is ubiquitous in the peripheral city (as is global culture and religion). Towers, roads, piped water, and the presence of the state in the form of police tend to be the product of a collusion of state and capital accumulation at a higher scale (see the classical cases in Eastern and Western Europe, as well as Canada, and the emerging modalities of governance, in particular the cases of Turkey and China, but also Egypt and Israel). This is also relevant for the non-residential structures and infrastructures that are put in place in the urban periphery, such as airports, ports, and industrial and commercial centres that now structure the "horizontalized" (Quinby, 2011) landscape of globalized capitalism in the spheres of both production and consumption. In all cases, there is not a clear-cut line of formal and informal forms of the production of massive suburban space. There are grey zones (following Yiftachel's [2009] notion of "gray cities"), in which identities and lives are assembled "in between" the modalities of state, market, and private authoritarianism. The dialectics of formal and informal is at work here, and spills the "peripheral" beyond its borders (a phenomenon on display in the Moroccan, South African, and Philippine chapters in this book). In Morocco, for example, state-market collusion produces massive formal suburban subdivisions; yet the clandestine slums are the flipside of this massive formality – they become massive informality. The pirate suburbs are a product of political exclusion, economic expulsion, and self-organization and autonomy (see Belarbi and Rousseau in this book). In South Africa, a regime of bricolage creates a grey space between the state, capital, and community through amenities and "backyarding" (see Charlton and Rubin in this book), while in Manila the state's attempts at regulation of peripheral housing both create and threaten "professional squatters" in poor neighbourhoods (see Coker in this book).

We can conclude, then, that massive suburbanization is by no means uniform suburbanization. There can be massive slums, government sponsored shacks with amenities, and so on, and all of that can be altered by activities such as backyarding, gecekondu building, etc. In the traditional world of state-sponsored housing (and ancillary urbanization) in the suburbs, in the classical North American suburbs (see the chapters on Chicago and Toronto), and in the state sponsored peripheries of Halle (Saale), Montpellier, Paris, Prague, and also Toronto, massivively built-up environments are coping with rapid demographic, economic, and morphological change that recompose massive suburbanisms in often unpredictable ways.

The Plan for the Book

The chapters in this book were first presented at a workshop on "Building and Rebuilding the Urban Periphery" in Istanbul at the Istanbul Kültür University, 10–13 December 2015. The workshop was part of the Major Collective Research Initiative (MCRI) on Global Suburbanisms: Governance, Land and Infrastructure in the 21st Century (http://suburbs.info.yorku.ca/). The event was rooted in the larger questions posed by the suburban redevelopment and tower renewal team in the MCRI,[2] which started from the assumption that many suburban developments of the twentieth century are now areas in social and physical decline. Policy-makers, community activists, planners, and architects seek to define key issues and implement policy interventions. This suggests the value of investigating suburban redevelopment from various perspectives: physical form and its adaptability; vulnerability and poverty in suburbs; and social housing as a key component of social infrastructure. In some cases, total redevelopment allows what private land markets consider higher and better uses. In others, emphasis is on social renewal, addressing education, poverty, and unemployment.

This work recognizes that the great variety among suburban neighbourhoods in terms of physical form (low-or high-density), housing tenure (privately owned or social housing), social class, and ethnocultural composition makes their renewal complex and contested. While the redevelopment of the physical, social, and environmental legacy of the massive suburban settlements of the twentieth century poses great challenges in urban areas in some parts of the world, the newly built massive peripheral zones in other global regions are only beginning to pose questions to planners, inhabitants, and students of suburban change.

This book, then, examines the political economy and anthropology of this massive suburbanization, its path-dependencies and disruptions. It engages the built, natural, and social environments of large-scale housing projects (and their adjacent commercial, infrastructural, and industrial estates). The chapters in this book deal with the politics, financial mechanisms, and technologies of building and rebuilding the periphery in this suburban century on a suburban planet (Keil 2013, 2018). The chapters also discuss emerging suburbanisms as particular ways of life. Some chapters also focus on the governmentalities and ideologies of homeownership in neoliberal suburbanism that prop up the material processes of building and rebuilding the periphery.

The book is divided into sections. This introduction is one of three conceptual articles that open the conversation on rethinking massive globalized suburbanization and the meaning of peripheries. Stefan Kipfer and Mustafa Dikeç in their chapter that follows the introduction discuss the ideological articulations that have characterized the notion of periphery and warn against spatial reification in the use of such terminology. Their argument is tightly linked to recent political developments in France and the United States during which the periphery – as a socio-spatial trope – has been related to the emergence of new political constellations such as (right-wing) populism. In another conceptual chapter, building on extant debates on planetary urbanization, Wilson, Boodram, and Smith discuss the dramatic changes Chicago's periphery has undergone in recent decades. The authors provide a "sketch" of planetary suburbanization in which they highlight the entrepreneurialization of suburban areas and in particular of specific places.

A second section focuses on the legacies of massive suburbanization in the twentieth century. The geographical emphasis here is on Eastern and Western Europe as well as Canada. Matthias Bernt details the building and rebuilding of the high-rise periphery of Halle-Neustadt in the former East Germany, and Steven Logan discusses the contemporary history of Prague's South City. Both chapters observe keenly the tremendous changes over the last generation in a post-socialist political and economic environment – with financialization and privatization of housing stock at the centre of attention – and demonstrate the path-dependencies, or in Keller Easterling's (2014), words their "reproducibility," which now get tested under fast-changing conditions. Douglas Young takes the reader to Toronto, Canada, a city with a similar modernist morphological legacy to the (post)socialist cases of massive suburbanization, but with a distinctly capitalist history of building and rebuilding. Young bemoans the "tragedy" of the allegedly progressive "Tower Renewal" strategy that is targeting the decaying and stigmatized modernist housing areas in that city. This North American example is followed by two chapters dealing with massive housing governance and (re)building in the French metropolitan periphery: Roza Tchoukaleyska investigates the interplay of national housing policy and local activism in Montpellier's La Paillade neighbourhood. Stefan Kipfer presents findings from a long-term research project on the periphery of Greater Paris, with an intrinsic comparative perspective on Toronto.

A third section of the book casts a spotlight on Istanbul. There are pragmatic reasons for this singling out of Turkey's largest city in this book on massive suburbanization. Istanbul was the site of the 2015 workshop that initiated this present book, and the city has been the subject of much interest among urbanists looking at the momentous shifts in urbanization patterns on the Bosporus (see, for example, Caldeira 2016). As K. Murat Güney explains in his chapter, Istanbul is a special case in the study of global suburbanisms for at least three reasons, including the unprecedented speed of population growth, the authoritarian governance of mass housing development through a distinctive institutional body created by the government, and the fast rise of housing prices fuelled by speculation. The contributions to this section approach the analysis of these shifts from different yet compatible methodological and thematic perspectives. Erbatur Çavuşoğlu and Julia Strutz tell the detailed story of the development in one part of Istanbul's western periphery from Kayabaşı to Kayaşehir, which signifies a semiotic change that involves the meaning of the location – in the mind of the beholder – of a once-marginal place into one of desire, significance, and value. K. Murat Güney, in his chapter on northern Istanbul, deepens this theme and provides systematic background to the understanding of the political economy and social anthropology of Turkish housing strategies under the AKP government over the past two decades. The section is concluded with a chapter by Murat Üçoğlu on the (sub)urban political ecologies of the massive building of Istanbul's periphery.

The fourth and final section of the book presents cases from the emerging global periphery where new-built massive suburbanization now mostly takes place. Examples from Africa and Asia present a wide-ranging, on-the-ground investigation of peripheral construction and reconstruction in the coordinates of state action, market pressures, and community activism. Wafae Belarbi and Max Rousseau present the so-called pirate suburbs of Casablanca as a case of informalization that is tightly linked to purposeful state action. Similarly, Sarah Charlton and Margot Rubin in their case study on Johannesburg, at the other edge of the African continent, look at the career of state housing policy and the effects on residents and entrepreneurs who add lodging space and marketable units to the envisioned settlements supported by the government. This in-betweenness of state and market action (accompanied and contested by community politics) is a pattern also visible in Manila. Abidemi Coker focuses on the interplay of housing policy and what some call "professional squatting" in the Filipino capital's sprawling housing areas. While Johannesburg and Manila represent distinct cases of massive suburbanization between the state, the market, and the community, they also provide interesting commonalities for further reflection. In his subsequent chapter on Israel, Oded Haas also looks at state-led development of housing in Israel. His case study of the Arab suburb of Tantour explains the ethno-colonial nature of massive suburbanization in Israel as a project of national population sorting. An interesting contrast is provided in Karl Schmid's look at Israel's neighbour Egypt through the lens of cultural assemblages that define Cairo's suburbanization. The final chapter of this section and of the book takes the reader to China. Fulong Wu and Tianke Zhu present a case study of the Jiangning district in Nanjing in eastern China and reveal a fascinating diversity of form and social structure in the new town periphery of that country.

NOTES

1 Nine per cent of investments into single-family homes in the suburban York region of Toronto comes from foreign investors, for example (Mahoney and Giovannetti 2017).
2 This group is led by Douglas Young at York University and includes collaborators Matthias Bernt and David Wilson, who have key contributions in this volume.

REFERENCES

Aalbers, M. (2009). *Place, Exclusion and Mortgage Markets*. Malden, MA: Blackwell
– (Ed.). (2012). *Subprime Cities: The Political Economy of Mortgage Markets*. Oxford, UK: Wiley-Blackwell.
– (2013). "Neoliberalism Is Dead ... Long Live Neoliberalism." *International Journal of Urban and Regional Research* 37, no. 3, 1083–90. https://doi.org/10.1111/1468-2427.12065
Aalbers, M., and Christophers, B. (2014). "Centring Housing in Political Economy." *Housing, Theory, and Society* 31, no. 4, 373–94. https://doi.org/10.1080/14036096.2014.947082
Abbruzzese, T. (2016). "'Build Toronto' (Not Social Housing) Neglecting the Social Housing Question in a Competitive City Region." In Keil, R., Hamel, P., Boudreau, J.A., and Kipfer, S. (eds), *Governing Cities Through Regions*, 143–72. Waterloo, ON: Wilfrid Laurier University Press.
Addie, J.P., and Keil, R. (2015). "Real Existing Regionalism: The Region between Talk, Territory and Technology." *International Journal of Urban and Regional Research* 39, no. 2, 407–17. https://doi.org/10.1111/1468-2427.12179
Allard, S.W. (2004). "Access to Social Services: The Changing Urban Geography of Poverty and Service Provision." Washington, DC: The Brookings Institution, Center on Urban and Metropolitan Policy, Survey Series.
Angel, S., Parent, J., Civco, D.L., and Blei, A. (2010). "The Persistent Decline in Urban Densities: Global and Historical Evidence of 'Sprawl.'" Working paper. Cambridge, MA: Lincoln Institute of Land Policy.
ARCH+, Zeitschrift für Architektur und Städtebau. (2011). "Sonderheft Planung und Realität: Strategien im Umgang mit den Großsiedlungen." ARCH+ 44, no. 203 (June).
Bain A., and Peake, L. (eds) (2017). *Urbanization in a Global Context*. Oxford, UK: Oxford University Press.

Balducci, A. (2012). "Planning the Post-metropolis." *disP – The Planning Review* 188, no. 1, 4–5. https://doi.org/10.1080/02513625.2012.702953

Beck, U. (1999). *World Risk Society*. Cambridge, UK: Polity Press.

Belina, B., and Lehrer, U. (2016). "The Global City Region: Constantly Emerging Scalar Fix, Terrain of Inter-Municipal Competition and Corporate Profit Strategy." In Keil, R., Hamel, P., Boudreau, J.A., and Kipfer, S. (eds), *Governing Cities Through Regions*, 83–97. Waterloo, ON: Wilfrid Laurier University Press.

Bloch, R. (2015). "Africa's New Suburbs." In Hamel, P., and Keil, R. (eds), *Suburban Governance: A Global View*, 253–77. Toronto, ON: University of Toronto Press

Brenner, N. (ed.). (2014). *Implosions/Explosions: Towards a Study of Planetary Urbanization*. Berlin, Germany: Jovis.

Brenner, N., and Schmid, C. (2014). "The 'Urban Age' in Question." *International Journal of Urban and Regional Research* 38, no. 3, 731–55. https://doi.org/10.1111/1468-2427.12115

– (2015). "Towards a New Epistemology of the Urban." *City* 19, no. 2–3, 151–82. https://doi.org/10.1080/13604813.2015.1014712

Caldeira, T.P.R. (2016). "Peripheral Urbanization: Autoconstruction, Transversal Logics, and Politics in Cities of the Global South." *Society and Space* 35, no. 1, 3–20. https://doi.org/10.1177/0263775816658479

Chédiac, S., and Bernié-Boissard, C. (2010). "La Résidentialisation." In J.P. Volle, L. Viala, E. Négrier, and C. Bernié-Boissard (eds), *Montpellier: La ville inventeé*, 92–109. Marseille: Edition Parathése/Gip Epau

Charmes, E., and Keil, R. (2015). "The Politics of Post-suburban Densification in Canada and France." *International Journal of Urban Research and Action* 39, no. 3, 581–602. https://doi.org/10.1111/1468-2427.12194

Charmes, E., and Rousseau, M. (2014). "Le pavillon et l'immeuble : Géopolitique de la densification dans la région métropolitaine de Lyon." *Géographie, Economie, Société* 16, no. 2, 155–81. https://doi.org/10.3166/ges.16.155-181

Conley, D., and Gifford, B. (2006). "Homeownership, Social Insurance, and the Welfare State."*Sociological Forum* 21, no. 1, 55–82. https://doi.org/10.1007/s11206-006-9003-9

Cooke, T.J., and Denton C. (2015). "The Suburbanization of Poverty? An Alternative Perspective." *Urban Geography* 36, no. 2, 300–13. https://doi.org/10.1080/02723638.2014.973224

Coutard, O., and Rutherford, J. (eds). (2015). *Beyond the Networked City Infrastructure: Reconfigurations and Urban Change in the North and South*. London, UK: Routledge.

Cowen, D., and Parlette, V. (2011). *Toronto's Inner Suburbs: Investing in Social Infrastructure in Scarborough.* Toronto, ON: Neighbourhoodchange.ca. http://neighbourhoodchange.ca/wp-content/uploads//2011/06/Cowen-2011-Social-Infrastructure-in-Scarborough-N-Change.pdf

Dardot, P., and Laval, C. (2014). *The New Way of the World: On Neoliberal Society*. London, UK: Verso.

De Meyer, D., et al. (1999). *The Urban Condition: Space, Community, and Self in the Contemporary Metropolis*. Rotterdam, the Netherlands: 010 Publishers.

Dikeç, M. (2007). *Badlands of the Republic: Space, Politics and Urban Policy*. Malden, MA: Wiley-Blackwell.

Doling, D., and Ronald, R. (2010). "Homeownership and Asset-Based Welfare." *Journal of Housing and the Built Environment* 25, no. 2, 165–73. https://doi.org/10.1007/s10901-009-9177-6

Drummond, L., and Labbé, D. (2013). "We're a Long Way from Levittown, Dorothy: Everyday Suburbanism as a Global Way of Life." In Keil, R. (ed.), *Suburban Constellations: Governance, Land and Infrastructure in the 21st Century*, 46–51. Berlin, Germany: Jovis.

Dunham-Jones, E., and Williamson, J. (2011). *Retrofitting Suburbia: Urban Design Solutions for Redesigning Suburbs* (updated paperback ed.). Hoboken, NJ: John Wiley & Sons.

Easterling, K. (2014). *Extrastatecraft: The Power of Infrastructure Space*. London, UK: Verso.

Edensor, T., and Jayne, M. (eds) (2012). *Urban Theory beyond the West: A World of Cities*. New York, NY: Routledge.

Ekers, M., Hamel, P., and Keil, R. (2012). "Governing Suburbia: Modalities and Mechanisms of Suburban Governance." *Regional Studies* 46, no. 3, 405–22. https://doi.org/10.1080/00343404.2012.658036

Enright, T. (2016). *The Making of Grand Paris: Metropolitan Urbanism in the 21st Century*. Cambridge, MA: MIT Press.

Erder, S. (1996). *İstanbul'a bir semt kondu: Ümraniye* (A neighborhood has landed in İstanbul: Ümraniye). İstanbul, Turkey: İletişim Yayınları.

Filion, P. (2015). "Suburban Inertia: The Entrenchment of Dispersed Suburbanism." *International Journal of Urban and Regional Research* 39, no. 3, 633–40. https://doi.org/10.1111/1468-2427.12198

Filion, P., and Keil, R. (2016). "Contested Infrastructures: Tension, Inequity and Innovation in the Global Suburb." *Urban Policy and Research* 35, no. 1, 7–19. https://doi.org/10.1080/08111146.2016.1187122

Filion P., and Pulver, N. (eds). (2019). *Global Suburban Infrastructure*. Toronto, ON: University of Toronto Press.

Ford, L.R. (1994). *Cities and Buildings: Skyscrapers, Skid Rows, and Suburbs.* Baltimore, MD: The Johns Hopkins University Press.

Foster, T.B., and Kleit, R.G. (2015). "The Changing Relationship between Housing and Inequality, 1980–2010." *Housing Policy Debate* 25, no. 1, 16–40. https://doi.org/10.1080/10511482.2014.933118

Graham, S. (2016). *Vertical: The City from Satellites to Bunkers*. London, UK: Verso.

Gururani, S. (2013). "On Capital's Edge: Gurgaon, India's Millennial City." In Keil, R. (Ed.), *Suburban Constellations: Governance, Land and Infrastructure in the 21st Century*, 182–9. Berlin, Germany: Jovis.

Gururani, S., and Kose, B. (2015). "Shifting Terrains: Questions of Governance in India's Cities and their Peripheries." In Hamel, P., and Keil, R. (eds), *Suburban Governance: A Global View*. Toronto, ON: University of Toronto Press.

Haila, A. (2015). *Urban Land Rent: Singapore as a Property State*. Malden, MA: Wiley-Blackwell.

Hamel, P., and Keil, R. (eds). (2015). *Suburban Governance: A Global View.* Toronto, ON: University of Toronto Press.

Harlander, T. (2011). "Die 'Modernität' der Boomjahre: Flächensanierung und Großsiedlungsbau." *ARCH+* 44, no. 203, 14–24.

Harris, R. (1996). *Unplanned Suburbs: Toronto's American Tragedy, 1900–1950.* Baltimore, MD: The John Hopkins University Press.

– (2004). *Creeping Conformity: How Canada Became Suburban, 1900–1960.* Toronto, ON: University of Toronto Press.

– (2010). "Meaningful Types in a World of Suburbs." In Clapson. M., and Hutchison, R. (eds), *Suburbanization in Global Society*, 15–47. Bingley, UK: Emerald.

– (2013). "Suburban Land." In Keil, R. (ed.), *Suburban Constellations: Governance, Land and Infrastructure in the 21st Century*. Berlin: Jovis.

Harris, R., and Keil, R. (2017). "Globalizing Cities and Suburbs." In Bain, A.L., and Peake, L. (eds), *Urbanization in Global Context*, 52–69. Don Mills, ON: Oxford University Press.

Harris, R., and Lehrer, U. (eds). (2018). *The Suburban Land Question: A Global Survey*. Toronto, ON: University of Toronto Press.

Harris, R., and Vorms, C. (eds). (2017). *What's in a Name? Talking about Suburbs.* Toronto, ON: University of Toronto Press.

Heeg, S. (2017). "Governing the Built Environment in European and Metropolitan Regions: Financialization, Resposibilization and Urban Competition." In Keil, R., Hamel, P., Boudreau, J.A., and Kipfer, S. (eds),

Governing Cities Through Regions, 65–82. Waterloo, ON: Wilfrid Laurier University Press.

Herzog, L. (2015). *Global Suburbs: Urban Sprawl from the Rio Grande to Rio de Janeiro*. New York, NY: Routledge.

Hirt S., and Kovachev, A. (2015). "Suburbia in Three Acts: The East European Story." In Hamel, P., and Keil, R. (eds), *Suburban Governance: A Global View*, 177–97. Toronto, ON: University of Toronto Press.

Hodson, M., and Marvin, S. (eds). (2016). *Retrofitting Cities: Priorities, Governance and Experimentation*. London, UK: Routledge.

Hulchanski, D. (2010). *The Three Cities within Toronto: Income Polarization among Toronto's Neighbourhoods, 1970–2005*. Toronto, ON: University of Toronto.

Jacobs, J. (2006) "A Geography of Big Things." *Cultural Geographies* 13: 1–27. https://doi.org/10.1191/1474474006eu354oa

Jessen, J., and Roost, F. (eds). (2015). *Refitting Suburbia: Erneuerung der Stadt des 20. Jahrhunderts in Deutschland und den USA*. Berlin, Germany: Jovis.

Kaika, M., and Luca, R. (2013). "Land Financialization as a 'Lived' Process: The Transformation of Milan's Bicocca by Pirelli." *European Urban and Regional Studies* 23, no. 1, 3–22. https://doi.org/10.1177/0969776413484166

Keil, R. (ed.). (2013). *Suburban Constellations: Governance, Land and Infrastructure in the 21st Century*. Berlin, Germany: Jovis.

– (2018). *Suburban Planet: Making the World Urban from the Outside In*. Cambridge, UK: Polity Press

King, A.D. (2004). *Spaces of Global Cultures: Architecture Urbanism Identity*. London, UK: Routledge.

Kipfer, S. (2016). "Demolition and Counterrevolution: La Rénovation Urbaine in Greater Paris." "Neocolonial Urbanism: La Rénovation Urbaine in Paris." *Antipode* 48, 3, 603–25. https://doi.org/10.1111/anti.12193.

Kneebone, E., and Berube, A. (eds). (2013). *Confronting Suburban Poverty in America*. Washington, DC: Brookings Institution.

Knox, P.L. (2008). *Metroburbia USA*. New Brunswick, NJ: Rutgers University Press.

– (2017). *Metroburbia: The Anatomy of Greater London*. London, UK: Merrell.

Kohn, M. (2016). *The Death and Life of the Urban Commonwealth*. New York, NY: Oxford University Press.

Konvitz, J.W. (2016). *Cities and Crisis*. Manchester, UK: Manchester University Press.

Kuyucu, T. (2014). "Law, Property and Ambiguity: The Uses and Abuses of Legal Ambiguity in Remaking Istanbul's Informal Settlements."

International Journal of Urban and Regional Research 38, no. 2, 609–27. https://doi.org/10.1111/1468-2427.12026
Labbé, D. (2014). *Land Politics and Livelihoods on the Margins of Hanoi, 1920–2010*. Vancouver: UBC Press.
García-Lamarca, M., and Kaika, M. (2016). "'Mortgaged Lives': The Biopolitics of Debt and Housing Financialisation." *Transactions of the Institute of British Geographers* 41, no. 3, 313–27. https://doi.org/10.1111/tran.12126
Lapavitsas, C. (2013). *Profiting Without Producing: How Finance Exploits Us All*. London, UK: Verso
Lefebvre, H. (2003). *The Urban Revolution*, Minneapolis, MN: University of Minnesota Press.
Le Goix, R. (2016). *Sur le front de la métropole: Une géographie suburbaine de Los Angeles*. Paris, France: Publications de la Sorbonne.
– (2017). "Some Reflections on Comparing (Post-)Suburbs in the United States and France." In Harris, R., and Vorms, C. (eds), *What's in a Name? Talking about Suburbs*, 255–81. Toronto, ON: University of Toronto Press.
Leyshon, A., and Thrift, N. (2007). "The Capitalization of Almost Everything: The Future of Finance and Capitalism." *Theory, Culture, and Society* 25, no. 7–8, 97–115. https://doi.org/10.1177/0263276407084699
Logan, S. (2015). "Modernist Urbanism in the Age of Automobility: Producing Space in the Suburbs of Toronto and Prague." PhD diss., York University.
Mabin, A., Butcher, S., and Bloch, R. (2013). Peripheries, Suburbanisms and Change in Sub-Saharan African Cities. *Social Dynamics* 39, no. 2, 167–90. https://doi.org/10.1080/02533952.2013.796124
Madden, D., and Marcuse, P. (2016). *In Defense of Housing*. London, UK: Verso.
Mahoney, J., and Giovannetti, J. (2017). "Foreigners Bought 9.1 Percent of Homes in Suburbs of North of Toronto." *The Globe and Mail*, 11 July 2017. https://www.theglobeandmail.com/real-estate/toronto/foreigners-bought-91-per-cent-of-homes-in-suburbs-north-of-toronto/article35662716/
McFarlane, C. (2010). "The Comparative City: Knowledge, Learning, Urbanism." *International Journal of Urban and Regional Research* 34, no. 4, 725–42. https://doi.org/10.1111/j.1468-2427.2010.00917.x
– (2016). "The Geographies of Urban Density: Topology, Politics and the City." *Progress in Human Geography* 40, no. 5, 629–48. https://doi.org/10.1177/0309132515608694

McGee, T. (2013). "Suburbanization in the Twenty-First-Century World." In Keil, R. (ed.), *Suburban Constellations: Governance, Land and Infrastructure in the 21st Century*, 18–25. Berlin, Germany: Jovis.

– (2015). "Deconstructing the Decentralized Urban Spaces of the Mega-Urban Regions in the Global South." In Hamel, P., and Keil, R. (eds), *Suburban Governance: A Global View*, 325–36. Toronto, ON: University of Toronto Press.

Monte-Mor, R.L. (2014a). "Extended Urbanization and Settlement Patterns in Brazil: An Environmental Approach." In Brenner, N. (ed.), *Implosions/Explosions: Towards a Study of Planetary Urbanization*, 109–20. Berlin, Germany: Jovis.

– (2014b). "What Is Urban in the Contemporary World?" In Brenner, N. (ed.), *Implosions/Explosions: Towards a Study of Planetary Urbanization*, 260–7. Berlin, Germany: Jovis.

Moos, M., and Mendez, P. (2014). "Suburban Ways of Living and the Geography of Income: How Homeownership, Single-Family Dwellings and Automobile Use Define the Metropolitan Social Space." *Urban Studies* 52, no. 10, 1864–82. https://doi.org/10.1177/0042098014538679

Moos, M., and Walter-Joseph, R. (2017). *Detached and Subdivided: An Atlas of Suburbanisms*. Berlin, Germany: Jovis

Muth, R.F. (1968). "Urban Residential Land and Housing Markets." In Perloff, H.S., and Wingo, L. (eds), *Issues in Urban Economics*, 285–334. Washington, DC: Resources for the Future.

O'Callaghan, C., Kelly, S., Boyle, M., and Kitchin, R. (2015). "Topologies and Topographies of Ireland's Neoliberal Crisis." *Space and Polity* 19, no. 1, 31–46. https://doi.org/10.1080/13562576.2014.991120

Palomera, J. (2014). "How Did Finance Capital Infiltrate the World of the Urban Poor? Homeownership and Social Fragmentation in a Spanish Neighborhood." *International Journal of Urban Regional Research* 38, no. 1, 218–35. https://doi.org/10.1111/1468-2427.12055

Peck, J. (2015). "Cities beyond Compare?" *Regional Studies* 49, no. 1, 160–82. https://doi.org/10.1080/00343404.2014.980801

Perloff, H.S., and Wingo, L. (1968). "Introduction." In Perloff, H.S., and Wingo, L. (eds), *Issues in Urban Economics*, 1–42. Washington, DC: Resources for the Future.

Phelps, N. (2015). *Sequel to Suburbia: Glimpses of America's Suburban Future*. Cambridge, MA: MIT Press.

Phelps N., and Vento, T. (2015). "Suburban Governance in Western Europe." In Hamel, P., and Keil, R. (eds), *Suburban Governance: A Global View*, 155–76. Toronto, ON: University of Toronto Press.

Poppe, W., and Young, D. (2015). "The Politics of Place: Place-Making versus Densification in Toronto's Tower Neighbourhoods." *International Journal of Urban and Regional Research* 39, no. 3, 613–21. https://doi.org/10.1111/1468-2427.12196

Quinby, R. (2011). *Time and the Suburbs: The Politics of Built Environments and the Future of Dissent*. Winnipeg, MB: Arbeiter Ring Publishing.

Robinson, J. (2015). "Comparative Urbanism: New Geographies and Culture of Theorizing Urban." *International Journal of Urban and Regional Research* 40, no. 1, 187–99. https://doi.org/10.1111/1468-2427.12273

– (2016). "Thinking Cities through Elsewhere: Comparative Tactics for a More Global Urban Studies." *Progress in Human Geography* 40, no. 1, 3–29. https://doi.org/10.1177/0309132515598025

Robinson, J., and Roy, A. (2016). "Debate on Global Urbanisms and the Nature of Urban Theory." *International Journal of Urban and Regional Research* 40, no. 1, 181–6. https://doi.org/10.1111/1468-2427.12272

Rolnik, R. (2013). "Late Neoliberalism: The Financialization of Homeownership and Housing Rights." *International Journal of Urban and Regional Research* 37, no. 3, 1067–74. https://doi.org/10.1111/1468-2427.12062

Ronald, R. (2008). *The Ideology of Home Ownership: Homeowner Societies and the Role of Housing*. New York, NY: Palgrave Macmillan.

Roy, A. (2009). "Why India Cannot Plan its Cities: Informality, Insurgence and the Idiom of Urbanization." *Planning Theory* 8, no. 1, 76–87. https://doi.org/10.1177/1473095208099299

– (2009). "The 21st-Century Metropolis: New Geographies of Theory." *Regional Studies* 43, no. 6, 819–30. https://doi.org/10.1080/00343400701809665

– (2015). "Governing the Postcolonial Suburbs." In Hamel, P., and Keil, R. (eds), *Suburban Governance: A Global View*, 337–47. Toronto, ON: University of Toronto Press.

Saunders, D. (2010). *Arrival City: The Final Migration and Our Next World*. Toronto, ON: Alfred A. Knopf.

Schipper, S. (2013). *Genealogie und Gegenwart der Unternehmerischen Stadt. Neoliberales Regieren in Frankfurt am Main zwischen 1960 und 2010*. Münster, Germany: Westfälisches Dampfboot.

– (2015). "Towards a 'Post-neoliberal' Mode of Housing Regulation? The Israeli Social Protest of Summer 2011." *International Journal of Urban and Regional Research* 39, no. 6, 1137–54. https://doi.org/10.1111/1468-2427.12318

– (2018). *Wohnraum dem Markt entziehen? Wohnungspolitik und städtische soziale Bewegungen in Frankfurt und Tel Aviv*. Wiesbaden, Germany: Springer.

Sevilla-Buitrago, A. (2013). "Urbs in Rure: Historical Enclosure and the Extended Urbanization of the Countryside." In Brenner, N. (ed.), *Implosions/ Explosions: Towards a Study of Planetary Urbanization*, 236–59. Berlin, Germany: Jovis.

Shen J., and Wu, F. (2016). "The Suburb as a Space of Capital Accumulation: The Development of New Towns in Shanghai, China." *Antipode* 49, no. 3, 761–80. https://doi.org/10.1111/anti.12302

Sieverts, T. (2003). *Cities without Cities: An Interpretation of the Zwischenstadt*. London, UK: Spon Press.

Simone, A. (2011). "Introduction to Urban Life Itself." *International Journal of Urban and Regional Research* 35, no. 2, 403–4. https://doi.org/10.1111/j.1468-2427.2010.01054.x

– (2016a). *Always Something Else: Urban Asia and Africa as Experiment*. Basel, Switzerland: Basler Afrika Bibliographien.

– (2016b). "It's Just the City After All!" *International Journal of Urban and Regional Research* 40, no. 1, 210–18. https://doi.org/10.1111/1468-2427.12275

Soederberg, S. (2013). "The Politics of Debt and Development in the New Millennium: An Introduction." *Third World Quarterly* 34, no. 4, 535–46. https://doi.org/10.1080/01436597.2013.786281

Stanilov, K. (ed.). (2007). *The Post-socialist City: Urban Form and Space Transformations in Central and Eastern Europe after Socialism*. Dordrecht, the Netherlands: Springer.

Stout, N. (2016). "Indebted: Disciplining the Moral Valence of Mortgage Debt Online." *Cultural Anthropology* 31, no. 1, 82–106. https://doi.org/10.14506/ca31.1.05

Strebel, I., and Jacobs, J.M. (2013). "Houses of Experiment: High-Rise Housing and the Will to Laboratorization." *International Journal of Urban and Regional Research* 38, no. 2, 450–70. https://doi.org/10.1111/1468-2427.12079

Thompson, W.R. (1968). "Internal and External Factors in the Development of Urban Economies." In Perloff, H.S., and Wingo L. (eds), *Issues in Urban Economics*, 43–80. Washington, DC: Resources for the Future.

Tonkiss, F. (2013). *Cities by Design: The Social Life of Urban Form*. Cambridge, UK: Polity Press.

Touati-Morel, A. (2015). "Hard and Soft Densification Policies in the Paris City-Region." *International Journal of Urban and Regional Research* 39, no. 3, 603–12. https://doi.org/10.1111/1468-2427.12195

Tuvikene, T. (2016). "Strategies for Comparative Urbanism: Post-socialism as a De-Territorialized Concept." *International Journal of Urban and Regional Research* 40, no. 1, 132–46. https://doi.org/10.1111/1468-2427.12333

Tzaninis, Y. (2016). *Building Utopias on Sand: The Production of Space in Almere and the Future of Suburbia*. Amsterdam, the Netherlands: University of Amsterdam.

Üçoğlu, M., and Güney K.M. (forthcoming). "City with no Boundary: Suburbanization as a Mode of Wealth Accumulation in Istanbul." In Hamel, P. (Ed.), *Suburban Governance under Scrutiny: Revisiting the Understanding of City Regions*. Toronto, ON: University of Toronto Press

Wacquant, L. (2008). *Urban Outcasts: A Comparative Sociology of Advanced Marginality*. Cambridge, UK: Pluto Press

Walker, R.A. (2016). "Why Cities? A Response." *International Journal of Urban and Regional Research* 40, no. 1, 164–80. https://doi.org/10.1111/1468-2427.12335

Walks, R.A., and Bourne, L.S. (2006). "Ghettos in Canadian Cities? Racial Segregation, Ethnic Enclaves, and Poverty Concentration in Canadian Urban Areas." *The Canadian Geographer* 50, no. 3, 273–97. https://doi.org/10.1111/j.1541-0064.2006.00142.x

Walks, A. (2012). "Suburbanism as a Way of Life, Slight Return." *Urban Studies* 50, no. 8, 1471–88. https://doi.org/10.1177/0042098012462610

– (2013). "Mapping the Urban Debtscape: The Geography of Household Debt in Canadian Cities." *Urban Geography* 34, no. 2, 153–87. https://doi.org/10.1080/02723638.2013.778647

– (2016a). "Our Mortgaged Future: Understanding the Overleveraged State of Housing Finance in Canada." *Alternatives Journal* 42, no. 2, 22–7. https://www.alternativesjournal.ca/our-mortgaged-future-understanding-overleveraged-state-housing-finance-canada

– (2016b). "Homeownership, Asset-based Welfare and the Neighbourhood Segregation of Wealth." *Housing Studies* 31, no. 7, 755–84. https://doi.org/10.1080/02673037.2015.1132685

Weber, R. (2010). "Selling City Futures: The Financialization of Urban Redevelopment Policy." *Economic Geography* 86, no. 3, 251–74. https://doi.org/10.1111/j.1944-8287.2010.01077.x

Weiss, H. (2014). "Homeownership in Israel: The Social Costs of Middle-Class Debt." *Cultural Anthropology* 29, no. 1, 128–49. https://doi.org/10.14506/ca29.1.08

Wiese, A. (2004). *Places of Their Own: African American Suburbanization in the Twentieth Century*. Chicago, IL: University of Chicago Press.

Wu, F. (2013). "Chinese Suburban Constellations: The Growth Machine, Urbanization, and Middle-Class Dreams." In Keil, R. (ed.), *Suburban Constellations: Governance, Land, and Infrastructure in the 21st Century*, 190–4. Berlin, Germany: Jovis.

Yiftachel, O. (2009). "Theoretical Notes on 'Gray Cities': The Coming of Urban Apartheid?" *Planning Theory* 8, no. 1, 88–100. https://doi.org/10.1177/1473095208099300

Zhan, Y. (2015). "'My Life Is Elsewhere': Social Exclusion and Rural Migrants' Consumption of Homeownership in Contemporary China." *Dialectical Anthropology* 39, no. 4, 405–22. https://doi.org/10.1007/s10624-015-9401-6

1 Peripheries against Peripheries? Against Spatial Reification

STEFAN KIPFER AND MUSTAFA DIKEÇ

Introduction: Who Trumps Whom?

During recent electoral and referendum campaigns in the United States, Britain, and France, debates raged over the nature of popular support for populist and neo-fascist politicians. Who voted for Brexit and why? Who supported candidates Trump and Le Pen? What explains the fact that working-class fractions vote for hard-right politicians and propositions? Is it class-based anger against neoliberal capitalism and its consequences? Should we focus on racist resentment against migrants, Blacks and Muslims? Could it be that voters accept invitations to live out virulent sexism and homophobia? Or, and here is our own opening question, should we not think about all these explanations together instead of treating them as (false) alternatives?

In the United States, Thomas Frank (2016a, 2016b, 2017) has long warned against a Liberal disdain towards members of the working class who cast their vote for right-wing candidates. He argued again during the recent presidential elections that this Liberal contempt buttressed the working-class vote for Trump, which for Frank is less about racism than about economic despair (see also Hochschild and Frank 2016; Lilla 2016). Analysing the recent election, critics have pointed out that Frank's argument minimizes the cross-class character of support for Trump as well as the centrality of racism and sexism in that support. Given the solid upper- and middle-class vote for Trump, and the strong working-class support for Clinton, the fact that segments of the working class voted for Trump does not make him the man of the US working class. In turn, Frank fails to investigate the racialized dynamics that connect the otherwise contradictory electoral bloc of disproportionately white men and women among Trump voters (see also Bouie 2016; Kelley 2017; Fassin 2017).

There are strong spatial dimensions to these debates about class, race, and the right-wing vote. For Frank (2016a, 2016b, 2017), Trump's America conjures up small-town agony, job loss, and poverty somewhere in the Midwest. Some spatial specialists in effect corroborate this image. Jacques Lévy (2016), who is well-known in France because of his electoral maps that try to show the inverse statistical relationship between the "extremist" vote (right or left) and "urbanity," thinks that the vote for Trump reflects a distance from central cities, and an inward-looking conception of the good life opposed to the supposed tolerance and cosmopolitan openness of metropolitan America. Even if Lévy is clear that the "gradient of urbanity" cuts across class and ethnic lines, his "ecological" explanation of the Trump vote paints an empirically misleading (Bouba-Olga 2016) mirror image of Frank's picture of Trumpist small-town USA: a socially diverse and worldly urbane America.

Lévy's (2016) intervention shows how electoral political geography, whatever its merits, can accentuate and solidify one-sided social interpretations of political trends (for previous arguments, see Kipfer and Saberi 2014, 2016). In France, Lévy's electoral imaginary of urbane spaces that resist the populist vote has a powerful rival: Christophe's Guilluy's (2014, 2016) idea of peripheral France. In a sense, Guilluy territorializes, in the French context, Frank's argument about the right-wing working-class voter. He does so by supplanting an already existing spatial imaginary of the "periphery" – the stigmatized social spaces inhabited to a large extent by non-white working class inhabitants (*banlieues*) – with another one: small and mid-sized towns, villages and select exurban zones between major urban centres, spaces that are presumed to be inhabited by the "native" working class. Pitting one "periphery" against another disarticulates race and class by virtue of a schematic spatial image of social and political life. In turn, and particularly importantly in contexts like France (and the United States), schematic oppositions between globalist, urbane (metropolitan) and national, home-grown (peripheral) social spaces obscure the imperial and neo-colonial relations that cut across both (Hart, forthcoming).

In the following, we interpret Guilluy's (2014, 2016) image of competing peripheries as a political intervention. This image constructs a model of the "periphery" whose critique of the "centre" (globalized metropolitan France) flirts with the self-image of the French extreme-right party, the Front National (FN). We take this image as an invitation to stage a brief return to the spatial turn in social research, as well as our respective work on space, politics, and social relations. The insights gained from the spatial turn urge us to be wary of spatial reification, and they provide an important resource with which to understand the one-sided and politically dangerous territorialization of social relations achieved by Guilluy and related versions of political geography. Such territorialization of social relations, we will argue, neglects the relationality of the social world, and thus risks a spatially determinist rendering of political action confined to a methodological individualism. We will conclude by suggesting that a critique of spatial reification leads us to a conception of politics not as a one-sided expression of demographic variables and territorial complexes but as an active force, one that is certainly shaped by sedimented socio-spatial realities, but one that also disrupts and re-organizes social relations and their territorial dimensions.

Space and Social Relations: A Reminder about the Dangers of Spatial Reification

Between the late 1970s and the 1990s, English-speaking debates raged over the geographical dimensions of history and social relations. Some took these debates to be a manifestation of a wider spatial turn in intellectual life. At its core, the spatial turn was a warning against the perils of reducing space to a given container and to fixity. It was a call to pay attention to processes, tensions, and struggles in the production of space, instead of seeing space as a pre-given entity, either an insignificant backdrop to history or an independent variable determining social life.

In this general sense, the spatial turn continues to help us problematize rigid and essentializing distinctions such as centre and periphery. It also suggests that we pay attention to multiple dynamics involved in the production of space, viewed not simply as a piece of land considered merely in its topographic aspect, but as something whose production and meaning cannot be grasped independently of its discursive articulations and the practices associated with it. We can also say, as we will below in different ways, that politics is integral to the shaping of the spatial relations between dominant and dominated territories.

The implication for the "peripheries" question is that periphery is not something that is already given. Such reification risks turning these territories into undifferentiated and timeless masses to be wary of. While the centre – or, better yet, what is perceived to be the centre – stands for order, normality, and progress, the periphery becomes the site of disorder, deviance, and backwardness – a threat to the centre. Thus, the periphery must be controlled, not only to protect the centre, but also to protect the periphery from itself, from its own chaos, backwardness, and deviancy (Said 1978).

Thinking about peripheries in reified territorialized terms has a governmental dimension. To use Rancière's (1999) terms, such partitioning has a policing effect. "The police" here does not refer to the uniformed forces of order, and the term is not necessarily pejorative. It is a form of symbolization that consolidates orders of space and time, hierarchies of places; and through these, it may lead to institutionalized and legitimized practices of domination. In other words, governing regimes consolidate and work through spatial and temporal orders that become naturalized, which then may become sources of domination and inequality. The political element here is to "resist the givenness of place," to not take reified territorialization or other forms of fixity as naturally given designations with innate attributes (Dikeç 2015).

What follows from Rancière's (1999) work is an emphasis on the processes by which centres and peripheries are established. For example, one may offer the term peripheralization to refer to the construction of an area, a group, and so on, as something as exterior, marginal, and deviant, as the part that does not fit in or indeed threatens the centre. An example here would be French urban policy that was conceived in the early 1980s to address the problems of social housing neighbourhoods located mainly in the peripheral areas of cities, the *banlieues*. By spatially designating these areas as spaces of intervention and associating them with discourses of deviance and unruliness, this policy contributed to the peripheralization of these areas as the "badlands" of the republic – places that somehow do not fit in the republican imaginary and threaten the republican order. The policing effect of this policy has been to establish territorial hierarchies and to consolidate institutionalized practices of domination towards such areas, constituted as undisciplined and unruly. The debate about "new peripheries" in France, as we will see, recasts the focus and content of dominant spatial designations.

For Henri Lefebvre, the language of centre and periphery played an increasingly central role in the period between his urban research and his work on space and the state (Kipfer and Goonewardena 2013).

In this context, centres and peripheries do not refer simply to existing places (where some places are peripheral given their geographical distance from central places – that is, spaces that concentrate human activity). Neither are they mere metaphors for power, the powerful and the powerless. Centre and periphery refer to relationships of domination strongly shaped by the state, that condensation of social life that is, among other things, "in the business" of organizing and establishing territorial hierarchies between various scales of activity while also shaping the boundaries and meanings of these local, regional, and national scales. Urban policy is one such way of organizing territorial hierarchies, as it helps consolidate a certain spatial order and encourages a certain way of thinking about it (through mappings, spatial designations, policy discourses, categorizations, etc.).

As with Rancière, a Lefebvrian entry point to centre and periphery leads us to emphasize the processes and strategies ("peripheralization" and "centralization," say) by which centre-periphery relationships are produced. Formal knowledge, including policy research and other university-based knowledge, participates in these processes and strategies. This insight allows us to understand the way in which visualized statistical knowledge based on some of the most quintessential forms of state knowledge (the census and the map) can nudge researchers who map electoral choices towards solidifying the very territorialized way of seeing that state action itself can produce.

Of course, even if one understands them as historically malleable creations, centre-periphery relations do not exhaust social totality. Insofar as they refer to interdependent, socially produced spatial forms, they mediate the broader social world instead of expressing (let alone determining) that social world. Lefebvre's (2003) dialectical language of mediation allows us to "see" that social relations of class, race, or gender cannot be mapped fully. The spatial dimensions of social relations are too complex and multiple to be translated into neatly equivalent geographical relations or forms. In turn, the struggles that help define space and social relations escape any easy geographical characterization, and politics may defy spatial fixity.

Politics against Fixity

A brief return to the "spatial turn" in social theory and the additional insights from the works of Jacques Rancière and Henri Lefebvre leaves us with important conceptual tools to use against reified ways of shrinking social relations to their spatial or territorial dimensions. As the earlier examples from electoral geographies in the United States and France indicate, such reified conceptions of space and politics borrow from and participate in a wider territorialization of social relations. Most starkly illustrated in French debates about "new peripheries," such reified visions of the social world threaten to disarticulate class and race by attaching each to separate territories. In light of the real threat of neo-fascism, which can only be countered with an alliance of a broad range of popular forces, such a disarticulation would mean nothing short of political suicide.

Our view is that politics cannot be reduced to a series of actions (from voting to revolting) that derive from socially and spatially fixed positions. It is about organizing, disorganizing, or transforming these positions. What follows is that political engagement should certainly accept where people "are at" in the here and now – but as a starting point, not a point of destination. Ultimately, politics is about reordering people's places and positionalities, not keeping things – and people – fixed in place.

As Rancière ([1983] 2004) suggests, accepting people's position as a naturally given starting point implies that they are mental prisoners of their situation. Politics is about resisting such givenness and fixity of place. As he showed in his account of nineteenth-century workers in *The Nights of Labour* (1989), the division between work and rest that the workers had to obey prevented them from engaging in other activities such as writing or discussing. At night, the worker had to sleep and rest to be ready again for work the following morning. Emancipation implied undoing this order, resisting what the given division imposed:

> This dividing line has been the object of my constant study. It was at the heart of *The Nights of Labor*, where the assertion of worker emancipation was first of all the upheaval of this division of temporalities that anticipated the redistribution of social and political shares by making night into the laborer's time of rest – by inscribing him within the cycle of production and reproduction that separated him from the leisure of thought. It was this that was at stake in *The Philosopher and His Poor*, the Platonic allocation transforming the worker's "absence of time" into the worker's very virtue. But this "absence of time" was itself only a symbolic division of times and spaces. What Plato had excluded was the slack time and empty space separating the artisan from his purely productive and reproductive destination: the space/time of meetings in the *agora* or the assembly where

> the power of the "people" is exerted, where the equality of anyone with anyone is affirmed. (Rancière [1983] 2004, 225–6)

According to this approach, politics is about the givens of a situation. It is, in other words, about the established order of things, including established practices of identification and articulation. It implies a disruption in the established order in the name of equality and in order to address violations of equality. This disruption, however, does not necessarily have to be revolutionary or heroic: "Just as denials of equality take mundane forms, so could affirmations of it" (Dikeç 2017a, 52). The emancipatory potential here is to draw attention to such denials and affirmations through disruptions that expose them.

The expository potential of disruption makes it a central theme in this understanding of politics. Think about, for example, how the 2014 Ferguson, Missouri uprising exposed the concerted effort of Ferguson's police department and its municipal court to generate revenue for the city by preying on Black citizens (Dikeç 2017b). There is, however, another reason why disruption is key to politics. Politics thus understood interrupts normalized structures and routinized practices that may be sources of injustices and denials of equality. Such disruptions force us to "stop-and-think" (Arendt 1978, 78), rather than getting caught up in the accepted givens, normalized and repetitive practices, and ordering principles of everyday life.

While Rancière developed his insights in part by breaking with certain, notably Althusserian, approaches to Marxism, his conception of politics resonates with all those revolutionary and radical traditions that insist that politics, while not indeterminate, is crucially a transitive force and an art of transformation. Not only Lefebvre but also Antonio Gramsci and Frantz Fanon, for example, hold that politics is practice that "moves" people and the spatial dimensions of their existence at the same time as it moves the world itself (Kipfer 2018). To put it differently, politics is about transforming the immediate (what both Fanon and Gramsci would call "spontaneous") interests and identities that form the starting point of people's political engagement through that very engagement.

Concretely, this also means that the territorial complexes and imaginaries that may co-define the starting points of struggle ("city," "country," "land," "*banlieue*," "*quartiers*," "ghetto") may also be subject to transformation through that very struggle. While differently situated, Gramsci's conception of the united front (Thomas 2009) and Fanon's (1963) notion of national liberation both assume that political practice can transform – instead of just aggregating – the social foundations of revolutionary alliances (between workers, agricultural labourers, and peasants, or peasants and the lumpenproletariat) and their spatial dimensions ("city" and "countryside," "shantytown"). Meanwhile, Lefebvre's notion of the right to "the city" (political centrality; Lefebvre 1996) implies that "peripheries" (peripheralized subaltern groups) make claims to political power by transforming each other through the very processes of mobilization that make them encounter and confront each other.

A political orientation that is open to reordering place and position alerts us to moments of fluidity that offer possibilities of transforming spatial relations (and the social relations they mediate and reflect) in struggle, or that point to efforts to turn established spatial relations/structures into matters of struggle/contestation. Examples need not be grandiose, they can be grand or small. Through ongoing research dealing with various forms of anti-fascist practice in France, one of this chapter's authors has encountered intellectuals (organizers, that is) whose engagements challenge the territorial and social fixation of race and class as separate entities; among these are members of a national anti-fascist labour union network who confront the revanchist drift among organized workers (also in towns run by the Front National) and activists in Greater Paris who build new political relationships by straddling "city" and "*banlieue*," moving across various social divides and colour lines, and linking distinct political milieus and mobilizations: politicized soccer fan clubs, high school blockades in east-central Paris, youth mobilizations against police violence in working-class suburbs, mass marches against labour-law reform, and anti-racist coalitions led by non-white organizers.

There is also the example of the recurrent urban uprisings of the past decade or so, many of which took place under liberal democratic regimes committed to the principle of equality (Dikeç 2017b). These uprisings exposed how, even under conditions of formal equality, inequality and discrimination are deeply entrenched parts of the everyday urban lives of certain groups. Institutionalized violence (such as police brutality) that has fuelled the anger leading to these uprisings was usually focused on spatially stigmatized areas articulated with markers of deficiency or threat. Although these were not premeditated events calculated to have long-lasting effects in negotiating with the powers that be, some uprisings, including the ones in France, led to the creation of more permanent oppositional forces to give voice to inhabitants, challenge essentializing discourses, and grapple with the constraints imposed by state-bound territorialization (Kipfer 2009, 2011; Hancock 2017). Even when such formations did not take place, however, such spontaneous insurrections were invitations to stop-and-think, as they exposed routinized violence visited upon certain groups living in stigmatized spaces.

We have seen the implications of an understanding of politics that resists fixity. Let us now go back to Guilluy's argument about "peripheral France" to see how spatial reification and fixity paves the way to various forms of essentialism that feed extreme-right sensibilities.

New Peripheries?

You see, there are two kinds of people in this world, my friend. Those with loaded guns, and those who dig. You dig.

Sergio Leone, *The Good, The Bad, and the Ugly* (1966)

In a series of publications encapsulated most succinctly in *La France périphérique: comment on a sacrifié les classes populaires*, Christophe Guilluy (2014, 9, 176) argues that globalization has divided France into two categories of people (winners and losers) and territories (metropolitan and peripheral France). According to Guilluy, peripheral France is a constellation of settlement forms outside the twenty-five largest urban agglomerations, urban areas, and metropoles.[1] In his texts, peripheral France is also a short form for the working classes (*classes populaires*: that is, workers and employees)[2] and a series of "vulnerabilities" (from unemployment and precarious employment to low income) that have a higher-than-average statistical occurrence there.

For Guilluy (2014), there is an unbridgeable, ultimately cultural gap between peripheral and metropolitan France. He acknowledges that metropolitan France is fractured; it houses not only the bourgeoisie, the managerial strata, and cultural elites but also a large proportion of the non-white working class, the one that is often described as living in stigmatized social spaces and housing estates (*banlieues*). Yet metropolitan France coheres insofar as it is an integral part and beneficiary of the economic and cultural dynamics of globalization. While peripheral France is relegated to a state of economic stagnation, socio-spatial immobility, and "cultural insecurity," metropolitan France is "nomadic," socially mobile, oriented to the world at large, and infused with multicultural sensibilities. For Guilluy (2014, 237), France's territorial divide has clear political results: if metropolitan France remains committed to economic and cultural liberalism, peripheral France, which is caught in what he calls, in analogy to the notion of the neighbourhood effect, an insidious territorial effect is hostile to both. It rejects politics (including the left-right cleavage) and abstains or votes for the FN, which does not shape but merely responds to the sensibilities of peripheral France. This image further peripheralizes this peripheral France by stripping it of all political capacity, as all it can do, in this constructed image, is to respond with reactionary impulses.

It is not difficult to poke holes into Guilluy's (2014) interpretation and visual representation of census data and surveys. The supposedly qualitative divide between metropolitan and peripheral France is drawn on the basis of strictly relative statistical thresholds. While glossing over the depths of class-based and racialized exploitation and exclusion in "metropolitan France," it paints a systematically homogenized and historically static picture of social life in "peripheral France," which, far from being only a space of relegation, combines *multiple* subaltern residential and employment zones with *other* spaces, including middle- and upper-class spaces of privilege, exclusion, and mobility (Charmes 2014; Girard 2017a, 2017b; Rivière 2013). Guilluy also wrongly interprets metropolitanization. Not restricted to a few places (defined by the settlement boundaries established by official demographers, as he suggests), metropolitanization reaches all parts of the hexagon. While highly uneven, it cuts across social divides and links up with political action, including the FN vote, in multiple ways (Veltz 2017).

Guilluy's (2014) argument about "peripheral France" is in many ways the opposite of Jacques Lévy's. Lévy (2003, 2007) imagines French political geography as a few large and round egg yolks (FN-less metropolitan centres) surrounded by stringy hard shells (demographically extensive but FN-rich hinterlands, towns, and exurban districts; see also Ravanel, Buléon, and Fourquet 2003, Lévy and Lussault, 2014). Lévy, too, has received criticism for interpreting the French electoral landscape in dualist and homogenizing ways, giving short shrift to the socio-spatial cleavages that riddle landscapes (Ripoll and Rivière 2007; Bacqué, Charmes, and Vermeersch 2016; Girard 2017a, 229–31; Rivière 2013; Batardy et al. 2017). The empirical limitations of Guilluy's and Lévy's research reveal deeper theoretical and methodological problems, however. These relate to methodological individualism and spatial reification. Both are vital for their respective invocations, affirmative or critical, of what they think are central spaces in France: urbane and cosmopolitan France. As we will discuss further below, we hold these invocations to be performative political acts, not just descriptions of social and geographical trends. In other words, these spatial invocations are not innocent representations of the world, but world-making practices with political implications.

Unless self-reflexively triangulated with the help of other sources and methods, quantitative research on electoral behaviour (in political geography and other disciplines) makes methodologically individualist assumptions. Insofar as it takes the unit of analysis in voting or polling data for granted and correlates it with other, decontextualized census data derived from individual responses, it shares with liberalism the assumption that politics is a punctual expression of individual preferences. These, the assumption goes, can be studied sufficiently by correlating them mechanically to other individual characteristics (from income to nationality) without the collective and contradictory relations within which voter preferences are situated (Pilon 2015). According to these methodological assumptions, politics becomes a reflex of demography.

The methodological limitations of polling analysis risk becoming invisible once they are visualized, paradoxically. To translate electoral data onto maps requires drawing boundaries between spatial segments, each of which is demarcated from other, differently coloured segments on the basis of statistical cut-offs that measure the relative spatial concentration of particular qualities. This very way of clustering data, without which visualization is impossible, *necessarily* risks giving a misleadingly uniform image of electoral zones. While these are only proportionally, not absolutely, distinct aggregates of individual votes, maps hide or visually minimize those votes that do not reach the chosen numerical benchmark in a particular zone.

The fixation of electoral maps on the proportional distribution of votes also hides the overall mass of voting data. Maps that suggest that the central city vote for the FN presidential candidate does not matter (because of the party's lower-than-average weight there) miss the fact that that very vote can be decisive because of the absolute numbers of voters involved (Veltz 2017). For example, Marine Le Pen would not have reached the second round of the 2017 presidential election without the almost 500,000 votes she garnered in the fifteen largest cities (Le Monde 2017). In short, even the most refined map must homogenize to be effective – that is, legible. Well-narrated and understood as part of a broader texture, they can be precious tools. But read in isolation, which is usually the case in public debates, electoral maps risk reifying social realities, or as Rivière (2013) put it, hiding "inhabitants underneath the map." Given the aesthetic primacy of the visual in capitalism, which makes maps spontaneously more convincing than texts or words, one can understand the social power of spatial reification.

Mapped or not, electoral data can provide useful first cuts for electoral analyses. But the qualitative meaning of the "FN vote" cannot be adequately grasped through such data analysis alone. It must be complemented with a range of qualitative approaches that manage to relate the contingency of individual voting behaviour to the spatially mediated social relations that give meaning to that behaviour. This is the case even more so now because the very expansion of electoral support for the FN has made this support more diffuse than it was in the 1980s. The Mediterranean and Northeastern bastions of the FN are still there. Yet, despite persistent spatial variation, the FN vote is a truly national phenomenon that cuts across unevenly urbanizing France (Veltz 2017; Gombin 2015). Socially, it is sustained by a largely white but cross-class and cross-geographical constellation of support (Gaxie 2017; Lehingue 2017; Fourquet 2015).

Facilitated by a sequence of right-leaning recompositions of French politics (including a destructuring of the left), working-class support for the FN has been strong since the 1990s. Yet, even though the FN draws crucial electoral support from working-class segments, the FN is not the party of the French working class as a whole. Studies show how the FN appeals to people in a range of positions. These include merchants and entrepreneurs in smaller towns; anti-republican elements of the haute bourgeoisie; French Algerians (*pieds-noirs*), notably in the Mediterranean South; more "established" but weakly unionized working-class segments in exurban districts; some workers and middle classes in the deindustrializing Northeast; public sector workers at the heart of state authority (policemen and firefighters, for instance); and owners of single-family bungalows adjacent to stigmatized housing estates (Sabéran 2014; Igounet and Jarousseau 2016; Girard 2017a, 2017b; Pierru and Vignon 2017; Bouron and Drouard 2017; Beaud and Pialoux 2017; Pudal 2017; Mischi 2017; Négrier 2012). In these largely white milieus, the FN's perennial themes, "immigration" and "security" (racialized identity and authority), vary in their precise class-specific, racialized, and gendered meanings; they make it easy for people to interpret their various socio-spatial situations as being threatened simultaneously by "those above" and by "those below."

Lévy and Guilluy eschew a relational conception of the social world to the benefit of a spatially determinist, even "ecological," reading of political action based on methodological individualism. Reifying the social spatially, they abstract from the contradictory situations that define the constellation of social spaces upon which the FN draws, as well as the authoritarian and racist ideologies, policies, and practices the FN has both defined and appropriated. This spatial ordering resonates with Jacques Rancière's (2003, 7) notion of "partition of the sensible" or "distribution of sensible evidences," which refers to "the way in which the abstract and arbitrary forms of symbolization of hierarchy are embodied as perceptive givens." Sensible evidences (such as mapping, for example) are available to the senses, and they make sense. Thus, by invoking a sense of obviousness, they not only consolidate, as in this case, a spatial order, but they also encourage a certain way of thinking about it, while discouraging others.

Politics, Spatial Imaginaries, and the Far Right

Abstracting from the relations and situations that shape politics is not just an academic affair. The weight of Guilluy's and Lévy's arguments exceeds their empirical rigour and coherence. While they both have an important public profile in the "centre" of French public life, so to speak, they are both silent about how they themselves contribute to the decisions that define the characteristics and boundaries of centre and periphery. Guilluy's and Lévy's effective contribution to struggles for hegemonic spatial imaginaries (those that are beyond questioning) makes their spatial essentialism politically potent.

Lévy's (2003, 2007, 2017) contrast between worldly urban centres and inward-looking landscapes and Guilluy's dualism of "metropolitan" and "peripheral" France ignore the imperial geographies of which France is a part, and relegate to secondary importance the distinction between the "Republic" and the "*banlieue*" that has framed so much political antagonism in the hexagon since the 1980s. They thus redirect the territorialization of social relations that also became entrenched in state repertoires (place-based urban policy focusing largely on *banlieues*) since that time. However, they do so in ways that resonate with differing political perspectives. Lévy's conventionally urbanist conception of centrality is physically and psychosocially distant from inward-looking, propertied, and family-centred social spaces that reject alterity. This imaginary dovetails with the social-liberal sensibilities with which a wing of the

French bourgeoisie and intelligentsia defends itself against what they call "populism," – that is, the neo-fascist and Sarkozyst right and the far left. These sensibilities are well captured by the dailies *Le Monde* and *Libération*, as well as Socialist and centrist (now "Macroniste") politicians.

At first sight, Guilluy's work is attractive because it develops an explicit critique of such social-liberal sensibilities and the ruling-class circles and social spaces that help produce them. At the time, his *France périphérique* joined others (Baumel and Kalfon 2011) who criticized the Socialist Party's 2012 electoral strategy of abandoning, on the grounds of divergent "values," workers (*ouvriers*) for the benefit of an ad hoc electoral coalition of urbanites, youth, women, minorities, and the educated (Ferrand, Prudent, and Jeanbart 2012; Bréville and Rimbert 2015). Guilluy's more recent *La crépuscule de la France d'en haut* (2016) takes to task the liberal bourgeoisie and its petty-bourgeois allies (gentrifiers included) for their hypocritical multiculturalism. He argues that the latter is lived out in self-segregated spaces that benefit from the labour of the non-white working class inhabitants, while staying socially distant from these same inhabitants and exhibiting indifference or outright disdain for the non-metropolitan working classes.

Guilluy's (2016) critique is only seemingly a class analysis of "top-down France," however. Going well beyond Frank's (2016a) critique of American Liberalism, Guilluy's critique not only inverts the Socialist Party's 2012 culturalist "liberation" of the left from class (Amable and Palombarini 2017). It also resonates with essentialist themes dear both to right-wing Socialists like Manuel Valls and leaders of the Front National like Bruno Gollnisch (Bréville and Rimbert 2015). The true cleavage in Guilluy's analysis is ethno-cultural. It runs between globalizers and multiculturals (including migrants), on the one hand, and immobilized "villagers" (whose supposed marginality and "cultural insecurity" inevitably pushes them to reject others; Guilluy 2014, 93–4; 2016b, 128, 190, 219–22; see also Bouvet 2015). Here, Guilluy (2014) approximates frontist ideology. He does so by uniting three elements against the reviled globalist centre: an anti-metropolitan imaginary (towns, villages, and exurbs), an ethnically cleansed working class (separated from their non-white fractions on cultural grounds), and an ethno-differentialist and naturalized conception of xenophobia (presumed to be expressive of a universal need to be amongst one's own).

Guilluy's (2014) critique of "metropolitan France" and his portrayal of the FN as the party of the "peripheral" French working class pits class against race while essentializing both (Gintrac and Mekdjian 2014;

Girard 2017b). Naturalizing the FN's basic strategy of translating a variety of social situations into racist resentment and violent authoritarianism, this critique runs counter to emancipatory imperatives. There is in fact a growing recognition in overlapping anti-racist, autonomist, and socialist circles of the French radical left that a combined challenge against the *social war* promised by neoliberalism (President Macron's party, La République en marche) and the *civil war* conjured up by the Front National is impossible without connecting non-white inhabitants to those segments of the white working class that are not captured fully by the frontist project.

Guilluy (2016, 171–7) goes so far as to say that anti-fascism is a weapon in the class struggle waged against peripheral France. He thinks that anti-fascism expresses a hatred of working-class sensibilities, of which the FN is a mere symptom. Indeed, Guilluy is one of several well-known French intellectuals (including Jean-Claude Michéa and Michel Onfray) who have effectively built red-brown bridges to the FN and its real if deceptive attempt to present its neo-fascist project as a (national) populist critique of (global) liberalism. Key to such red-brown bridging efforts is the determination to separate a critique of the bourgeoisie, liberalism, and capitalism from a critique of nationalism, racism, and patriarchy in order to divide the world into ethno-culturally essentialized cross-class camps: globalizers and nationals, metropolitans and provincials. As Gill Hart (forthcoming) points out, such spatial dichotomies also ignore how "globalism" and "nationalism" are intertwined in world order, for example through the imperial projects and memories that are advanced and mobilized by the far right itself.

Our main point is to say that that spatially reified conceptions of centre and periphery can underwrite the construction of such collaborationist camps. Examples abound. During the 2017 presidential campaign, Marine Le Pen's campaign worked hard to supplant existing social and political divides (including the left-right polarity, that old fascist bugbear) with cleavages between exactly such camps: the *patriotes* (the FN) and the *mondialistes* (Emmanuel Macron's liberal En Marche). During that same campaign, Guilluy (2017) and Lévy (2007) provided arguments and maps that seemed to corroborate the schisms conjured up by the top two candidates. In so doing, they sidestepped decisive evidence that pointed out, once more, that the votes for Macron and Le Pen defied a spatial metropole-periphery model: each articulated a distinct constellation of multiple socio-spatial realities situated within the uneven dynamics of urbanizing France (Gilli 2017; Le Bras 2017; Giblin 2017; Gilli, Jeanbart, Pech, and Veltz 2017).

As the latest round of public debate about French electoral geography indicates, the struggle over spatial imaginary is undoubtedly at the heart of politics. It is, therefore, important to expose the methodological assumptions of such spatial imaginaries, which, as we argued, are not objective representations of immutable social worlds, but situated world-making practices with political implications. To challenge reified spatial imaginaries does, in turn, require perspectives that understand social spaces as relational, contradictory, politically co-constituted, and, thus, subject in principle to political contestation and transformation. In our view, perspectives that highlight the role of politics in the production of space belong in any anti-fascist toolkit.

ACKNOWLEDGMENTS

We would like to thank the editors and reviewers, Anne Clerval, Gillian Hart, Arnaud Huc, Parastou Saberi, and Karen Wirsig, for their helpful comments on earlier versions of this chapter.

NOTES

1 Although Guilluy (2014) does not technically exclude the offshore territories of France from his term "periphery," these territories matter even less to his analysis than the *banlieues* in France proper. In his work, the distinction between metropole and periphery thus refers not to the difference between mainland France and its still-existing colonies but to different social spaces in the hexagon itself.

2 Guilluy (2014) refers to the technical terms used by the French statistical service: *ouvriers* (industrial, artisanal, warehouse, transport, and agricultural workers) and *employés* (workers in services, administration, and the public sector, including police and military personnel).

REFERENCES

Amable, B., and Palmobarini, S. (2017). *L'illusion du bloc bourgeois: Alliances sociales et avenir du modèle français*. Paris, France: Raison d'agir.

Arendt, H. (1978). *The Life of the Mind*. Volume 1: *Thinking*. New York, NY: Harcourt Brace Jovanovich.

Bacqué, M.-H., Charmes, E., and Vermeersch S. (2016). "Des territoires entre ascension et déclin: Trajectoires dans la mobilité périurbaine." *Revue française de sociologie* 57, no. 4, 681–710. https://doi.org/10.3917/rfs.574.0681

Batardy, C., Bellanger, E., Gilbert, P., and Rivière J. (2017) "Présidentielle 2017: Les votes des grandes villes au microscope." *Métropolitiques.eu*, 9 May. https://www.metropolitiques.eu/Presidentielle-2017-Les-votes-des-grandes-villes-au-microscope.html

Baumel, L., and Kalfon, F. (eds). (2011). *Plaidoyer pour une gauche populaire: La gauche face à ses électeurs*. Lormont, France: Le bord de l'eau.

Beaud, S., and Pialoux, M. (2017). "Les ouvriers et le FN. L'exacerbation des luttes de concurrence." In Mauger, E., and Pelletier, W. (eds), *Les classes populaires et le FN: Explications de vote*, 133–48. Paris, France: Editions du Croquant.

Bouba-Olga, O. (2016). "Non, le vote des Américains n'oppose pas le centre et la péripherie." *Le Monde*, 19 November. https://www.lemonde.fr/idees/article/2016/11/19/olivier-bouba-olga-non-le-vote-des-americains-n-oppose-pas-le-centre-et-la-peripherie_5033995_3232.html

Bouie, J. (2016). "How Trump Happened." *Slate*, 13 March. http://www.slate.com/articles/news_and_politics/cover_story/2016/03/how_donald_trump_happened_racism_against_barack_obama.html

Bouron, S., and Drouard, M. (2017). "Voter Front National au château." In Mauger, E., and Pelletier, W. (eds), *Les classes populaires et le FN: Explications de vote*, 227–39. Paris, France: Editions du Croquant.

Bouvet, L. (2015). *L'Insécurité culturelle*. Paris, France: Fayard.

Bréville, B., and Rimbert, P. (2015) "Une gauche assise à la droite du peuple." *Le Monde Diplomatique*, 8–9 March. https://www.monde-diplomatique.fr/2015/03/BREVILLE/52741

Charmes, E. (2014). "Une France contre l'autre?" *La vie des idées*, 5 November. https://laviedesidees.fr/Une-France-contre-l-autre.html

Dikeç, M. (2015). *Space, Politics, and Aesthetics*. Edinburgh, UK: Edinburgh University Press.

– (2017a) "Disruptive Politics." *Urban Studies* 54, no. 1, 49–54. https://doi.org/10.1177/0042098016671476

– (2017b) *Urban Rage: The Revolt of the Excluded*. New Haven, CT: Yale University Press.

Fanon, F. (1963). *The Wretched of the Earth*. New York, NY: Grove Press.

Fassin, E. (2017). *Populisme, le grand ressentiment*. Paris, France: Textuel.

Ferrand, O., Prudent, R., and Jeanbart, B. (2012). "Gauche, quelle majorité électorale pour 2012?" *Terra Nova: Le Think Tank Progressiste*, 10 May. http://tnova.fr/rapports/gauche-quelle-majorite-electorale-pour-2012

Fourquet, J. (2015). *Karim vote à gauche et son voisin vote FN*. Paris, France: Éditions de l'Aube.
Frank, T. (2016a). "Donald Trump Is Moving to the White House, and Liberals Put Him There." *The Guardian*, 9 November. https://www.theguardian.com/commentisfree/2016/nov/09/donald-trump-white-house-hillary-clinton-liberals
– (2016b). "Millions of Ordinary Americans Support Donald Trump. Here's Why." *The Guardian*, 8 March. https://www.theguardian.com/commentisfree/2016/mar/07/donald-trump-why-americans-support
– (2017). "The Intolerance of the Left: Trump's Win as Seen from Walt Disney's Hometown." *The Guardian*, 27 January. https://www.theguardian.com/us-news/2017/jan/27/why-donald-trump-win-walt-disney
Gaxie, D. (2017). "Front National: Les contradictions d'une résistible ascension." In Mauger, E., and Pelletier, W. (eds), *Les classes populaires et le FN: Explications de vote*, 43–66. Paris, France: Éditions du Croquant.
Giblin, B. (2017). "Ce que disent les cartes de la 'fracture.'" *Le Monde*, 27 April, 20.
Gilli, F. (2017). "Non, il n'y a pas deux France qui s'opposent." *Le Monde*, 28 April, 25.
Gilli, F., Jeanbart, J., Pech T., and Veltz, P. (2017). *Elections 2017: Pourquoi l'opposition métropole-périphérie n'est pas la clé*. Paris, France: Terra Nova.
Gintrac, C., and Mekdjian, S. (2014). "Le peuple et la 'France périphérique': La géographie au service d'une version culturaliste et essentialisée des classes populaires." *Espaces et Sociétés* 1, no. 156–7, 233–9. doi:10.3917/esp.156.0233
Girard, V. (2017a). *Le vote FN au village: Trajectoires de ménages populaires du périurbain*. Paris, France: Éditions du Croquant.
– (2017b). "Classes populaires, habitat périurbain, et votes FN: Retour sur quelques fausses évidences." In Mauger, E., and Pelletier, W. (eds), *Les classes populaires et le FN: Explications de vote*, 101–13. Paris, France: Éditions du Croquant.
Gombin, J. (2015) "Le changement dans la continuité: Geographies électorales du Front National depuis 1992." In Crépon, S., Dézé, A., and Mayer, N. (eds), *Les faux-semblants du Front National: Sociologie d'un parti politique*, 395–416. Paris, France: Les Presses Sciences Po.
Guilluy, C. (2014). *La France périphérique: Comment on a sacrifié les classes populaires*. Paris, France: Champs actuel.
– (2016). *La crépuscule de la France d'en haut*. Paris, France: Flammarion.
– (2017) "Macron, le candidat de la France d'en haut." *Le Monde*, 27 April, 20.
Hancock, C. (2017). "Feminism from the Margin: Challenging the Paris/*Banlieues* Divide." *Antipode* 49, no. 3, 636–56. https://doi.org/10.1111/anti.12303

Hart, G. (forthcoming) "Why Did it Take so Long? Trump-Bannonism through Southern Lenses." *Historical Materialism.*

Hochschild, A.R., and Frank, T. (2016). "Cette Amérique profonde qui s'estime oubliée: Entretien croisée." *Le Monde,* 2 November. https://www.lemonde.fr/elections-americaines/article/2016/11/02/cette-amerique-profonde-qui-s-estime-oubliee_5023875_829254.html

Igounet, V., and Jarousseau, V. (2016). *L'illusion nationale: Deux ans d'enquêtes dans les villes FN.* Paris, France: Les Arènes.

Kelley, R.D.G. (2016). "After Trump." *Boston Review,* 15 November. http://bostonreview.net/forum/after-trump/robin-d-g-kelley-trump-says-go-back-we-say-fight-back

Kipfer, S. (2009). "Tackling Urban Apartheid: The Social Forum of Popular Neighbourhoods in Paris." *International Journal for Urban and Regional Research* 33, no. 4, 1058–66. https://doi.org/10.1111/j.1468-2427.2009.00929.x

– (2011). "Decolonization in the Heart of Empire: Some Fanonian Echoes in France Today." *Antipode* 43, no. 4, 1155–80. https://doi.org/10.1111/j.1467-8330.2011.00851.x

– (2018). "Quel Gramsci décolonial? Plaidoyer pour une piste Fanon-Gramsci." In Caloz-Tschopp, M.-C., Felli, R., and Chollet, A. (eds), *Rosa Luxemburg, Antonio Gramsci actuels,* 359–74. Paris, France: Kimé.

Kipfer, S., and Goonwardena, G. (2013). "Urban Marxism and the Post-colonial Question: Henri Lefebvre and 'Colonization.'" *Historical Materialism* 21, no. 2., 76–117. https://doi.org/10.1163/1569206X-12341297

Kipfer, S., and Saberi, P. (2014). "From 'Revolution' to Farce? Hard-Right Populism in the Making of Toronto." *Studies in Political Economy* 93, no. 1, 127–52. https://doi.org/10.1080/19187033.2014.11674967

– (2016). "Times and Spaces of Right Populism: Notes from Paris and Toronto." *Socialist Register* 52, 312–32.

Le Bras, H. (2017). "Le malaise social n'est pas la seule cause du vote Le Pen." *Le Monde,* 26 April, 18. https://www.lemonde.fr/idees/article/2017/04/26/herve-le-bras-le-malaise-social-n-est-pas-la-seule-cause-du-vote-le-pen_5117919_3232.html

Lefebvre, H. (2003). *The Urban Revolution.* Minneapolis: University of Minnesota Press.

– (1996). *Writings on Cities.* Cambridge, MA: Blackwell.

Lehingue, P. (2017). "L'électorat du Front National: Retour sur deux ou trois 'idées reçues.'" In Mauger, E., and Pelletier, W. (eds), *Les classes populaires et le FN: Explications de vote,* 19–42. Paris, France: Éditions du Croquant.

Le Monde. (2017). "Résultats, présidentielle – second tour." *Le Monde,* 9 May, 1–25.

Lévy, J. (2003). "Vote et gradient d'urbanité." *EspacesTemps.net*, 5 June. https://www.espacestemps.net/articles/vote-et-gradient-urbanite/
– (2007). "Regarder, voir: Un discours informé par la cartographie." *Annales de la Recherche Urbaine* 42, no. 1, 17–20.
– (2016). "Les riches ont voté Trump, les villes Clinton." *Le Monde*, 17 November. https://www.lemonde.fr/idees/article/2016/11/16/les-riches-ont-vote-trump-les-villes-clinton_5031798_3232.html
– (2017). "Deux France face à face: 'Maastricht a amorcé le recul du clivage droite-gauche.'" *Le Monde*, 27 April, 18–19.
Lévy, J., and Lussault, M. (2014). "Périphérisation de l'urbain." *EspacesTemps.net*, 15 July. https://www.espacestemps.net/articles/peripherisation-de-lurbain/
Lilla, M. (2016). "The End of Identity Liberalism." *New York Times*, 18 November. https://www.nytimes.com/2016/11/20/opinion/sunday/the-end-of-identity-liberalism.html
Mischi, J. (2017). "Essor du FN et décomposition de la gauche en milieu populaire." In Mauger, E., and Pelletier, W. (eds), *Les classes populaires et le FN: Explications de vote*, 117–32. Paris, France: Éditions du Croquant.
Négrier, E. (2012). "Le Pen et le peuple: Géopolitiques du vote FN en Languedoc-Roussillon." *Pole Sud* 37, no. 2, 153–66. https://www.cairn.info/revue-pole-sud-2012-2-page-153.htm
Pierru, E., and Vignon, S. (2017). "Comprendre les frontistes dans les mondes ruraux." In Mauger, E., and Pelletier, W. (eds). *Les classes populaires et le FN: Explications de vote*, 77–100. Paris, France: Éditions du Croquant.
Pilon, D. (2015). "Researching Voter Turnout and the Electoral Subaltern: Utilizing 'Class' as Identity." *Studies in Political Economy* 96, no. 1, 69–92. https://doi.org/10.1080/19187033.2015.11674938
Pudal, R. (2017). "L'attrait du FN sur les sapeurs-pompiers." In Mauger, E., and Pelletier, W. (eds), *Les classes populaires et le FN: explications de vote*, 183–94. Paris, France: Éditions du Croquant.
Rancière, J. (1989) *The Nights of Labour: The Workers' Dream in Nineteenth-Century France*. Philadelphia, PA: Temple University Press.
– (1999) *Disagreement: Politics and Philosophy*. Minneapolis: University of Minnesota Press.
– (2003). "The Thinking of Dissensus: Politics and Aesthetics." Response paper presented at the Fidelity to the Disagreement: Jacques Rancière and the Political Conference, Goldsmiths College, London, UK, 16–17 September.

– (1983) 2004. *The Philosopher and His Poor*. Durham, NC: Duke University Press.

Ravanell, L., Buléon, P., and Fourquet, J. (2003). "Vote et gradient d'urbanité: Les nouveaux territoires des élections présidentielles de 2002." *Espaces, Populations, Sociétés* 3, 469–82. https://doi.org/10.3406/espos.2003.2099

Ripoll, F., and Rivière, J. (2007). "La ville dense comme seul espace légitime?" *Annales de la Recherche Urbaine* 102, 120–30. https://doi.org/10.3406/aru.2007.2701

Rivière, J. (2013). "Sous les cartes, les habitants: La diversité du vote des périurbains en 2012." *Esprit* 3, 34–44. doi:10.3917/espri.1303.0034

Sabéran, H. (2014). *Bienvenue à Hénin-Beaumont: Reportage sur un laboratoire du Front National*. Paris, France: La Découverte.

Said, E. (1978). *Orientalism*. London, UK: Routledge and Kegan Paul.

Thomas, P. (2009). *The Gramscian Moment*. Leiden, The Netherlands: Brill.

Veltz, P. (2017). "Fractures sociales, fractures territoriales?" *Metis: Correspondances européennes du travail*, 20 February. http://www.metiseurope.eu/fractures-sociales-fractures-territoriales_fr_70_art_30504.html

2 Public Housing, Heroin Addiction, and America's Industrial Suburbs: A Planetary Urbanist Perspective

DAVID WILSON, BASMATTEE BOODRAM, AND JASMINE SMITH

Introduction

What is going on in the most stigmatized zones of America's rapidly changing industrial suburbs, its inglorious public housing? As negligence has shaded into vengefulness in this housing's provision, and new economic and political processes afflict these fabricated islands of isolation, social problems have recently ramped up, particularly a nationally publicized scourge of heroin addiction that is the focus of this chapter (Cicero, Ellis, Surratt, and Kurtz 2014; Kuehn 2014). Suburban projects in Chicago, our empirical focus, now experience an epidemic of heroin addiction that leaves in its wake thousands of near-dead, disproportionally Black people (Abc.chicago, 2016). The new rag-tag health care frontline – schools, churches, mosques, needle exchange sites, social service providers – is perplexed about what to do. Piecemeal and shadowy responses, glaringly ineffectual, reflect a combination of scant state resources, ad hoc ameliorative strategies, and shifting, unclear interventionist modes (Strang et al. 2012).

This chapter reinterprets this proliferating drug addiction, centring something new: the sensibility of planetary (sub)urbanization. Planetary urbanization, in brief, was originally referenced by French philosopher Henri Lefebvre, who provocatively noted that there is now no longer anything in the world outside the urban. In *The Urban Revolution* (2003), Lefebvre anticipated the rise of a planetary "fabric" from an ever-restless, expanding capitalist urbanization that would profoundly and intimately touch vast spaces. Whether Lefebvre was practising messy speculation or concrete projection is open to debate. But this futuristic vision is now being taken up by urbanists as one potentially innovative way to frame understandings of the world and its crises.

We interpret the planetary urbanization notion as an ethos that recognizes that an ever-changing urban (or its fragments) pervasively moves into places and their many spheres across the globe with often profound effects on people and places. Neither a perspective, a conceptual framework, nor a theory of everything, this sensibility nuances the dissecting of capitalist urbanization in all its variations and varieties. The urban, informed by this sensibility, becomes less a kind of place than a set of wide ranging processes found in cities and beyond that are provided form by connections, relations, extensions, and tentacles of interdependence. City events, following Brenner (2017), become merely the starting-point for understanding the urban. Amidst recent criticisms of this vision (see Walker 2015; Roy 2016; Ruddick et al. 2017) that challenge it especially as being overly totalizing, economically mechanistic, and epistemologically reductionist, we believe that there are important ideas here worthy of refining and deploying. In this vein, we re-work the urban concept by providing it on-the-ground areal and process specificity in current times. We address two issues: what precisely urbanization and the forces that power it today are, and the role that on-the-ground institutions and people play in constituting and mediating messy, current urbanization.

Why apply a refined planetary urbanist notion to the issue of increased heroin addiction in these suburban housing projects? Because policy and media explanations for this addiction are overwhelmingly simple and reductionist. These voices notably borrow from the resilient detritus of poverty subculture theory to rehearse well-worn, barren notions of destructive, encroaching city cultures. Invoking an atomistic ontology and a simple city-suburb distinction, Brenner's (2014, 15) "methodological city-ism," a destructive, ensnaring "social city" moves seamlessly into suburbs especially via outflight of the poor from cities (Hall 2015, discussion with author; Reitz 2015, discussion with author). Suburban youth, engaging the city creep, are before long infected and seamlessly slide into turbulent peer groups, problematic moral and social principles, and new tumultuous lifestyles. This seamless "city-ization" – as youth negotiate the life-cycle stage of rebelliousness – makes them ripe for falling into the core trap: the abyss of heroin abuse (see O'Leary 2016). City culture, an established social construct in the common imaginary, once more takes a beating.

Our empirical focus is eight public housing projects in Aurora and Cicero, Illinois (figure 2.1).[1] These projects, like much of their suburban

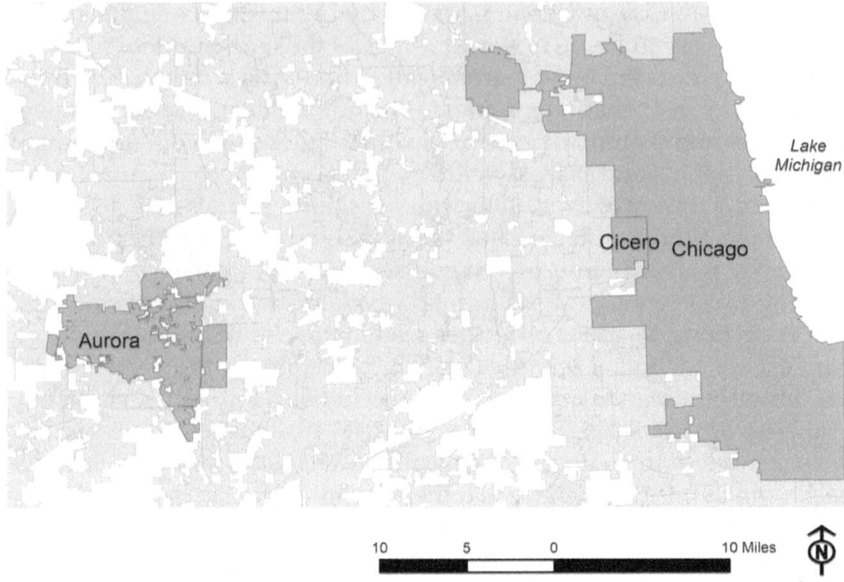

2.1 Condition of housing estates in Aurora and Cicero in Chicagoland

counterparts in the 1950s and 1960s, have a rancid history (table 2.1). The story is distinctly American capitalist: As suburbs like these experienced explosive growth at the time, they quickly became bourgeois escape zones for the relatively wealthy (Walker 1981). Amid escalating land values, governances soon desired to find storehouses for "ocular-toxic" black, brown, and white poor bodies that found themselves here: public housing fit the bill (Anderson 2010). Before long, neglected and disparaged housing projects appeared in the shadows of teeming middle-class suburban subdivisions in a process that continues today. Even as gritty, industrial suburbs across America like Aurora and Cicero experienced a stepped-up deindustrialization and decline in recent decades, these projects persisted as race-class storehouses in these areas. That is where, today, the eight public housing projects that we study in this chapter find themselves.

Two brief caveats are in order before we proceed further. First, we forge links between current urbanization and increased heroin addiction, zeroing in on the North American reality. We see current urbanization in North America as distinctive – that is, as "structurated grids of influences" – whose places share key properties and processes (Brenner 2017).

2.1 Public housing projects in Aurora and Cicero

Name	Housing Quality Designation
Association Homes	Fair
Constitution House	Fair
Fox Shore Apartments	Poor
Sage Crest Homes	Poor
Fox Point Apartments	Fair
Coulter Court Residences	Fair
North Island Apartments	Fair
Marywood Homes	Poor

Compiled by authors. Housing quality designation from windshield survey.

We thus recognize a conjuncturally crafted, immensely rich constellation of urbanizations across the globe: North America is the site for one such urbanization. Second, our approach to urbanization focuses on the human usage of heroin and its precipitating conditions rather than the globalized raising, harvesting, and distribution of heroin. This focus does important work for us: it privileges the political-economic realities of vulnerable populations being situated within at-risk social worlds over what we see as something less important, the contextual dimensions of the issue. Our story, then, should not be seen as capturing the complete circuitry of this drug-planetary urbanist relationship, but rather as interrogating the core of this relation.

Planetary Urbanization: Its Recent Codifying

A handful of urbanists have engaged in theorizing the planetary urbanist sensibility recognize a core package of tenets. They posit a temporally specific, powerfully concentrating, and far-reaching urbanization that infiltrates the hermeneutic, the material, the physical, the social, and the political of places (Schmid 2017; Robinson 2017; Wilson and Jonas, 2018). Current urbanization, insinuating itself across the globe in life's everydayness, shapes many things, including the day's thoughts, desires, meanings, structuring conditions for human materiality, opportunity structures for political mobilization, everyday actions, and physical morphologies. It is Abdul Maliq Simone's (2017) "phantom of changed everywhereness," which percolates into the vastness of social life and landscape form as fragments and traces that re-make the established. This urbanization, in its sly insinuating, ultimately muddies the world and our taxonomies of it by moving across the boundaries of our traditional categories: the local, the regional, the global, the social, the political, the economic, the suburban, and the urban (Keil 2016, discussion with author).

Andy Merrifield, Neil Brenner, and Christian Schmid have to date provided the most incisive concreteness to the notion of planetary urbanization. Merrifield (2014, 2016) gives us an immensely variegated urbanization that retains both capitalism's relentless expansion and its dialectical problematics. Merrifield highlights an unsteady, performative, and expanding urbanization that now both changes the earth's character (fabricating constellations of more hybridized, capital-plundering but unstable places) and emboldens and sharpens people's hermeneutic and political gazes (people deciding to engage and struggle against a stepped-up everyday commodification). Urbanization, on a place-by-place basis, has the contradictory effect of wreaking new destructiveness but also awakening people to its perversities. People are transformed: it "create[s] a changed ontological reality inside us ... something immanent rather than extrinsic ... another way of seeing" (Merrifield 2014, 6). A rapidly transforming socio-spatial world of places, each distinctively navigating tension-ridden capitalist forces, assumes new features, textures, and meanings that reverberate to the very heart of how people think, feel, and act.

Neil Brenner (2014, 2017) and Christian Schmid (2014a, 2014b) similarly posit planetary urbanization as emergent sets of flows, fluxes, and processes in a relentlessly expanding capitalist world. Especially interested in the mechanics of this urbanization, they borrow from Lefebvre to anchor and build on his notion of the ever-changing links between capitalist forms of agglomeration ("implosions") and broader transformations of territory and landscape ("explosions") set within evolving, multi-scalar uneven development. Infusing this notion with variability and non-determinacy, their identification of urbanization as "structurated grids of influence" (Brenner, 2017) refuses density and agglomeration as the sole traits of urbanization while positioning this as ever-mediated, geographically conditioned, and structuring the conditions for change (rather than changing things directly). Implosions and explosions near and far from sites of targeted restructuring spearhead erupting urbanization that forges the conditions for the production of new, unpredictable geographies.

Our project follows from this work to provide a focused planetary urbanist sensibility in the North American context that draws upon this notion's rich possibilities. We show how the central activating attribute of current urbanization – an improvisational re-entrepreneurialization of places – sets in motion kinds of change to sharpen and deepen current inequalities. Why the positing of the entrepreneurialization of space as a core aspect of urbanization? Because we believe it drives a complex urbanization that currently unfolds in North America and in our two study areas. Entrepreneurialism in North America today, David Harvey (2012) notes, is the current lynchpin of urbanization, the temporally refined and powerfully delivered process that infuses places with a new round of socio-physical change, meanings, and understandings (under the auspices of providing "growth"). Yet we suggest that in current neoliberal times, this is not the blunt, brutal, monolithic force that neoliberal governance analysts often portray it as. Entrepreneurialism is seen to generate the structurated conditions – as something insinuative – within which institutional actions give rise to a rich amalgam of entrepreneurial-infused changes (in the material and discursive of places).

A Focused Planetary (Sub)Urbanization: A Sketch

Our "focus project" frames current planetary urbanization in North America in four important ways. First, it is seen to involve a central catalyst, governances, which strive to re-make places especially by re-entrepreneurializing them in current neoliberal days. Our focus, this analytic object and its driving of this process, links to national and international economies as it improvisationally strives to "business-up" places via transforming civic life on the ground. To make this work, governances are responsive (reading and adjusting strategies and discourses) to changing political currents and social trends. Schools, work places, social relations, social spaces, and community and city physical forms, in idiographic settings, are tinged by the new competitive ethos that grips these places; no one or thing seems immune from it. Taking variegated forms, actions from Montreal to Chicago are tainted by this unmistakable drive.

Second, this urbanization is seen to cultivate two fundamental things in places: a hyper-"businessing-up" of places and the conditions for this future unfolding. Current urbanization, distinguished from "development" and "growth," reverberates through places as institutions seek to create both current place re-entrepreneurializing and the discursive ground for its continuation into the distant future. Soja (2014), studying Los Angeles, hints at this current "double-nature" of urbanization in a nuanced rendition of this city's ongoing transformation. Changing LA, to Soja, experiences economic change as much in the long-term (through attempts to naturalize "entrepreneurial traces" in the discursive) as the short-term (through attempts to restructure city morphology).

To Soja and us, governance actions are amazingly intricate. Current urbanization, it follows, is as much about seeding an entrepreneurial telos in the recesses of common thought as it is about facilitating immediate re-entrepreneurialization.

Third, this urbanization is seen as both remarkably expansive and as deeply reaching into places. On these two fronts, urbanization bores deeply into as well as expands across the entirety of a place's spheres (the economic, the political, the social, and the socio-spatial). Place re-entrepreneurialization, in this sense, moves beyond the narrow domain of enhancing economic practices and supporting physical morphologies, as governances seek to manufacture and manage the social and political dimensions that give legitimacy, flavour, and framing to common economic belief. Inducing change is seen as achievable, and this involves constituting enabling social and political arenas for its current and future unfolding. In this setting, urbanization unleashes projects that plumb the core "habitus" of places. Everyday social, political, and social life become its realms of attention in Swyngedouw's (2012) "era of deeply penetrating everyday restructuring in global times."

Fourth, it follows that we see urbanization as something different from how many define development and growth. Whereas development and growth, in all their current forms (e.g., "smart growth," "urban sustainability," "urban resilience"), fundamentally seek to induce physical and social change in places, our urbanization involves a deeper reach into transforming places and has a different objective. Development and redevelopment seek to change the present, and strive to primarily transform the socio-physical form of places – the content and arrangement of neighbourhoods, districts, physical infrastructure, and patterns of land use (see Stone and Sanders 1987; Wheeler and Beatley 2014). Alternatively, current urbanization seeks to change the present and future, and reaches out to manage and control the additional domains of the hermeneutic, the imagined, and the habitus in the everyday. Urbanization today, we suggest, is as much about building the social base to establish "the end of place history" (a timeless normalizing of maximum entrepreneurialism in places) as driving short-term socio-physical restructuring.

Moving Urbanization Along

But how do North American governances propel this urbanization today? We suggest the importance of two powerful forces that governances work through and draw on: (1) a double-barrelled globalization, and (2) neoliberal transnational dynamics.[2] These two forces, taking many faces on the ground as they anchor current urbanization, are less originated and invented by governances than they are emergences seized, re-worked, and mobilized (using place-specific fears, and place-based stocks of understandings and symbolic ensembles) to advance place re-making. As appropriated "opportunity structures," these forces, extending Lefebvre, are negotiated by governances and exert influence through rounds of institutional practices and actions wrapped in regimes of truth and formal planning. As Lefebvre would have recognized, this urbanization is immensely complex. Crucial to this chapter, we now briefly amplify this re-entrepreneurialization's two dominant forces (the new doubled-barrelled globalization and transnational dynamics) and their current repercussions for places.

(1) The New Double-Barrelled Globalization and Its Outcomes

Globalization, as a double influence today, is, first, something many are familiar with: governance responses to the reality of a damaging place transformation spun out by an internationally extending capital. Documented by urbanists, governances now feel they must cajole and attract capital, which can now freely operate across vast continents and use and abuse places. In this context, place-bound governances fervently strive to re-make the political climate, social milieu, and morphology of places. Massive economic giveaways (e.g., governances doling out tax abatements, or land write-downs), reforms of investment conditions and possibilities, and creations of innovative public-private partnerships become crucial parts of current urbanization. Such strivings for economic retention fundamentally shape the form and fabric of cities.

But second, current globalization now takes on a new dimension: a political trope as an elaborate, place-based invoking of a powerful, frightening global reality (taking advantage of "new fear days," especially due to current economic insecurities from the global economy). This fearful global invocation – Ekman's (2015) new seducing, psychosocial haunt – moves through the fabric of places as a second-wave rhetoric of globalization that bolsters place-entrepreneurializing efforts. Recognize global capital's ferocity and destructiveness, people in places across North America are told, and support efforts to tame this force. Populations are to respond rationally and appropriately – that is, to embrace "progressive" urbanization, pro-gentrification redevelopment, and austerity measures. The drive to garner support for potentially controversial restructuring greases the wheels to reculturalize places, reform political cultures, build bold and glitzy downtowns, produce gentrified neighbourhoods, entrepreneurialize public spaces, re-make investment climates, and neuter "business-unfriendly" labour unions.

This double-barrelled globalization propels an afflicting urbanization that leaves in its wake three well-documented things important to this study: eclipsed welfare provision, burgeoning low-wage economies, and more fractured places. Especially important for us, certain places experience more flagrant and ostentatious exclusion of the poor and their neighbourhoods (Wilson 2004, 2007; Jefferson 2016). Exclusion from this is now frequently bold and audacious (Sassen 2015). This is not a behind-the-scenes pushing of the poor to back-alley neighbourhoods, as has marked so many places in the United States, Canada, and Mexico for decades (see Davis 2003). It is now a spotlighted visual drama of re-making places which supposedly must banish poverty and near-poverty populations to established areas of exclusion and containment. In this context, "ghetto" neighbourhoods are provided a more flagrant "inside the place" structuralist logic: they are to both house mobilizable, able-bodied labourers and contain them as potential human contaminants to the new entrepreneurial place re-making.

(2) Neoliberal Transnational Dynamics and Their Outcomes

Neoliberal transnationalism is the second dominant, recently changed force now being worked through to entrepreneurialize places. Far from being a simple in-movement of "foreigners" into places, governances manage this reality as an opportunity to grow economies and re-business places. Governances now obsess with cultivating growth in two ways: growing low-wage service sectors and renewing real-estate accumulation; massive immigrant in-movement can help this agenda. In this context, governances play a double hand: they simultaneously welcome new immigrants into places to secure their much-needed labour power but also stigmatize them which helps forge re-culturalized downtowns, gentrified enclaves, and creative-class play spaces. People who are desired and needed in places to toil in horrific low-wage economies are also socio-spatially situated – upon arrival – to be invisible at key sites to the bourgeois gaze. Through these two contradictory steps, immigrants become as central to current place urbanization as any population in these places.

At one level, there is a generic welcoming for these recent arrivals. Narratives of multi-culturalism and cultural diversity infuse places with an aesthetic of tolerance and warmth. If low-wage economies are to flourish, this labour is needed. But at another level, banishing these recent entrants to corners and crevices is stark. Governances across the United States, Canada, and Mexico stigmatize and isolate them through new and revised programs and policies: isolated council (public) housing, immigrant holding and detention centres, police crackdown on loitering and public gathering, and charter schools replacing public schools. Spaces of the city, relentlessly produced and re-produced by these mechanisms, become naturalized as the proper terrain for a particular socio-racial class (Ahmed 2015). The outcome: the daily round of immigrants – journeys to work, journeys to shop, the work experience, common travel – become afflicting via a normalized marginality (Ghertner 2011). This current process, in context-specific settings, expands the pool of shadow neighbourhoods and shadow citizenships in places.

The extent of the migratory in-movement? Since 2000, US metropolitan areas have received more than 1.2 million people from East Asia and South Asia (many of them poor, working-class, and stigmatized); Canadian metropolitan areas have received more than 250,000 people from Mexico and Latin America (many of them desperately poor); and Mexican metropolitan areas have received more than 200,000 people from other continents (many struggling to find their place in rapidly growing, balkanizing places; Miraftab, Wilson, and Salo 2015). At the core of this, the globe now experiences unprecedented levels and geographies of residential dislocation (Harvey 2012; Dear 2013). Strife, wars, and political instabilities driven by afflicting nation-state interventions and the unceasing process of accumulation by dispossession ravage and dislocate vast populations across the globe (Harvey 2012).

Heroin Addiction and Youth in Suburban Public Housing

The Haunt

Heroin addiction among youth flourishes in "the projects" of Chicago's industrial suburbs (abcchicago, 2015). Its complexities root in a something central: a history of youth ensnared in long-term debilitating trauma (Jones 2015, discussion with author). Our discussions with youth corroborated this reality, and found the trauma to be less an outcome of isolated sets of acts (e.g., a beating, a moment of sexual abuse) than of a sustained pattern of deeply disturbing activities (e.g., living in deep poverty for years or decades, being homeless for a sustained period, living in severe stigma for years). Yet these hauntings arise out of a complex double consciousness of this youth, which shows both an acceptance of and active struggle with trauma (see Agar and Wilson 2002). A two-ness (a dual vision) in outlook structures a simultaneity of despondency and hope for the future, as both visions fleetingly move in and out of complex human thought, as we will now discuss.

Self-Affliction and the Haunt

In the first level of consciousness, this youth feels persistent despondency, discord, and a sense of being caught in a powerfully destructive past and present. Many are sceptical that there are ways out of heroin addiction or their current realities; pathways to other ways of living seem distant, remote, and foreclosed. A constant rehearsal of trauma in the everyday, from the past and present, activates this thought as a dominant everyday "structure of feeling." Heroin is identified either as the bane of it all (what one young man called "the driving force in my life at the moment") or a condition brought on by oppressive life circumstances. Heroin is seen in multiple ways: as an unequivocal evil, a crutch to get by, or a substitute for short-term happiness, but it is always an albatross around their neck, whose shedding seems nearly impossible. "It's pretty obvious I can't stay addicted to this stuff [heroin]," one twenty-three-year-old man told us, "but frankly I don't know what to do ... It's scary, it's hard man, but getting out of this and starting a better life is not easy."

Their current plight, moreover, is widely seen in the framing of themselves as, in their own words, "being responsible, choosing persons ... to be what we are." A powerful sense of self-blame – for bad decision-making and poor choice of lifestyle – often scars common thought. Bearing mainstream beliefs about social culpability, this youth often turn on themselves and villainize their own pattern of decisions. As one seventeen-year-old girl told us, "The biggest mistake in my life has been the choices that I have made ... it has afflicted me, and I just can't even think seriously of how to turn this craziness around." In this way, an already fragile identity takes a beating. Not surprisingly, a sense of blaming others for their circumstances was seldom brought up with us. In the context of our interviews, there tended to be no recurrently identified targets, enemies, villains, or causes for their plight outside of their own decision-making.

Hopefulness and the Haunt

Yet, set against this haunting, this youth constructs something else: a positive identity placed in a conception of a still shapeable and malleable reality. This youth strikes out and codes the world in ways that dull fears, enrich identities, and enable pleasure. For example, seemingly stark and foreboding events – copping drugs, seeing junkies lying on corners, huddling in abandoned buildings to avoid the cold and catch some sleep – also become social engagements with positive connotations. One women noted that one widely seen afflicting event, sleeping in abandoned buildings, is "also, through the pain, a kind of bonding thing, a moving along together of kindred spirits ... among friends ... like when we were younger ... that helps us feel okay about ourselves." Earlier in the discussion, ironically, this was something very different: "A kind of living that was barely tolerable." Similarly, copping drugs, to a male who quickly turned fanciful, is a kind of "thrill-of-the-hunt, a sometimes weird but always adventurous trip into the business world." Moments earlier, in a different flash of consciousness, this drug obtaining was described as "awful and unsettling."

At the same time, the numerous people this youth negotiates in daily life are understood through this hope: fellow addicts, halfway-house workers, needle-exchange site workers, policeman, strangers, drug dealers, and casual acquaintances. In one breath, oppressive people become re-cast as actors in an unfolding drama of potential possibility, what one women termed to us "the other side of how I and we can see this crazy bunch of people that we must see and deal with every day." Similarly, a day's scary and uncertain spaces – streets, parks, back alleys, needle-exchange sites – become re-cast as exotic places for making contact with others (friends, acquaintances). As one struggling youth told us: "I'm out here on my own and struggle," she noted, "and it's gonna be this way for a while. But I think about going to school, getting married, and having a family. I have been thinking lately about kickin' this thing and getting a good job. Sometime soon I will do it ... I have no doubt." This turn of reality may be delusional, but it fulfils an important function: it ultimately converts a bitter pill (the trace of the haunt) into a short-term digestible concern.

Planetary Urbanization Structuring the Haunt

Today, two governance urbanization strategies are accelerating in Aurora (population 204,000) and Cicero (population 84,000), afflicting this youth's turbulent emotional state and shaping their susceptibility to heroin use. In both cases, governances work through the new globalization (its reality and tropes) and transnational dynamics. First, they seek to expand the sense of opportunity for economic growth in places, exploding low-wage service sectors (i.e., hotels, banks, restaurants, department stores, hotels and motels, auto repair shops, medical service stores); and second, they strive to restructure their downtowns to enhance capabilities, especially to attract capital and jobs. This is the boldly advertised, two-pronged panacea for place resuscitation that supposedly holds the key to the futures of Aurora and Cicero. These actions cut resoundingly through their socio-spatial fabrics.

The first drive, bolstering the low-wage economic sectors, involves both a deepened city agglomeration (implosion) and an expansive spread of social networks across the globe (explosion). On the agglomeration (implosion) side, Aurora and Cicero strive to replenish and reinvigorate the job bases of especially retail corridors and job nodes. Both places recently instituted a number of massive tax abatement zones along these corridors, wrapped in decisive narratives of "needing to attract more business that is so essential to city growth that will help all" (Fox 2015, discussion with author). Use of tax abatements here, like in other places in Illinois and beyond, has a rich history. Moreover, initial business improvement district (BID) designations in the two dominant retail corridors of the places seek to bolster and retain what planning worker Fox (2015, discussion with author) calls a "critical mass of businesses" (the technique pools private funds to provide businesses enhanced security, road fix-up, and signage upgrade). BIDS are widely seen by governance actors as the latest innovative technical fix that could help this drive to restructure.

On the far-flung reach (explosion) side, governance actors strive to transform far-away people's social relations with these Illinois communities to snatch them out of international migration fields. Thus, both Chambers of Commerce and local governments aggressively recruit workers, notably by mobilizing World Relief and Catholic Charities (Hall 2015, discussion with author). Two techniques weave together these long-distance logistics networks. First, colourful websites intended for global populations extoll immigrant living in Aurora and Cicero, and second, World Relief and Catholic Charities periodically visit communities and court potential relocatees in North Africa (Algeria, Morocco), the Middle East (Iran, Lebanon), and South Asia (India, Pakistan, Bangladesh). In both cases the goal is to "bond with potential in-movers and have them see Illinois [and these places] for what it is – a great destination for them to come to" (Hall 2015, discussion with author).

Much subtlety exists here. Beneath these features, this urbanization relies on a core tactic: intimately (but subtly) moving into every life's diverse spheres – the economic, the cultural, the emotive, the political. The core of this is the drive to sculpt a new desired place-subject against which immigrants can be seen, measured, and responded to (to enable them to be brought into the new dead-end service economy, but to isolate their presence in the places). Borrowing from stocks of sensibilities about the ideal American citizen seared into common thought (see Anderson 2010), governance officials speak of the need for industrious, productive, place-committed subjects to re-entrepreneurialize place in the present and future (pushing the logic of a timeless re-entrepreneurial ethos). Here is Planner C. Hall's (2015, discussion with author) "important person who would lead the way to a resuscitating of Aurora in current times."

In this context, immigrants are paradoxically served up as a source of local possibility and anxiety: they are to be both accepted and marginalized subjects in everyday community life. Turbulent narratives in Aurora and Cicero thus fluctuate between an abstract celebrating of ethnic diversity ("Black heritage," "Latino heritage") and problematizing the same people for their habits and ways. Place re-making both establishes climates of social tolerance and instills sensibilities of intolerance. The offer of new immigration seeks resolution to a vexing political project: how a place can be projected as both accepting of otherness and suspicious of it. A logic is forged to accept people in these places, but also to carefully manage their places of residence and activity spaces. Thus, in Aurora and Cicero, the business sector boldly identifies "communities of tolerance and acceptance where all kinds of people can flourish ... but we need to identify kinds of citizens to help lead our community into a healthy future ... and know who is not being sufficiently contributory" (Harris, 2015, discussion with author).

In the second mode of urbanization in these places, governances work one more time through global and transnational dynamics to re-make downtowns as more upper-middle class in consumption and aesthetic. Downtowns are to be sites for and objects of consumption. Pleading economic necessity in global times, these cores are to concentrate non-poor, white-collar jobs and culturally discerning activities while new immigrants and the poor are to be marginalized and repulsed. Selling this has been nothing short of fervent and aggressive. Governances position themselves as economic and cultural warriors fervently defending place health. All are to see the governances as involved in important performative skirmishes (re-making downtowns, controlling and managing immigrants and the racialized poor, strengthening the character of neighbourhoods). Again, deepening city agglomeration and globally extending city networks proceed simultaneously.

On the implosive side of inducing change, Aurora has begun a massive restructuring of downtown following its recently unfurled "Seizing the Future Plan." The goal is to make the area "second to none among Chicagoland places ... create a potential hotspot for investors, speculators, and creative-class individuals and households ... nothing else will do" (Lee, 2015, discussion with author). Three blocks of touted blight clearance begins to re-green, de-grime, and re-brand the core. There is also current discussion among planners and city councillors of using tax increment financing to induce change. In Cicero, the 2014 "Change the Downtown" plan heralds a new era of re-making the core to stimulate municipal growth. Cicero's very future, it seems, is reliant on this. Once more, governances strive to forge a critical cluster of economically propulsive businesses and people.

The explosive side of inducing this change is equally important: both downtown re-makes also reach across the globe's entirety as thinly disguised attempts to seize and steal businesses and investment from other places across the globe. Economic agglomeration seemingly requires this outward extending: Aurora and Cicero purportedly cannot re-build themselves without poaching from other places. Planning worker E. Lee (2015, discussion with author) in Aurora notes this: "Our new downtown will be a deeper jump into global competitiveness, a concerted stroke to change us ... the message, we want your business and business leaders ... it's a battle in the present, that's just the way it is and what we have to do today." Again, a global engagement far beyond the immediate confines of Aurora and Cicero drives a kind of place transformation. In the final analysis, a simultaneity of concentration and stretching, areal focus and areal extension, underpins economic and political projects in these two places.

Urbanization once more moves deeply into every life's fabric: the goal is to re-entrepreneurialize place and establish its long-term logic. First, the governances borrow and deploy for public consumption the previously discussed central character, the new ideal civic subject, to engineer the influx of affluent creatives and new investment downtown. Oratory serves up an emergent civic person in these places who exudes the ideal cultural values (civic-mindedness, love for history in built environments, love for nature and open space) and political beliefs (a deep respect for private markets, a love of ethnic and cultural diversity) for these places to flourish economically and socially. Symbolism that snuggles into local life's intimacies casts this being in bold terms as the hope and future of the places. Tapping a now established iconography labels them "the new creatives who will help to change things in Aurora all around" (Lee 2015, discussion with author).

Punishing the Poor

Sinister outcomes follow from this urbanizing that are crucial to this paper. First, Aurora's and Cicero's poorest areas, the eight public housing projects on the East Side and North Side, respectively, have become poorer. Efforts to build low-wage service sectors and not a base of decent-waged jobs in these places have devastated workers and families. Poverty in the projects and their neighbourhoods has increased by an estimated 25 per cent in the last four years (Smith 2015, discussion with author). The more than 400 adults in the projects, most living below the poverty level, work jobs (they are not unemployed) that fetch non-living wages. Yet planners and politicians, seemingly oblivious to what happens to citizens when municipal policy promotes growth in dead-end jobs, continue to push for generic job creation in their municipalities. In Cicero, as planning worker P. Harris (2015, discussion with author) noted, "Yes, we have an aggressive job creation strategy that seeks to bring in employment opportunities to help our residents ... and it's working, our job numbers have grown." In Aurora, city councilperson A. Abes (2015, discussion with author) applauds his governance's "aggressive drive to create jobs, a push that in our community enlarges the number of jobs we have."

Second, re-culturalizing and re-glamorizing the two downtowns and socio-spatially marginalizing the new immigrants and poor afflicts the well-being of these housing project subalterns. A deeper stigma as to their worth and character permeates the everyday fabric of these places and their social spaces. As a youth in the Wells projects in Aurora told me: "Yes it is hard to keep your head up sometimes in my life ... I struggle, ya know, have to fight through the way I know I'm being seen by others ... on the streets, in the store, how I feel about myself." Yet, there is little formal recognition of this heightened affliction. Planners and public officials in Aurora and Cicero speak only of economic and cultural gain from re-making the downtowns. Failing to see in their actions the simultaneity of creating inclusions and exclusions, acceptances and banishments, there is only "civic improvement" in governance efforts (Abes 2015, discussion with author).

And what about the youth in these housing projects and their realities of drug usage? Too many now live in stifling, eviscerating realities. Amid the damage wrought by the destructive combination of implosive and explosive place remaking, living conditions too often become near intolerable. Aurora psychologist B. Rett (2015, discussion with author) notes this in discussing her clinical practice: "Many of my patients are from this area [the East Side] of Aurora." To Rett, "families here feel hardship, lots of kids and teenagers are being raised in toxic families – alchoholism, drug abuse, broken and unstable families ... social issues have been frequent on Aurora's East Side." Comments from youth bear this out. One twenty-three-year-old sums this up in conversation: "I grew up on the East Side of Aurora, it was tough, man, nothing but trouble and depressed times all over ... I moved out of my father's place two years ago, he always worked shit jobs and was angry about it. I ain't going back, stuff went down that would kill you, I mean literally ... it is a miracle I survived."

Not surprisingly, heroin usage and addiction too often follow. Current urbanization, moving through these places, institutionalizes in the everyday fabric a harsh, punishing social and economic reality that further downgrades the lives of the most vulnerable. For youth in the projects, there is no ducking this reality, with often devastating consequences. One youth, reflecting upon this, noted his plummet into pain, hopelessness, and drug dependency:

> My family life and everything else with my life, it was just hard ... later, flippin' burgers, man, was too much. It hurt and was demeaning; I couldn't do it for long. And lots of shit happening ... Soon, I quit and went home, nothing to do here but hang out and try and avoid my father. Before, long, man, I tried heroin and liked it; it took me to a new place, a new space.

Another youth, a two-year addict, recounts a similar story:

> There was little positive in my upbringing, lots of fighting and drinking, my father left when I was young. I ended up leaving home four years ago. Did it affect me? Of course, and I am now on my own, I had no place to

go, drugs and heroin came pretty easy to me after that ... it's a scary way to live, this hole I am now in, it's day-to-day.

Discussion and Conclusion

This chapter has sought to understand the connections between category-collapsing, complex urbanization and the perplexing issue of proliferating heroin addiction in America's suburban housing projects. To accomplish this, our focus has shaken the bones of a central concern of radical political economy, urbanization, to demonstrate that this today differs from common notions of "growth" and "development." Its dynamic: urbanization seizes the latest dominant forces on the current scene, globalization and transnationalism, and, through rounds of implosions and explosions that move through places in expansive and deep ways, re-sculpts two important things: the current conditions and future flooring for the re-entrepreneurialization of places. Urbanization today, in all its faces, processes, and features, is as much about the drive to cultivate the social and political grounds for an endless future of place re-entrepreneurialization as the current desire to establish a current re-entrepreneurialization. Thus, a range of spheres in places are targeted for re-engineering: the social, political, economic, cultural, and socio-spatial. Through the simultaneity of implosions and explosions, a fiery urbanization erupts everywhere to re-shape the everyday of places.

What is at the heart of the drug upsurge in suburban public housing? We suggest it is the human impacts that a punishing urbanization afflicts on an acutely vulnerable, confined population. This urbanization, offering punishing programs, policies, invectives, and reasoning, moves through these places to widen material and symbolic divides across races-classes as choreographed recipes to transform places. With growing inequality, stigma, and economic insecurity arising from this, expanding numbers of youth experience and feel tumultuousness, heightened suffering, and diminished hopes. Reflecting the current neo-revanchist and global-exaggeration times, then, this is not a simple denigration, but a purposeful assault that positions the afflicted to be outside any serious sense of acceptable deprivation and core citizenship in community and society. Here in common thought is the rapidly growing Other America, in thick, gory detail. Amid horrific oppression, heroic and often poignantly reflexive struggles by youth unfold, but drug abuse too often proliferates in these circumstances.

To be sure, a planetary urbanist world today has not produced something totally new here in its punishing, ghettoizing, and situating of people in lifeworlds of marginality that foster drug use and drug addiction. America has seen this before, for example, in the crack epidemic in Harlem and Bedford-Stuyvesant in the 1980s, and the heroin addiction epidemic in America's Black communities in the 1940s. Yet today this process unfolds through a temporally distinctive set of forces in a planetary urbanist world: a double-barrelled globalization and neoliberal transnationalism enabled by the construction of new programs and policies, new rhetorical formations, and new governance actions. These forces, recent in origin, collectively define a new urbanization that pushes socio-spatial divides and human marginality to new depths. More than before, a humanly produced marginality is seen as acceptable given the supposed need for this new kind of urbanization in purported place-eviscerating global times.

The basic connections between current urbanization and this proliferating drug affliction in the two suburban communities is made clear in Rett's (2015, discussion with author) comments: "Without doubt families in East Aurora, and in the projects, where I get so many of my patients, are being torn up by the new economy in America and Aurora." To Rett, "The struggle is real, and it can be deadly ... clearly too many kids and teenagers are coming of age in neighborhoods whacked by the craziness of current economic and political realities." Punctuating this, the youth we talked to repeatedly identified this reality. In our discussions, one twenty-three-year-old reflected this: "I grew up on the East Side of Aurora, it was tough, man, I moved out of my father's place two years ago. I ain't going back, the neighborhood has collapsed, I am not into being hit or abused ... it was beyond difficult, man." As many youth in these projects live through a punishing every day best described as historically unprecedented and human-numbing (Edidin et al., 2012), they and drugs more tightly interconnect.

With this focused planetary urbanist sensibility set down, we need to re-work it. At issue, for us us, is to recognize and place at this notion's centre fluid, structurating processes whose forms and relations are set in specific times and places. The challenge is to extend the planetary urbanist sensibility into the realm of human problems and to make it about historically specific processes, relations, and forces that give rise to structurated grids of influence. Using this vision we do not seek to foreclose the world's array of empirical possibilities, but instead to open this up by excavating mixtures of interconnecting forces layered onto sites. Because this is a seminal analysis, what one colleague called (when reading this chapter) a "first-shot stretching paper," we hope it will open up new vistas for urbanists to contemplate and investigate. In this spirit, we believe, there is much work ahead that needs to be done.

ACKNOWLEDGMENTS

This chapter has benefited enormously from discussions and communications with Neil Brenner, Daniel Gonzales, Melissa Heil, Brian Jefferson, Roger Keil, Robert Lake, Andy Merrifield, Priyam Tripathy, and Murat Ucoğlu. To all we are very grateful.

NOTES

1 We talked to thirty-four heroin-addicted youth at a needle exchange site in West Chicago over twelve months in 2015 to learn of their lives and predicaments. All respondents either lived in public housing at one point in their lives or had intimate knowledge of the realities here via social networks. They and other interviewees in the study (interviews were conducted in 2015, 2016, and 2017) were provided anonymity (names were changed) to enable more credible data extraction.

2 By conjunctural forces we mean the convergence of constellations of elements in particular settings that make places distinctive for the unfolding of urbanization.

REFERENCES

Abcchicago.com. (2016) City to fight opioid epidemic in Chicago. 6 October 2016. https://abc7chicago.com/health/city-to-fight-opioid-epidemic-in-chicago/1541880/

Abes, A. (2015). City Councilperson, City of Cicero, Discussion with Author, 9 May, Cicero, IL.

Agar, M.H., and Wilson, D. (2002). "Drugmart: Heroin Epidemics as Complex Adaptive Systems." *Complexity* 7, no. 5, 44–52. https://doi.org/10.1002/cplx.10040

Ahmed, S. (2015). "The 'Emotionalization of the "War on Terror"': Counter-Terrorism, Fear, Risk, Insecurity and Helplessness." *Criminology and Criminal Justice* 15, no. 5, 545–60. https://doi.org/10.1177/1748895815572161

Anderson, M. (2010). "The Discursive Regime of the 'American Dream' and the New Suburban Frontier: The Case of Kendall County, Illinois." *Urban Geography* 31, no. 8, 1080–99. https://doi.org/10.2747/0272-3638.31.8.1080
– (2017). Preface to N. Brenner, *Critique of Urbanization: Selected Essays*, 9–13. Basel, Switzerland: Birkhauser.
Brenner, N. (ed.). (2014). *Implosions/Explosions: Towards a Study of Planetary Urbanization*. Berlin, Germany: Jovis.
– (2017). *Critique of Urbanization: Selected Essays*. Basel, Switzerland: Birkhauser.
Cicero, T.J., Ellis, M.S., Surratt, H.L., and Kurtz, S.P. (2014). "The Changing Face of Heroin Use in the United States: A Retrospective Analysis of the Past 50 Years." *JAMA Psychiatry*71, no. 7, 821–6. https://doi.org/10.1001/jamapsychiatry.2014.366
Davis, M. (2003). "Planet of Slums." *New Left Review* 11, 12–19.
Dear, M. (2013). *Why Walls Won't Work: Repairing the US-Mexico Divide*. Oxford, UK: Oxford University Press.
Edidin, J.P., Ganim, Z., Hunter, S.J., and Karnick, N.S. (2012). "The Mental and Physical Health of Homeless Youth: A Literature Review." *Child Psychiatry & Human Development* 43, no. 3, 354–75. https://doi.org/10.1007/s10578-011-0270-1
Ekman, M. (2015). "Online Islamophobia and the Politics of Fear: Manufacturing the Green Scare." *Ethnic and Racial Studies* 38, no. 11, 1986–2002. https://doi.org/10.1080/01419870.2015.1021264
Fox, B. (2015). Worker, Building Department, Town of Cicero, Discussion with Author, 3 August, Cicero, IL.
Ghertner, D.A. (2011). *Rule by Aesthetics: World-Class City Making in Delhi*. New York, NY: Wiley-Blackwell.
Harris, P. (2015). Operative, Cook County Commission on Social Innovation, Discussion with Author, 6 August, Cicero, IL.
Hall, C. (2015). Planner, City of Aurora, Discussion with Author, 10 July, Aurora, IL.
Harvey, D. (2012). *Rebel Cities: From the Right to the City to the Urban Revolution*. New York, NY: Verso.
– (2014). "Cities or Urbanization?" In Brenner, N. (ed.), *Implosions/Explosions: Towards a Study of Planetary Urbanization*, 52–66. Berlin, Germany: Jovis.
Jefferson, B.J. (2016). "Broken Windows Policing and Constructions of Space and Crime: Flatbush, Brooklyn." *Antipode* 48, no. 5, 1207–91. https://doi.org/10.1111/anti.12240
Jones, E. (2015). Worker, Department of Community Affairs and Special Projects, Town of Cicero, Discussion with Author, 4 August, Cicero, IL.

Keil, R. (2016). Professor, York University, Discussion with Author, October, Istanbul, Turkey.

Kuehn, B. (2014). "Driven by Prescription Drug Abuse, Heroin Use Increases among Suburban and Rural Whites." *Journal of the American Medical Association* 312, no. 2, 118–19. https://doi.org/10.1001/jama.2014.7404

Lefebvre, H. (2003). *The Urban Revolution*. Minneapolis, MN: University of Minnesota Press.

Lee, E. (2015). Operative, Cook County Commission on Social Innovation, Discussion with Author, 12 August, Chicago, IL.

Merrifield, A. (2014). *The Politics of the Encounter: Urban Theory and Protest under Planetary Urbanization*. Athens: University of Georgia Press.

– (2016). "Planetary Urbanization in a Changing World." Plenary Lecture at the Eighth International Conference on Spaces and Flows, Philadelphia, PA, April.

Miraftab, F., Wilson, D., and Salo, K. (2015). *Cities and Inequalities in a Global and Neoliberal World*. London, UK: Routledge.

O'Leary, J. (2016). "5 Socio-Cultural Factors that Cultivate Addiction." *Rehabs.com*, 15 April. http://www.rehabs.com/5-socio-cultural-factors-that-cultivate-addiction/

Reitz, B. (2015). Planner, City of Aurora, Discusion with Author, 12 April, Aurora, IL.

Rett, B. (2015). Child Psychologist, City of Aurora, Discussion with Author, 1 August, Aurora, IL.

Richards, A. (2016). Social Worker, City of Aurora, Discussion with Author, 9 August, Aurora, IL.

Robinson, J. (2017). Panel Presentation at New Trends in Planetary Urbanization Panel. Boston, MA, April.

Roy, A. (2016). "What Is Urban about Critical Urban Theory?" *Urban Geography* 37, no. 6, 810–23. https://doi.org/10.1080/02723638.2015.1105485

Ruddick, S., Peake, L., Tanyildi, G., and Patrick. D. (2017). "Planetary Urbanization: An Urban Theory for Our Time?" *Urban Geography* 36, no. 3, 387–404. https://doi.org/10.1177/0263775817721489

Sassen, S. (2015). "Neither Global nor National." Unpublished technical report available from author.

Schmid, C. (2014a). "Patterns and Pathways of Global Urbanization: Towards Comparative Analysis." In Brenner, N. (ed.), *Implosions/Explosions: Towards a Study of Planetary Urbanization*, 203–18. Berlin, Germany: Jovis.

– (2014b)."The Urbanization of Switzerland."In Brenner, N. (ed.), *Implosions/Explosions: Towards a Study of Planetary Urbanization*, 268–76. Berlin, Germany: Jovis.

– (2017). Panel presentation at New Trends in Planetary Urbanization Panel. Boston, MA, April.
Simone, A.M., (2017). Panel presentation at New Trends in Planetary Urbanization Panel. Boston, MA, April.
Smith, A. (2015). City Official, City of Aurora, Discussion with Author, 17 April, Aurora, IL.
Soja, E. (2014). *My Los Angeles: From Urban Restructuring to Regional Urbanization.* Los Angeles: University of California Press.
Stone, C., and Sanders, H. (1987). *The Politics of Urban Development.* Lawrence: University of Kansas Press.
Strang, J., et al. (2012). "Drug Policy and the Public Good: Evidence for Effective Interventions." *The Lancet*, 379:9810, 71–83. https://doi.org/10.1016/s0140-6736(11)61674-7
Swyngedouw, E. (2012). "The Politics of the Square." Millercom Presentation, University of Illinois at Urbana-Champaign, 14 March.
Walker R. (1981). "A Theory of Suburbanization: Capitalism and the Construction of Urban Space." In Dear, M. and Scott, A.J. (eds). *Urbanization and Urban Planning in Capitalist Society*. London, UK: Methuen.
– (2015). "Building a Better Theory of the Urban: A Response to 'Towards a New Epistemology of the Urban?'" *City*, 19:2–3, 183–91. https://doi.org/10.1080/13604813.2015.1024073
Wheeler, S.M., and Beatley, T. (2014). *Sustainable Urban Development Reader*. London, UK: Routledge.
Wilson, D. (2004). "Toward a Contingent Urban Neoliberalism." *Urban Geography*, 25:8, 771–83. https://doi.org/10.2747/0272-3638.25.8.771
– (2007). *Cities and Race: America's New Black Ghetto*. London, UK: Routledge.
Wilson, D. and Jonas., A (2018). "Planetary Urbanization: New Perspectives on the Debate." *Urban Geography* 39, 1576–80. https://doi.org/10.1080/02723638.2018.1481603

PART TWO

Legacies

3 Estates under Pressure: Financialization, Shrinkage, and State Restructuring in East Germany

MATTHIAS BERNT

Massive housing estates located at the peripheries of cities are characteristic of most cities in East Germany. They make up around one-fifth of the housing stock so that, in 1990, more than 23 per cent of all East German households lived in one of 150 housing estates, each consisting of not less than 2,500 units. When these estates were built in the 1960s to 1980s, they signified progress. Their construction was based on modernist planning principles and included the extensive use of prefabricated elements. Housing standards here were high, and far better than in the deteriorating inner cities (where modern heating and other amenities were an exception) and rents were fixed at a very low level. Production, consumption, and management of these complexes rested on Fordist ideas of standardization and economies of scale and was controlled by public administrations or quasi-public cooperatives. The apartments built were rented to a broad variety of social groups. In sum, massive prefabricated high-rise suburbs stood as a paradigm for socialist policies, and the political goal until 1990 was to offer a satisfactory flat to every household and "solve the housing question."

Since then, the environment in which these estates need to function has changed completely, and large-scale housing estates in East Germany have come under increasing pressure. While "prefab" housing estates have always been subject to very specific policies since reunification, the last decade has provided particularly difficult challenges for the future development of these settlements.

In this chapter, I use the estate Am Südpark as an example to illustrate these changes.[1] Am Südpark is one of nine neighbourhoods in Halle-Neustadt, formerly the third-biggest estate in the German Democratic Republic (GDR) – and, in fact, for some time even an independent city. This westernmost part of Halle (Saale) was built between 1964 and 1989, for a projected population of 100,000 inhabitants. When it was built, Neustadt enjoyed a fairly favourable reputation as a "model city." After the reunification of Germany, the estate was hit hard by deindustrialization and population decline, as the collapse of the existing chemical industry in the region resulted in excessive unemployment, impoverishment, and outmigration, which considerably exceeded the already problematic average in East Germany. Today, Neustadt faces four problematic developments, making it a paradigmatic, yet extreme, example of the crisis of this form of "massive" suburbanization.

Two Rounds of Privatization

While housing had been organized as a part of a planned economy and disconnected from financial markets under state socialism, its finance, management, and ownership structures were successively marketized after the reunification of Germany. The first step was the reorganization of the existing municipal housing administrations and the cooperatives as commercial companies, which would be owned by the municipalities but act as commercial businesses; this was taken right after reunification in 1990. Very soon, these newly founded companies came under immense pressure and were forced to sell at least 15 per cent of their stock. The background for this was the Existing Debts Assistance Act (Altschuldenhilfegesetz), a federal law of 1993 which regulated the "debts" of municipal and cooperative companies in the territory of the Ex-GDR. In a nutshell, these debts resulted from funds which had been allocated to municipal housing administrations and cooperatives through the GDR state bank in order to provide the means for constructing new estates. When municipal housing administrations and cooperatives were transformed into free enterprises and the state bank of the GDR was sold to West German commercial banks, operation funds, which were appropriated in a planned socialist economy without interest rates, became "debts" of on average 15,000 DM (7,500 euro) per flat (Borst 1996; Bernt 2006). As it soon turned out, the newly founded companies hardly had the capital base to deal with these claims. Thus, the state intervened and took on the lion's share of the debts in 1993. The condition set by the German state was, however, a legally binding acceptance of the claims of the debtor and the requirement to privatize 15 per cent of the housing stock owned by 1999, ideally to their sitting tenants.

For the East German municipal housing companies and cooperatives this created problems, as sitting tenants (who in most cases did not have the necessary equity to engage in this operation) showed next to no interest in buying their apartments. As a consequence, the law was changed in May 1995 to allow for privatization of the entities to so-called in-between purchasers (Zwischenerwerber). In practice, this turned out to be a field for "prefab-gamblers" as a German magazine called them, who would buy huge numbers of prefabs at discounted rates, take high credits, and hope for a lucky resale (Der Spiegel 2000). When the East German housing markets plunged into crisis in the second half of the 1990s, these calculations often proved to be unrealistic and many privatization projects ran into insolvencies, re-sales, and even legal trials. The Aubis group, a subsidiary to the state-owned Berliner Bankgesellschaft, provides an illuminating example of the bizarre constellations which emerged at this time. Backed by securities provided by the state of Berlin, Aubis acquired more than 16,000 prefab apartments all across the former German Democratic Republic in the second half of the 1990s. Already at that time, experts raised concerns about the high prices paid and expressed doubts about the ability to pay down resulting loans costs. At around 2000 the whole construction fell apart: Aubis went into insolvency, the stocks were taken by the banks, and Berlin had to provide enormous means to compensate for the defaults in repaying mortgage debts. While there has never been a comprehensive study of this period, most experts today tend to see this period as a complete failure. Retrospectively, however, its main importance consists in laying the foundations on which much of a second round of privatizations was built in the 2000s.

This occurred in the mid-2000s and captured much of the stock that was sold under the Existing Debts Assistance Act in the 1990s. While it thus rested on a supply of already privatized flats, the economic background of the purchasers differed widely from what had been known before. As in other Western economies, this round of privatization was dominated by institutional investors, such as real estate investment trusts (REITs), real estate equity funds (REPE), and other property funds, often with branch offices around the world, who would seek undervalued assets to invest in. Here, the background was a bizarre mix of the consequences of local population decline (see below) and new global financial schemes. Instead of local demand, it was overaccumulation and increased liquidity after the stock market crash in 2000, historically low interest rates, and dramatic price deterioration in East German real estate markets that worked together to make large-scale housing estates in East Germany an attractive investment. This resulted in an increased relevance of financial investors, whose business strategies have vividly been described as follows:

> "They (the funds) purchased the cheapest, often problematic, housing estates, financed them with high debt rates and resold them as quickly as possible for a higher price. In a market situation, where demand was only increasing for housing in the better neighbourhoods, the opportunistic funds applied aggressive letting strategies with the imperative to serve their credit lines through decreasing vacancy rates. This fostered a concentration of socially disadvantaged tenants that have no access to the housing stock of better neighbourhoods" (Uffer 2013, 169).

In Neustadt, these privatizations had the strongest impact on the youngest and southernmost neighbourhood, Am Südpark. Here, the municipal company GWG Gesellschaft für Wohn-und Gewerbeimmobilien Halle-Neustadt mbH sold all of the properties it had in this area to Schultze & Partner, an "in-between purchaser" real estate company from Dortmund (which eventually went bankrupt in 1999), in the course of the Existing Debts Assistance Act. The Halle-Neustädter Wohnungsgenossenschaft e.G., a local cooperative founded in the 1970s, by contrast, tried to meet the obligations from the Existing Debts Assistance Act by founding a new cooperative named WG Am Südpark, to which it transferred 756 of its housing units in the area. In order to gain critical mass, it acquired another 552 apartments from the Bauverein cooperative. After only two years, when house prices plummeted and banks re-evaluated the value of the mortgaged properties and demanded new securities, WG Am Südpark had to declare bankruptcy in 2003 and the stock was taken over by a receiver and successively sold.

After this, the property ownership structures in the Am Südpark neighbourhood became increasingly complex (see figure 3.1). The insolvent stocks were taken over by the lending banks, which partitioned them into smaller portfolios. These were often put together with other portfolios and then sold to different purchasers (who again often resold). Eventually, most of these stocks were taken over by two private companies, WVB Centuria GmbH and Westminster Immobilien GmbH.

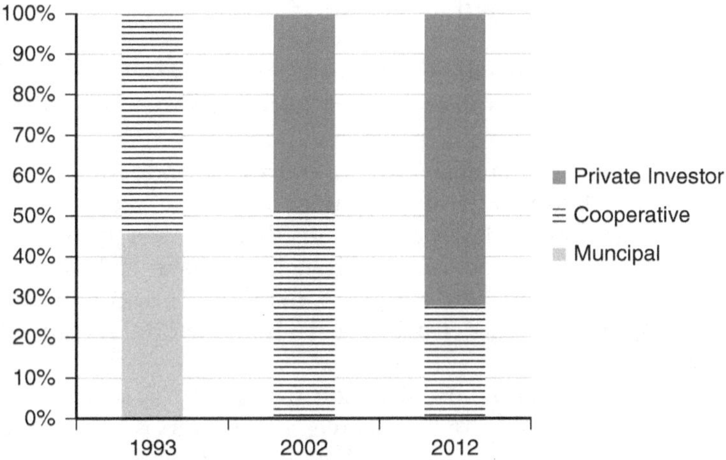

3.1 Percentage of property owned by municipality, cooperative, and private investors in Am Südpark. Source: City of Halle, author's calculations.

WVB Centuria GmbH is, simply put, a proprietary of the Paris-based global investment group Centuria Capital, which pools the money of international clients (often from Arab oil states), buys up underpriced properties for them, and makes sure that these are managed in a profitable way so that investors can gain an attractive return. Westminster, by contrast, is a family-owned company specialized in asset management. The company is a prime partner of banks who need to deal with distressed debts and has continuously expanded in recent years. What Westminster and Centuria have in common is a business model directed on the acquisition, development, and resale of underpriced properties for clients. In order to achieve full occupancy, both companies rent their apartments for very competitive prices. An interviewed manager of Centuria thus declared:

> "We consequently set the rent of the apartments at Am Südpark at a level defined by the 'costs for accommodation.' There is no point in offering apartments above this level here, as the majority of tenants do not pay their rent on their own. If we get tenants interested in renting here, they are welfare recipients as a rule. Whoever pays for his flat by himself would not move to Südpark."

State Withdrawal and Welfare Reforms

While private investors entered the local housing market, federal and state subsidies for urban development have been severely reduced in the last years.

Thus, reacting on financial distress and the introduction of a zero-new-debt quota by the federal government, the state of Saxony-Anhalt (in which Halle is located) has cut all subsidy programs for social housing. As a result, the number of social housing units in Halle (Saale) has already decreased from 7,061 (2008) to 4,718 (2012), and will continue to decline in the next years. The last social housing unit in Halle (Saale) will be released in 2019.

In addition, the budgetary situation of the city is so difficult that the city has dramatically reduced investments into its infrastructure. As a recent comparison reveals, Halle (Saale) is among the ten municipalities with the lowest per capita investments in Germany, achieving less than the one-seventh of the investment in Southern German cities (Arnold, Frier, Geissler, and Schrauth 2015, 1035ff.). The city has clearly embarked on a disinvestment policy, and expenses for green spaces, playgrounds, road surfaces, and other publicly maintained facilities have been severely reduced. While there is no publicly declared "red-lining," there are hints that cuts have been particularly massive in peripheral estates. This is also supported by field observations, which reveal that the public infrastructure is in a comparably worse state of maintenance here.

At the same time, welfare reforms introduced by Germany in 2004 (Hartz IV) have led to a substantial expansion of the demand for low-price apartments and made renting out apartments to welfare recipients a new commercial strategy for private landlords and speculators. Using the "welfare to work" approach of the UK Labour Government at the time as a blueprint, the reforms combined the former unemployment benefits for long-term unemployed (Arbeitslosenhilfe) and welfare benefits (Sozialhilfe), leaving them both at approximately the level of the lower of the two, the former Sozialhilfe (social assistance). They thus reduced the level of basic social security. At the same time, they made it possible to claim basic benefits as a supplement to a low income, thus supporting the dynamic growth of a low-wage sector. Furthermore, the new social welfare law distinguished between "cost of everyday living," paid by the national government, and housing costs, compensated by municipalities. Whereas the first is paid directly to the welfare recipient, the latter, called "cost for accommodation" (Kosten der Unterkunft – KdU), is in most cases directly assigned to the landlord. The point here is that there is a maximum rate for the "cost for accommodation" (KdU), which is set in accordance with the average rent per square metre in the municipality concerned. In other words, as long as rents are kept at the KdU level, landlords who rent out to welfare recipients can count on a stable income practically guaranteed by the state. Together, these conditions dramatically expanded the demand for low-price apartments and made renting out apartments to welfare recipients a new commercial strategy, which has been classified as the "Hartz IV-Geschäftsmodell" (i.e., welfare aid business model) by German planners.

Shrinkage and "Stadtumbau Ost"

Adding to the growing relevance of financial investors and the rise of a Hartz IV business model, local planning policies were hardly supportive. Here, the city administration of Halle (Saale) completely remodelled its planning strategies in 2001 and made "rightsizing" the city to a reduced population size the most important goal. This was done in line with the newly emerging federal subsidy program, Urban Regeneration East (Stadtumbau Ost; Bernt 2009). The background was the population loss and housing vacancies discussed above, which peaked in the late 1990s and became a serious threat to the functioning of the local housing market. With the aim of preventing an acceleration of vacancies and in line with a newly emerging Stadtumbau Ost federal subsidy program, Halle (Saale) planned to demolish thousands of flats. Demolitions were, however, to be implemented in a way that would help to strengthen what remained, instead of perforating the urban fabric. The basic planning idea for Urban Regeneration East in Halle-Neustadt was very plausible and easy to understand at first glance: it aimed to shrink from the outside in; that is, to concentrate demolitions in the southern and westernmost parts of the district and to strengthen the centre. The advantages of this plan were seen as self-evident. Through "building backwards" (rückbauen), the new bureaucratic term for demolishing, the peripheries would help increase population density, thus helping to maintain the population density necessary for the operation of social, technical, and commercial infrastructures. In addition, demolishing the outer parts of the estate would allow for the creation of new green spaces and, finally, provide opportunities to shrink redundant technical infrastructure (like water pipes or tram lines), making the grid more efficient.

In practice, however, this perspicuous idea met unforeseen difficulties. The plans to focus demolitions in the southern and western peripheries of Neustadt were objected to by those housing companies who had a major part of their stocks in these areas. To make things even more complicated, the then two major private owners in the Am Südpark neighbourhood – the in-between purchaser (Zwischenerwerber) Schultze & Partner and the newly founded cooperative WG Am Südpark – went bankrupt exactly at the time when the new plan was to be implemented and, therefore, they were incapable of taking any action. Even when the insolvent stock was later sold to new owners this did not help. As both Centuria and Westminster had just acquired their stock with the aim of developing them for financial investors, a demolition of their investment would not have made any sense. The outcome of this was a situation in which plans for demolishing the peripheries of Neustadt were made the official policy agenda, but hardly implemented.

While Stadtumbau Ost thus ended in a state of limbo in the peripheries of Halle-Neustadt, it had a number of unforeseen side-effects. The declaration of Am Südpark a restructuring zone thus significantly complicated the raising of credit for local owners. Interviewed experts reported that the announcement of the restructuring plans completely altered the lending policies of mortgage providers, who stopped providing further credit for any projects in the area and de facto introduced a form of "soft red-lining" (Aalbers 2011). The second side effect of the restructuring plans was that the neighbourhood Am Südpark would ultimately be excluded from all existing subsidy programmes for urban development projects. This is particularly important because in Germany stabilizing or upgrading disadvantaged communities is – as a general rule – very much dependent on the provision of public subsidies. The calculation not to waste money on areas that were deemed unworthy of retention thus effectively left one of the most problematic areas of the city without public support. Third, the declaration of the area as a "restructuring zone" accelerated the deterioration of the already stressed house prices, and interviewed experts have estimated that prices went down to 75 euros per square metre for prefab multifamily houses in Halle-Neustadt at around the turn of the millennium. This, however, reduced the costs for new investors, and facilitated market entry for the businesses like Westminster or Centuria.

3.1 Welfare dependency rates compared

	Unemployment	Child Poverty	Working Poor	Youth Unemployment
Südpark	22.4	70.5	30.2	11.3
(Halle)	8.7	34.7	10.8	3.8
Germany	5.0	14	n.a.	n.a.

Source: Federal Statistical Office, City of Halle, author's calculations.

Increased Impoverishment

The outcome of these simultaneous developments is an intensified social downgrading of the area. Local data on welfare recipients, poverty, and disadvantage clearly show that the expansion of privatized housing stocks at Am Südpark, which were rented at the lowest rates, triggered an increasing concentration of households that simply could not afford higher rents. While unemployment rates have dropped constantly in Halle (and even more so in Germany), the Am Südpark area has showed a steady increase (see table 3.1).

Studying the statistics, the picture is fairly dramatic: At Am Südpark, the level of unemployment is more than double that of Halle and nearly five times higher than in Germany as a whole. While in Halle only one out of three children depends on welfare aid, in Am Südpark close to three out of four children grow up in poverty. There are more working poor and higher youth unemployment. Household incomes are low and there is an immense need for support through charity organizations.

In short, planned shrinkage, state withdrawal from the housing sector, and welfare reforms have effectively worked together to pave the way for new investors who would capitalize on low sales price, low interest rates, and low maintenance and gain their profit on the basis of a Hartz-IV business model based on renting apartments to welfare recipients. The result of these simultaneous developments has been a growing level of inequality, an increased impoverishment of the Am Südpark neighbourhood, and a rise in the concentration of low-income households.

Are large-scale housing estates built under socialism then about to become "ghettos of the 21st century," as the Hungarian geographer Ivan Szelenyi (1996) forecasted two and a half decades ago? Given the sheer size of these estates, this is rather unlikely. Yet, while heterogeneity needs to be emphasized, there are a number of developments underway that have the potential to transform large parts of these neighbourhoods from fairly "normal" urban areas with widely mixed social groups into warehouses for the urban poor. While there has not been much research yet, and it still remains to be seen how much this constellation is to be found in other large-scale estates in East Germany, it is clear that the developments described here are at work in many other cities, too. Recently, they have been joined by the "refugee crisis" and the need to accommodate around a million immigrants who have come to Germany in the last two years. Numerous reports show that this is over-proportionally done in modernist prefab estates at the peripheries of the cities, where it is easier and less expensive to find accommodation. It does not take a crystal ball to see that this will add to the difficulties these neighbourhoods are already facing. In sum: the massive suburbs inherited from the socialist era in East Germany are about to face massive problems.

What can be learnt from this story for massive suburbs built elsewhere? Of course, there are numerous particularities which make East German estates a very peculiar story. Nowhere in the post-socialist world have privatizations taken place in the same way as in East Germany; nowhere has an Existing Debts Assistance Act resulted in such bizarre distortions within the housing sector; and nowhere has shrinkage been responded to with a subsidy programme comparable to Stadtumbau Ost. Against this background, the case described here seems to be rather unique. Yet, so are all cases; and taking a step back, a couple of issues emerge which make Halle-Neustadt a case capable of providing lessons for the future of similar neighbourhoods elsewhere.

The first of these is "urban shrinkage." In many cities, not only in the former Eastern Bloc, population numbers have decreased in the last decades and continue to do so. "Shrinkage" has become a normality in many settlements of the world (Oswalt 2005; Pallagst, Martinez-Fernandez, and Wiechmann 2013; Haase et al. 2016), and is experienced in cities as different as Osaka, Detroit, or Vorkuta. In this context, Halle-Neustadt is an extreme case, but not an atypical one. The point here is that with reduced population numbers, "rightsizing" the existing built environment becomes the order of the day. In Halle-Neustadt (as in many other East German estates) this sort of restructuring has not remained a planner's vision, but has actually been implemented. In this respect, Halle-Neustadt is a frontrunner for developments which are also relevant for many other "shrinking" cities. The bitter experiences made with "governing" shrinkage here demonstrate very clearly that "rightsizing" cities is hardly a merely technical issue, but a matter of power and interests. In Halle, state strategies for "planned shrinkage" were effectively counteracted by private interests, thus making public policy goals unfeasible. Learning from Halle thus suggests a need for pushing the discussion about future strategies for "rightsizing" beyond matters of design and coming to grips with matters of interests, power, and justice.

Second, policy fields which are usually analysed within very different communities (e.g., spatial planning, housing studies, and welfare research) have coalesced in Halle-Neustadt and this has brought about fairly problematic developments. While the interplay of different developments in generating urban change is hardly big news to urban studies in general, in practice many studies have difficulties with integrating the bewildering complexity of influential factors into their conceptual framework. In this sense, the case once again points towards the intricate interdependencies, uncertainties, shifts, and feedback loops influencing the production of specific spatial formations (see also Grossmann and Haase 2015). This complexity needs to be taken seriously. What is needed, then, are very complex and multi-faceted understandings of the constellations in which the future of massive housing estates will be figured out. Beyond ubiquitous statements, this implies directions for research designs, team compositions, and curricula, which need to be organized in a way that allows thinking beyond traditional disciplinary borders and including different fields of specializations.

Third, the case vividly demonstrates that ownership relationships in today's estates are defined in an increasingly global, interconnected, and liquid economic environment. The development of an estate is therefore not only an issue in the hands of local actors, but it gets increasingly intermingled with international financial markets and global asset strategies. Planning for a sustainable perspective thus demands coming to grips with the financialization of the real estate sector and fostering ownership structures which have both an interest in and the capacity to engage in the long-term development of the place where they own properties. The ongoing financialization of housing needs to be countered by the strengthening and/or the development of alternative forms of ownership and housing management at the local scale.

Up to now, Halle-Neustadt is clearly far from living up to all these challenges. Yet it can work as a reminder of some of the crucial issues future development strategies may likely face elsewhere.

NOTE

1 I draw on data that has been collected in a study on peripherization and urban politics, which I conducted together with Daniel Förste and Laura Colini at the Leibniz Institute for Regional Development and Structural Planning in Erkner from 2012 to 2014. Major findings of this study are published in Bernt, Colini, and Förste 2017.

REFERENCES

Aalbers, M.B. (2011). *Place, Exclusion, and Mortgage Markets*. Chichester, UK: John Wiley & Sons.

Arnold, F. Frier, R., Geissler, R., and Schrauth, P. (2015). "Große regional Disparitäten bei den kommunalen Investitionen" [Large and lasting regional disparities in municipal investments]. *DIW Wochenbericht* 43, 1031–40. http://hdl.handle.net/10419/121413

Bernt, M. (2006). "Fictitious Values, Imaginary Markets." In Oswalt, P. (ed.), *Shrinking Cities*, Vol. 2, *Interventions*, 592–6. Ostfildern, Germany: Hatje Cantz.

– (2009). "Partnerships for Demolition: The Governance of Urban Renewal in East Germany's Shrinking Cities." *International Journal of Urban and Regional Research* 33, no. 3, 754–69. https://doi.org/10.1111/j.1468-2427.2009.00856.x

Bernt, M., Colini, L., and Förste, D. (2017). "Privatization, Financialization and State Restructuring in Eastern Germany: The Case of Am Südpark." International Journal of Urban and Regional Research 41, 555–71. https://doi.org/10.1111/1468-2427.12521

Borst, R. (1996). "Volkswohnungsbestand in Spekulantenhand?" In Häußermann, H., and Neef, R. (eds), *Stadtentwicklung in Ostdeutschland*, 107–28. Opladen, Germany: Westdeutscher Verlag.

Der Spiegel. (2000). "Plattenbauhazardeure" [Prefab gamblers]. Accessed 23 August 2017. http://www.spiegel.de/spiegel/print/d15930875.html

Grossmann, K., and Haase, A. (2015). "Neighborhood Change beyond Clear Storylines: What Can Assemblage and Complexity Theories Contribute to Understandings of Seemingly Paradoxical Neighborhood Development?" *Urban Geography* 37, no. 5, 727–47. https://doi.org/10.1080/02723638.2015.1113807

Haase, A., Bernt, M., Grossmann, K., Mykhnenko, V., and Rink, D. (2016). "Varieties of Urban Shrinkage in European Cities." *European Urban and*

Regional Studies 23, no. 1, 86–102. https://doi.org/10.1177/0969776413481985

Oswalt, P. (ed.). (2005). *Shrinking Cities*. Ostfildern, Germany: Hatje Cantz.

Pallagst, K., Martinez-Fernandez, C., and Wiechmann, T. (eds). (2013). *Shrinking Cities – International Perspectives and Policy Implications*. New York, NY: Routledge.

Szelenyi, I. (1996). "Cities under Socialism – and after." In Andrusz, G., Harloe, M., and Szelenyi, I. (eds), *Cities after Socialism: Urban and Regional Change and Conflict in Post-socialist Societies*, 286–317. Oxford, UK: Blackwell.

Uffer, S. (2013). "The Uneven Development of Berlin's Housing Provision: Institutional Investment and Its Consequences on the City and Its Tenants." In Bernt, M., Holm, A., and Grell, B. (eds), *The Berlin Reader: A Compendium on Urban Change and Activism*, 155–70. Berlin, Germany: Transcript.

4 Learning from the Socialist Suburb

STEVEN LOGAN

In the opening pages to *Extrastatecraft* (2014), architectural theorist Keller Easterling places infrastructure space at the forefront of her thinking about architecture and cities. "Buildings are often no longer singularly crafted enclosures, uniquely imagined by an architect," she writes, "but reproducible products set within similar urban arrangements" (14). This process does not operate only at the level of the single building; entire cities are "constructed according to a formula." Although every global city has its signature building(s), Easterling argues that "the most prevalent formula replicates Shenzhen or Dubai anywhere in the world with a drumbeat of generic skyscrapers" (15). Easterling attaches this formulaic way of building cities to "free zones," golf courses, container ports, and, historically, to the mass-produced suburbs of places like Levittown. Critiques by Lewis Mumford (1961) or Humphrey Carver (1962) deride the mass-produced suburbs for their lack of architecture, but as Easterling (1999, 113) notes, "The organizational protocol was not merely that which facilitated the architecture: it was the architecture."

Easterling (1999) establishes a connection between the organization of the postwar suburb and the increasing importance of logistics and organization to contemporary global urbanism. This was an argument that had already been occupying Henri Lefebvre in *The Production of Space* ([1974] 1991, 75), where he contrasted urban space as a work (oeuvre) secreted slowly over time by its inhabitants, to urban space as a work of art by architects and planners, and to the city as a product of the "repetitive gestures" of bulldozers, cranes, machines and workers. The "modern towns, their outskirts and new buildings"to which Lefebvre was referring were both "reproducible" and the "result of repetitive actions."

The tension between the city as product of technologies of mass production and the city as a work of art planned in advance by architects and urbanists was central to the building of the massive socialist suburbs, the genesis of which stretch back to the architectural avant-garde of the 1920s. In a 1929 debate with Le Corbusier, Karel Teige, one of the key figures in the vibrant Czechoslovak avant-garde, claimed that instead of monuments, architecture should create instruments. In the 1920s, Teige critiqued the fetishization of modern architecture, in which design elements and a building's appearance came to be taken for a modern architecture that was increasingly emptied of what Teige saw as its fundamental reason for being: its function as an instrument of change, an architecture that challenged the status quo in the interests of society as a whole. Modern architecture was not simply about challenging old designs, but challenging architects who built villas for the rich instead of housing for the working class, hospitals, and schools. Teige ([1930] 1977) called these instruments "architecture without architecture." Instead of choosing between "architecture or revolution," as Le Corbusier posed the question in 1923 in his *Vers une architecture*, Teige ([1930] 2000, 298) wanted to "revolutionize architecture" and make it above all an "organizational process," a process in which architects would collaborate with sociologists, economists, artists, and others on a new and specifically socialist organization of society, and a complete remaking of the housing system. For Teige, architecture and construction were inseparable, and both were instruments for social change, while for Le Corbusier, architecture as a monumental work of art could only begin when construction ends and the engineers have finished their work.

This debate occupies an important place in the history that both Easterling and Lefebvre sketch out, and it is also an important aspect of the state socialist building programs of the 1960s, where architects and urbanists were charged with designing new towns from scratch. These developments were dominated by the construction of prefabricated concrete apartment blocks. These blocks, the lasting image of socialist urbanism, are the result of a repeatable spatial formula that came to define the postwar building machine. Czech architects even came up with a moniker for this kind of building: crane urbanism. The construction cranes and the tracks upon which they moved allowed for the quick and generic construction of cities throughout the Eastern Bloc and beyond.

This chapter makes the argument that the socialist suburb occupies a key moment in the history of postwar suburbanization, which too often focuses on the history of places like Levittown, rather than

4.1 A forgotten public space. Jižní Město and other socialist suburbs are dotted with sculptures commissioned during the socialist era, particularly in the 1970s and 1980s. Photo by Steven Logan.

their lesser-known socialist counterparts. This chapter has three parts: The first considers the socialist suburb, or sídliště in Czech, as an important typology in the history of massive suburbanization. Like the mass-produced suburbs of single-family houses, the sídliště were subject to persistent criticism in the 1960s that criticized the lack of architecture and urbanity in these mass-produced suburbs. The second part considers crane urbanism in more detail, and how it clashed with architects and planners' representations of space. In the third part of the chapter, I look more specifically at the tension between the city produced as a work of art in response to the supposed deficiencies of the sídliště, using the Prague suburb of Jižní Město (South City) as an example.

A Short History of the Sídliště

Like the postwar suburban landscapes of North America, the massive suburbs of the postwar periphery of Prague were derided from the moment of their inception. Although the sídliště were usually situated within rather than beyond the city limits, they were criticized for their lack of urbanity. Architects, planners, and sociologists described the existing sídliště as "grey, monotonous, dull, [and] lifeless" (Gottlieb and Todlová 1969, 211), "parasitic" (Nový 1971, n.p.), and "dormitory suburbs" lacking in public, social space (Hrůza 1967, 1). Sídliště is a notoriously difficult word to translate. It is essentially an archaeological term that predates its usage in the postwar context and can refer to any permanent settlement for a group of people, and more generally a settlement (osídlení) of any kind (Zadražilová 2007, 41). Architect and urban planner Karel Maier (Maier and Gerbert 1998, 2) calls the sídliště a "mass housing complex," which does not refer only to its postwar incarnation: since the industrial revolution, the sídliště have been about how to house a lot of workers in one place.

The term as it is used today, however, has come invariably to refer to the postwar peripheral housing developments built almost exclusively using prefabricated concrete panels, referred to in Czech colloquially as paneláky. In his 1985 seminal sociological study *Lidé a sídliště* (People and the sídliště), Jiří Musil (1985, 13) writes that the sídliště are "one of the most distinct markers of socialist-city building and socialist architecture." The Czech online dictionary Slovnik.cz offers sixteen different English translations. These include housing estate, housing development, neighbourhood unit, commuterland, commuterdom, and, very simply, blocks. Although the usual translation is "housing estate," I find this term does not resonate outside the UK, and as such does not capture both its positive and negative connotations, similar to such heavily loaded terms like "suburbia" or the French "grandes ensembles" (the sídliště are sometimes referred to as "large residential complexes").

The many definitions are themselves illuminating, particularly the association with Clarence Perry's neighbourhood unit, a transnational planning model (Ward 1999) that sought to define a neighbourhood – for Perry organized around a public school – that would include shops, parks, and community facilities within walking distance of every house and roads that discouraged car through traffic. In this sense, a sídliště would be made up of multiple neighbourhood units. Like their North American counterparts, the sídliště's association with commuting – commuterland, commuterdom – emphasizes how, although they were initially planned as self-sustaining settlements, the sídliště often turned into the socialist version of commuter suburbs. They too had their stereotypical depiction of the morning commute: instead of people all pulling out of their driveways at the same time, in the sídliště it was the long lineups at the bus stop (or in the best case, lines of people walking to the metro).

Although suburbs imply settlements outside of or near to the city, Häussermann (1996, 218) argues that if there was a form of socialist suburbanization, it was restricted to the "new high-rises on the city border, but within the city limits." That they were within the city limits led some scholars to suggest that these places are not a socialist version of suburbs, but a "direct application of functionalist city planning principles" (Sýkora and Mulíček 2014, 136), and that even if they look different from the nineteenth or early twentieth-century city, they are still part of the city-building project. Stanilov and Sýkora (2014a, 19) argue that the sídliště have specifically urban traits: they were built as extensions of the city, rather than as self-contained communities; they were "contiguous with the urban fabric"; and their "high urban densities" distinguished them from the surrounding countryside, especially since their growth boundaries were very clearly delimited, unlike the uncontrolled sprawl of some single-family-house suburbanization.

And yet in that same volume, Stanilov and Sýkora (2014b, 259–60) suggest that the postwar developments also exhibit "typically suburban characteristics," such as their "peripheral location" and the exclusive focus on residences. They were seen as "dormitory communities" because the residents most often had to commute elsewhere, either to the city centre or to another sídliště, for work. Musil (1985, 19) notes that the sídliště were generally located far from work, the city centre, cultural destinations like the cinema and theatre, and, in some cases, schools.

Although they were technically part of the city, inhabitants still perceived the sídliště as separate from it. The new style of building that the sídliště entailed – in keeping with modernist design principles of uniform towers, rejection of the traditional street and its mix of functions, and abundance of green, open spaces – made it difficult for people to believe that the sídliště were actually part of the city. In his study, Musil compares life in the sídliště with life in the older inner-city neighbourhoods of a number of different cities in Czechoslovakia. Unlike traditional forms of suburbanization where people flee the city to the quieter preserves of the periphery, many of the people who moved to the sídliště came from surrounding villages and towns to work in Prague (see Stanilov and Sýkora 2014b, 257). Musil writes that the participants in the study had difficulties with the semantics of the sídliště:

the majority of the people interviewed were unable to categorize the sídliště in terms of traditional concepts like city, suburb, or small town. The sídliště, he concludes, "is a phenomenon that cannot be classified with the help of old terms" (Musil 1985, 319).

Jiří Hrůza, an influential urban planner both during socialism and after, places what he called the "crisis of the sídliště" firmly within the social spaces of the sídliště: "community spaces," which in the old city are the street and the square, are missing from the sídliště. Hrůza (1967, 1) claims that these spaces traditionally serve both a social and a transiting function. Following a long line of modernists, he argued that in the building of the modern city, traditional streets, where different functions co-exist, are "unworkable." Modern forms of traffic had made that mixing untenable, so it was necessary to find a new solution. But in the sídliště, where separation of pedestrian and car traffic had become the norm, all attention was given to the function of transportation – usually, car, bus, and metro – while the task of replacing the social function of the traditional street had, regrettably, been "forgotten." Hrůza argues that the endless pedestrian pathways through wide open spaces that permeate the sídliště, and many developments like it worldwide, do not make up for the social space of the street because they are above all about getting from one place to another as quickly as possible. It was in this period, culminating in the 1964 urban plan, that Prague's three biggest postwar developments were envisioned: Severní Město (North City), Jihozápadní Město (Southwest City), and Jižní Město. These developments signalled a "necessary transition" from the smaller and more isolated sídliště to much larger settlements: "cities" (Borovička and Hrůza 1983, 89).

Cranes, Not Visions

The lasting image of the postwar socialist suburb is the row upon row of prefabricated apartment blocks in a landscape wiped clean by their construction. On 7 December 1954, Soviet President Nikita Khrushchev gave a speech to the All-Union Conference of Builders, Architects, and Building Industry Workers extolling the virtues of the industrialization of building. "Given concrete, electric motors, and lifting cranes, and other machinery," claimed Khrushchev ([1954] 1963, 161), "it is impossible to continue to work in ancient ways." Paralleling Teige's critique of Le Corbusier twenty-five years earlier, Khrushchev called for standardized designs in building, and he critiqued those architects who would rather "build monuments to themselves" (165). No apartment building should be a "replica of a church" – the architect might need "beautiful silhouettes," argued Khrushchev (170), but "the people" need apartments. The speech brought the consolidation of the building industry in Czechoslovakia and would completely change how buildings and cities were constructed. In order to take advantage of economies of scale, developments had to be on a large scale. The construction companies and state investors wanted projects of 5,000 to 15,000 apartments and not "the smaller housing projects of low-rise, detached buildings in order reap (supposed) economies of scale" (Szelenyi 1996, 305). In 1960s Prague, planners like Jiří Hrůza (2006, 38) insisted that maintenance and reconstruction of the existing building stock were necessary; however, the state had different priorities: "We had to fulfill the demand: find land for around 11,000 apartments annually." As a result, the dilapidated housing in the older, more central neighbourhoods contrasted with the new housing built on the periphery; in this context, it is not surprising that when comparing responses from his study, Musil found that people in the sídliště were more satisfied with their apartments than people living in the older, ill-maintained apartments in the city. The whole building apparatus was geared to such development: the pre-fabricated panels were constructed off-site in "house factories" (Szelenyi 1996, 305), and transported at great expense on the unprepared road system to the construction site, where workers simply had to assemble the panels into apartments with the help of cranes, which themselves required vast swaths of open space to operate.

The privileged actor in the industrial construction of these housing developments was neither the worker, reduced to stacking boxes one on top of the other, nor the architect, sidelined by the construction, but the crane and the tracks along which it moved – one of the central images of the industrialization of dwelling in socialist countries. Hrůza (2006, 38) describes crane urbanism thus: "For them [the state building organization] it was easiest if everything was made with one kind of panel, brought to the site and assembled." The crane also could not function without the rail tracks upon which it moved. Jiří Lasovský, Jižní Město's chief architect, points out that the unbelievably long buildings that characterize the sídliště were often a result of economizing on costs in what was otherwise a very expensive and complex process: it was cheaper to keep building, or to build in the immediate vicinity of the crane, than to have to move the track upon which the crane moved.[1] Significantly, the only place where there was enough room for the tracks, and where the economies of scale of crane urbanism and the state building quotas could be met, was on the open, flat spaces of the city's periphery (the tracks of the crane could not be built on Prague's hillsides). Crane urbanism might be better thought of as crane suburbanism. In neglecting the creative impulses of the architects and the planners, crane urbanism is the ultimate expression of anonymous mass production that characterizes twentieth-century industrialization, aptly named by Sigfried Giedion ([1948] 1969) "Mechanization Takes Command."

The attention to building dwellings to the exclusion of anything else save for the most essential infrastructure was typical of the sídliště of the postwar period. The cost of building meant that only the most necessary social services – like hospitals, clinics, and schools – would be constructed. Residents in Sofia call the large dwelling complexes "un-complex complexes" – that is, although the large dwelling complexes were imagined to be harmoniously composed ensembles with shops, cafes, cinemas, sport facilities, and so on, they ended up as a collection of individual residential buildings within an unfinished, and unforgiving, landscape (Hirt 2012, 87). Socialist city scholar Sonia Hirt (40) writes that although the sídliště will sometimes have overt, grand public spaces like "friendship parks," or large cultural centres, they often suffer from a lack of small retail, and "in-between" public spaces like cafes, small shops, or yards. Within the context of the codification of modern architecture and urbanism as an instrument of the state, as Tafuri writes (1998, 22), "the only task the architect can have is to organize that cycle." In his building speech, Khrushchev rejected the individualist caprices of the architect, which would in the post-1954 period turn the architect as artist into a "technical expert" (Beyer 2011, 89) and "technician" (Zarecor 2011, 295). In his history of concrete, Forty (2012, 249) notes that the biggest threat to architects in the postwar period was the system building of the paneláky, where the architect became simply a technician whose main role was the site arrangement of the buildings. The birth of the postwar socialist suburb took place in a space already coded by technologies of industrialization and state power, and expressed a tension between the city to be born and the technologies that would induce labour, to bring about the birth as quickly as possible. When the industrial overwhelms the urban, the socialist suburb becomes "a product strictu sensu: it is reproducible, and it is the result of repetitive actions" (Lefebvre [1974] 1991, 75). This production process mobilizes materials, be they concrete or steel or what Lefebvre calls the matériel (tools, instructions, know-how, etc.). The product is not a totality, but a result of conceiving space as both homogeneous

4.2 Jižní Město seen from the east just beyond the artificial lake. Jižní Město's row houses as well as houses from the pre-sídliště settlement can be seen just beyond the trees. Photo by Steven Logan.

and fragmented. Drawing on the work of Lefebvrian Polish geographer Bohdan Jałowiecki, Stanek (2011, 67) writes that when "industry became the main, if not the sole, factor in the construction of cities," the result was "spatial segregation, loss of urbanity, and ecological damage."

How a Suburb Is Born

"Jižní Město does not want to be a sídliště." So read the headline in the daily newspaper *Lidová demokracie* in July 1970, a headline that sums up the contrasting architectures at work in the socialist suburb: the architecture of crane urbanism increasingly dominating city building, and the architectural and urban plans for Jižní Město, which would draw on the utopian energies of 1960s modernism.

The plan was to build 20,000 apartments for 80,000 new residents on 1,200 hectares (the largest but least dense of the existing sídliště). Jižní

Město was built on the site of two already-existing independent communities, Chodov and Haje, in the south-east part of Prague in 1967. Prague widened its borders for the first time since 1921, annexing twenty-one independent municipalities, including Chodov and Haje (Borovička and Hrůza 1983, 90). Although some houses would need to be demolished, for the most part the old neighbourhoods were left intact.

Although Hirt (2012, 85) argues that the socialist suburbs "followed the tenets of modernist urbanism precisely as they were outlined in the Athens Charter," the role of modernist urbanism in the planning of Jižní Město was more complex. Although the separation of pedestrian and car traffic enshrined in the Athens Charter was a key aspect of Jižní Město's design, there was a significant movement in the postwar period critiquing the separation of functions, in particular the importance attached to automobile circulation. The goal was rather multi-functional centres, work close to dwellings, and shops and cafes easily reachable on foot on what Lasovský called the obytná ulice, or habitable street. Reflecting Lefebvre's contrast between dwelling and habiting (Lefebvre [1970] 2003), Lasovský's (1975, n.p.) aim was "to help create a typical urban environment in which people would not just inhabit [bydlet], but where they would be at home [doma]."

All through-traffic in the residential areas would be pedestrian, as residents would leave their cars in parking lots or garages before entering the car-free neighbourhood. An extensive pedestrian network would link Jižní Město's four neighbourhoods – Opatov, Litochleby, Haje, and Chodov (the "garden" neighbourhood) – each of which would have 20,000 residents. These streets would be complemented by a system of recreational pedestrian paths that would wind from the periphery to the individual centres. These recreational pedestrian paths would follow strips of greenery that extended from the peripheral landscape, allowing pedestrians to walk around the city or to the park out into the surrounding forests "without conflict with traffic and from most parts without any interruptions in the path" (Lasovský 1975, n.p.). In an attempt to create a diverse environment, the four neighbourhoods of Jižní Město would each be designed by a different collective of architects and would each use different building materials and technologies, not just the prefabricated concrete panels. The neighbourhoods were composed in such a way that the tallest buildings were clustered around the local centres, and the further from the centre one got, the less dense and tall the buildings became, until one reached the open spaces and the one- and two-storey row housing on the periphery, where the density would be around 100 people per hectare in contrast to the 350 people per hectare in the most dense neighbourhood. Despite the equation of the postwar housing estate with high-rise housing, Jižní Město was one of the few Prague sídliště where single-family houses were built; in the 1970s and 1980s, single-family housing made up 7 per cent of new housing construction in Prague (Maier, Hexner, and Kibic 1988, 60).

When the first residents moved into Jižní Město in 1976, much had changed. The hope and optimism of the 1960s both to transform the sídliště and draw on the energies of global modernism dissipated following the Russian occupation of Czechoslovakia in August 1968 and the political purges in the subsequent eighteen months. It put an end to the Prague Spring, and with the purges also affecting the Office of the Chief Architect in Prague, it soon thereafter put an end to the architects' plans for Jižní Město. The focus was on building apartments as quickly and efficiently as possible. Not only was Jižní Město the largest of the peripheral housing developments to date, it was also built in near-record speed. Lasovský put it thus: "The Prague Fathers want to set a record in Jižní Město. As opposed to the roughly 2,000 apartments delivered in 1969, they now want to build 5000 apartments yearly" ("Jižní Město Pražské" 1970, 3). In the ensuing drive to build apartments, many of the plans of the architects were sacrificed, including the multi-leveled local centres, the numerous parking garages that were to keep the streets within the neighbourhood relatively free of cars, and most notably, Jižní Město's city centre, the key aspect to bringing urbanity to Jižní Město.

As the industrial logic of crane urbanism became ever more obscene, Lasovský, along with the artists he brought on to work with him, attempted to construct the ideal urban scene in ever more fantastical ways, a kind of phantasmagorical counterpart to the damage inflicted by crane urbanism. However, he knew the city centre was never going to get built, because there was only money for the models. Despite that – or because of it – they created really elaborate plans. In her work on city centres in the GDR and the USSR, Elke Beyer (2011, 89) writes that architects and planners were "painfully aware of the limitations of their position within the existing system of state socialism," and yet at the same time they were coming up with elaborate models drawn from the tenets of the utopian modernist urbanism of the 1960s. This was especially the case with Jižní Město, as the infrastructure costs alone were immense, in particular building the metro out to Jižní Město: the further from the city, the more expensive the new sídliště became. The metro had to be extended five kilometres out to South City, and although the state claimed it was the most inexpensive

4.3 People gathering on the pedestrian walkways in the spaces between buildings. Photo by Steven Logan.

option, "this was an illusion," recalled Hrůza (2006, 37). Like its capitalist counterparts, the socialist economy was based on "extensive growth."

Lasovský was forced to step down as head architect and his atelier was disbanded. Although political purges were widespread, Hrůza (2006, 38) believed that in accounting for the gap between the many interesting, but unrealized designs, "the worst lobby was the state building organization – our biggest partner and the biggest antagonist." Lasovský places the blame squarely on the crane, his "successor" as chief architect. Although he had imagined different building methods and material for each of the neighbourhoods, only the concrete pre-fabricated panels were used, and the only difference was length – the ideal was the infinitely long building with almost no variation from one section to the next – and height – 4, 6, 8, or 12 storeys. And the crane works best when the tracks are straight, so the ideal building in a city of crane urbanism was the seemingly endless apartment building (one of which Jižní Město residents would dub the Great Wall of China). Lasovský had been interested in building low-rise apartments, which "should only be as tall as the tallest tree." Because the buildings could be closer together, recalls Lasovský, the density would be the same as the tall buildings, and the costs would be less as no cranes would be necessary. Lasovský's final remark on the building of Jižní Město was thus: "a total absence of architecture." Although he may have more accurately said, architecture without architecture.

Conclusion: Signs of Life?

"Learning from." The term comes from the well-known study of the architects Robert Venturi and Denise Brown, "Learning from Las Vegas," in which they undertook an analysis of the universally derided landscape of Las Vegas. A similar thrust governed their subsequent study "Learning from Levittown," which also emerged because architects were universally deriding the vernacular landscapes of postwar suburbia, built not by architects but by merchant builders, and the ways in which inhabitants modified their houses both on the outside and inside. Their inspiration was the sociologist Herbert Gans, who lived in Levittown. Venturi summed up their approach thus: "I want to look at what I hate because I am going to learn from that" (Colomina 2008, 49). The study culminated in an exhibition entitled *Signs of Life: Symbols in the American City* (1976). In her history of housing in Czechoslovakia from 1945 to 1960, Kimberley Zarecor (2011) points to the tension between the architecture of the unique building and an architecture of manufacturing, mass production, and materials. Offering a response to Lasovský's claim that there is no architecture in Jižní Město, Zarecor writes that when the architect ceases to practice his or her art, then architecture becomes "mere 'building' instead of architecture with a capital A" and is written off by historians as simply "economic and technological determinism" (296). From Lasovský's perspective, there is no "Architecture" because there were none of the great, monumental works of architecture that emerge from the mind of the heroic architect.

"Signs of Life." The seeming failure of places like Jižní Město has been now contrasted with their seeming success: Jižní Město, although a sea of mud and construction debris in the 1970s and 1980s, has grown into a garden suburb, surrounded by forests and even an artificial lake. This is not just a case of capitalism saving the socialist suburb, as many of the aspects which attract people – the good public transportation connections, the surrounding forests, the green spaces between buildings, and the separation of cars and pedestrians – were aspects of the socialist and modernist urbanism of the 1960s. In *Militant Modernism*, Hatherley (2008) argues that returning to left modernisms does not mean treating modernism as heritage, the architecture of which should be preserved, but looking to the visions and ideas that were behind their building, so much more than an architectural style of an individual building. Hatherley finds in the "concrete walkways and windswept precincts" of the British 1960s megastructures of Cumbernauld or the Barbican in London, as Walter Benjamin did in the arcades of the nineteenth century, a nostalgia for a modern future that did not quite happen. Given that these structures were produced at the height of British postwar social democracy, Hatherley argues that they offer a critique of present-day inequalities, particularly around architecture. The socialist suburb offered housing for everyone, so doctors lived next to bricklayers, actors next to taxi drivers. The modernism of the 1920s and the 1960s was "immersed in the quotidian," while today's architecture that still calls itself modernist is based on the spectacular and a "distance between itself and everyday life" (Hatherley 2008, 8, 12). It is only recently that Czech architectural theorists and historians have begun to treat these spaces as worthy of attention, whether for the utopian impulse behind their building or the particular ways that inhabitants negotiate what in many cases are complex, unfinished projects in urban building. Crane urbanism is not simply a concept to describe what is now a moment in the past. As Easterling (2014) and others have shown, infrastructure is not simply the invisible background to urban space, but is becoming increasingly visible as "the urban structure itself" (15). The disappointing beginnings of Jižní Město lies in the two extremes of architecture it presented. Those same extremes – star architects designing signature buildings and the mass production of cities in rapidly urbanizing countries like China – are still present. Learning from the socialist suburb suggests a productive meeting ground, as proposed by Lefebvre, between the city as product and the city as work of art. As architectural historian Lloyd Alter (2014, n.p.) writes, "Buildings are not isolated Frank Gehry sculptures; they exist to house people and give them places to work ... They should serve a societal need." The massive socialist suburb, for all of its failures, occupies an important place in the history of architecture and urbanism on the periphery.

NOTE

1 All reference to Lasovský, unless otherwise noted, are from an interview conducted on 24 June 2012 in Prague. The interview was conducted in Czech.

REFERENCES

Alter, L. (2014). "It's Time to Dump the Tired Argument That Density and Height Are Green and Sustainable." *TreeHugger*, 3 January. https://www.treehugger.com/urban-design/its-time-dump-tired-argument-density-and-height-are-green-and-sustainable.html

Beyer, E. (2011). "Planning for Mobility: Designing City Centres and New Towns in the USSR and the GDR in the 1960s." In Siegelbaum, L.H. (ed.), *The Socialist Car: Automobility in the Eastern Bloc*, 71–91. Ithaca, NY: Cornell University Press.

Borovička, B., and Hrůza, J. (1983). *Praha: 1000 let stavby města*. Prague, Czech Republic: Panorama.

Carver, H. (1962). *Cities in the Suburbs*. Toronto, ON: University of Toronto Press.

Colomina, B. (2008). "Learning from Levittown: A Conversation with Robert Venturi and Denise Scott Brown." In Blauvelt, A. (ed.), *Worlds Away: New Suburban Landscapes*, 49–69. Minneapolis, MN: Walker Art Center.

Easterling, K. (1999). "Interchange and Container: The New Orgman." *Perspecta* no. 30, 112–21.

– (2014). *Extrastatecraft: The Power of Infrastructure Space*. Brooklyn, NY: Verso.

Forty, A. (2012). *Concrete and Culture: A Material History*. London, UK: Reaktion.

Giedion, S. (1948) 1969. *Mechanization Takes Command*. New York, NY: W.W. Norton & Co.

Gottlieb, M., and Todlová, M. (1969). "Sociologické poznámky k jižnímu městě" [Sociological Notes on Jižní Město]. *Architektura ČSR* 15, no. 4, 211–14.

Hatherley, O. (2008). *Militant Modernism*. Winchester, UK: O Books.

Häussermann, H. (1996). "From the Socialist to the Capitalist City: Experiences from Germany." In Andrusz, G.D., Harloe, M., and Szelényi, I. (eds), *Cities after Socialism: Urban and Regional Change and Conflict in Post-socialist Societies*, 214–31. Cambridge, MA: Blackwell.

Hirt, S. 2012. *Iron Curtains: Gates, Suburbs, and Privatization of Space in the Post-socialist City*. Hoboken, NJ: Wiley & Sons.

Hrůza, J. (1967). "Krize sídliště" [Crisis of the Sídliště]. *Architektura ČSR* 13, no. 4 4, 1.

– (2006). "Jiří Hrůza." Interview by Ševčík, O., and Popelová, L. In Ulrich, P. (ed.), *Šedesátá léta v architektuře očima pamětníků*, 40–51. Prague, Czech Republic: Nakladatelství ČVUT.

"Jižní Město Pražské: Únorová diskuse u magnetofonu" [Prague's Jižní Město: February Discussion by the Tape Recorder] (1970). *Československý architekt* 16, no. 6–7, 1–3.

Khrushchev, N. (1954) 1963. "On Wide-Scale Introduction of Industrial Methods, Improving the Quality of and Reducing the Cost of Construction." In Whitney, T. (ed.), *Khrushchev Speaks: Selected Speeches, Articles, and Press Conferences, 1949–1961*, 153–92. Ann Arbor, MI: University of Michigan Press.

Lasovský, J . (1975). "Naděje a skutečnost" [Hope and Reality]. *Československý architekt* 28, no. 15–16, n.p.

Lefebvre, H. (1974) 1991. *The Production of Space* (Nicholson-Smith, D., trans.). Cambridge, MA: Basil Blackwell.

– (1970) 2003. *The Urban Revolution*. Minneapolis, MN: University of Minnesota Press.

Maier, K., and Gerbert, J.G. (1998). "Sídliště – díl první: Strašidla, mýty, realita" [Sídliště: First Volume: Specters, Myths, Reality]. *Umění a řemesla* 4, 2–6.

Maier, K., Hexner, M., and Kibic, K. (1998). *Urban Development of Prague: History and Present Issues*. Prague, Czech Republic: ČVUT.

Mumford, L. (1961). *The City in History*. New York, NY: Harvest Books.

Musil, J. (1985). *Lidé a sídliště* [People and the Sídliště]. Prague, Czech Republic: Nakladatelství Svoboda.

Nový, Otakar. 1971. Introduction to *Architekti Praze* [Architects for Prague], n.p. Prague, Czech Republic: HDKN Praha.

Stanek, Łukasz. 2011. *Henri Lefebvre on Space: Architecture, Urban Research, and the Production of Theory*. Minneapolis: University of Minnesota Press.

Stanilov, K., and Sýkora, L. (2014a). "Postsocialist Suburbanization Patterns and Dynamics: A Comparative Perspective." In Stanilov, K., and Sykorá, L. (eds), *Confronting Suburbanization: Urban Decentralization in Postsocialist Central and Eastern Europe*, 256–95. Malden, MA: Wiley-Blackwell.

– (2014b). "The Challenge of Postsocialist Suburbanization." In Stanilov, K., and Sykorá, L. (eds), *Confronting Suburbanization: Urban Decentralization in Postsocialist Central and Eastern Europe*, 1–32. Malden, MA: Wiley-Blackwell.

Sýkora, L., and Mulíček, O. (2014). "Prague: Urban Growth and Regional Sprawl." In Stanilov, K., and Sykorá, L. (eds), *Confronting Suburbanization: Urban Decentralization in Postsocialist Central and Eastern Europe*, 133–62. Malden, MA: Wiley-Blackwell.

Szelenyi, I. (1996). "Cities under Socialism – and after." In Andrusz, G.D., Harloe, M., and Szelényi, I. (eds), *Cities after Socialism: Urban and Regional Change and Conflict in Post-socialist Societies*, 286–317. Cambridge, MA: Blackwell.

Tafuri, M. (1969) 1998. "Toward a Critique of Architectural Ideology." In Hayes, K.M. (ed.), *Architecture Theory since 1968*, 3–35. Cambridge, MA: The MIT Press.

Teige, K. (1930) 1977. "K sociologii architektury" [Towards a Sociology of Architecture]. In *RED: měsíčník pro moderní kukturu* [RED: Monthly for Modern Culture], 163–223. Wurzburg, Germany: jal-reprint.
– (1930) 2000. "Modern Architecture in Czechoslovakia." In Teige, K., *Modern Architecture in Czechoslovakia and Other Writings* (Murray, I.Z., and Britt, D., trans.), 59–308. Los Angeles, CA: Getty Research Institute.
Ward, S.V. (1999). "The International Diffusion of Planning: A Review and a Canadian Case Study." *International Planning Studies* 4, no. 1, 53–77. https://doi.org/10.1080/13563479908721726
Zadražilová, L. (2007). "Domov na sídlišti: mytus nebo realita?" [Home in the Sídliště: Myth or Reality?]. In Hubatová-Vacková, L., and Říha, C. (eds), *Husákova 3 + 1: Bytová kutura 70. let* [Husákova 3 + 1: Apartment Culture I the 1970s], 39–56. Prague, Czech Republic: VŠUP.
Zarecor, K.E. (2011). *Manufacturing a Socialist Modernity: Housing in Czechoslovakia, 1945–1960*. Pittsburgh, PA: University of Pittsburgh Press.

5 Decline and Renewal in Toronto's High-Rise Suburbs: The Tragedy of Progressive Neoliberalism

DOUGLAS YOUNG

Introduction

The urban landscape of Toronto is unusual in the North American context given the large number of high-rise apartment buildings that were constructed in what were, in the 1950s, 1960s, and 1970s, peripheral suburban locations. Today, these now older or inner suburbs are seen as problem districts of worn-out rental housing and concentrations of racialized poverty. At the same time, in a city that has seen almost no new social housing built since the mid-1990s, the suburban rental towers represent a large stock of relatively affordable housing. One of the responses to "the problem of the inner suburb" is a City of Toronto program called Tower Renewal that was launched in 2008 with the goal of encouraging rental tower owners to rehabilitate their buildings.

In this chapter, I consider Tower Renewal as a case study of ever-evolving modes of neoliberal urban governance. In Toronto, such modes contribute to new understandings of "common sense" (Harvey 2005) by reshaping the discursive and material dimensions of processes of suburban decline and renewal. The strategy of Tower Renewal is to weave a progressive thread through an otherwise explicitly neoliberal project. This strategy appears to validate the potential of everyday life in postwar suburban districts, while at the same time supporting the idea that the market can be a good provider of affordable housing. The consequences of this "progressive" mode of neoliberal governance are tragic. Several years into Tower Renewal's mandate, it has contributed very little to improving the material well-being of inner suburban residents. Yet, in that tragedy, we can also see its success. The state limits its direct participation in non-profit housing provision and dampens any calls for new state-funded non-profit housing programs by appearing to take progressive action on housing for low-income households as part of a broader program of suburban renewal.

Toronto's High-Rise Suburbs

Toronto's high-rise suburbs are the result of planning policy from the 1960s that encouraged the development of more urban suburbs than had been built in the 1940s and 1950s. District plans of the 1960s called for a substantial number of dwelling units to be in the form of high-rise buildings. The result is the remarkable landscape of what are now referred to as the inner or older suburbs, with high-rise rental apartment buildings lining arterial roads and sometimes grouped into clusters of tall buildings. This type of landscape is what German urbanist Tom Sieverts (Sieverts 2003) and others (see Young, Wood, and Keil 2011) have described as an in-between city. It is a landscape that mixes elements of "typical" North American suburbia (wide roads, shopping centres, parking lots, neighbourhoods of detached bungalows, and land uses segregated by primary function) with elements of "typical" central city features (large numbers of immigrants, low household incomes, clusters of public housing, and notably, many high-rise buildings). Between roughly 1950 and 1980, more than 1,100 high-rise rental buildings were built across the city, many of them in what are now considered the inner suburbs. They contain more than 300,000 dwellings (about 30 per cent of the city's total) and are home to almost 20 per cent of the city's population (at least 500,000 people in a city of 2.7 million). Most of the towers are privately owned for-profit buildings with a minority in the non-profit sector. Ownership of the private sector buildings is quite varied and ranges from families with one building to large property management companies owning large portfolios to Real Estate Investment Trusts (REITS) that, likewise, have extensive holdings. Almost all the non-profit high-rise buildings are owned and operated by the Toronto Community Housing Company (TCHC), the city's non-profit housing authority. A much smaller number of the non-profit buildings are owned and operated as cooperatives.

Typically, the rental towers took the form of a slab building between fifteen and twenty-five storeys in height with a poured-in-place concrete structure and exterior walls built of concrete block faced with

5.1 Typical slab form private sector rental apartment building in Toronto's inner suburbs. Photo by Douglas Young.

brick (figures 5.1 and 5.2). A single main building entrance leads to a lobby and bank of two or three elevators shared by all residents of the building. Tenant mailboxes are located off of the lobby as is a property management office. On each floor, a double-loaded corridor provides access to, usually, twelve apartments (six on each side of the corridor) containing one, two, or three bedrooms in addition to the living room, dining room, kitchen, and bathroom. In most cases, each apartment has a balcony. Thus, a typical building contains 180 to 300 apartment units and houses in the range of 350 to 700 residents. Generally, one parking space per dwelling unit is provided in a single level underground parking garage that spreads quite far, limiting the potential for tree-planting above. Most towers are surrounded by extensive lawns with minimal additional landscaping. Often the only common space provided for residents is a laundry room located on the ground floor or in the basement. Most privately owned towers have a building superintendent who lives on site. The non-profit buildings generally have a property manager onsite during regular office hours.

5.2 Typical slab building in Toronto's inner suburbs, this one owned and operated as public housing. Photo by Douglas Young.

Many towers were built at what was, in the 1960s and 1970s, the geographical periphery of the city, some ten to twenty kilometres from the historic downtown. Today, they are no longer peripheral geographically, but they do occupy a social and political periphery. In a city region experiencing a several-years long condominium boom that is "Manhattanizing" large parts of it (the historic downtown, segments of the lakefront, and other locations next to high-order transit) into very dense neighbourhoods of all glass, forty- to eighty-storey point towers, the aging slab-form rental towers in the inner suburbs are increasingly perceived as shabby, badly sited in neighbourhoods with poor walkability and generally representative of the "failings" of modernist-era planning. Suburban tower neighbourhoods are widely considered today to be a problem in need of a solution. But, in the current era of austerity-driven municipal budget-squeezing, the solution to their social and physical problems is not readily apparent.

This "problem of the inner suburb" and its residential towers raises the broad question of how to govern the legacies of a previous era's urbanism, and how to do that in the specific context of twenty-first-century neoliberal Toronto. Theodore and Peck (2011, 21) provide direction to analysing neoliberal governance by telling us "to denaturalize neoliberal urbanism as a policy paradigm by exploring its origins, its evolution, and its variegated form." We should, they argue, challenge an understanding of neoliberalism as "an authorless, omnipresent, and monolithic phenomenon." We should see it, instead, as "a *constructed* project" (21; emphasis in original). In this regard, the Tower Renewal program can be explored in terms of its contribution to the project of constructing neoliberal governance in Toronto.

Underlying the Toronto-specific nature of the suburban governance challenge is a local crisis in housing affordability caused by several coincidental factors. One of the most important of those factors is the fact that funding for new social housing construction was cancelled by the federal government in the early 1990s and the Ontario provincial government in 1995, which has resulted in a current waiting list for rent-geared-to-income housing in the City of Toronto of 80,000+ households (Ontario Nonprofit Housing Association 2016). In the private market, there has been almost no new purpose-built rental housing created in the past thirty years and the rental vacancy rate in the city was a very low 1.3 per cent in 2016 (Canada Mortgage and Housing Corporation 2016). While rent control has been in place in Ontario since the 1970s, changes made to the legislation in the 1990s weakened it significantly: since 1998 vacancy decontrol applies to any rent-controlled unit when a tenant moves out (the rent can rise to whatever the market will bear; when the unit is reoccupied the new and higher rent is then subject to rent control). As well, new rental units (any occupied for the first time from 1991 on) were exempted from rent control; however, in the face of increasing evidence of the housing affordability crisis in Toronto, the provincial government introduced legislation in May 2017 that would eliminate the rent control exemption on new units. Adding tremendous pressure to all types of housing is continued population growth in the region of about 100,000 people per year.

Suburban Decline and Renewal in Three Frames: Neoliberalism, Urban Politics, and Conceptualizations of Suburbs

Decline and renewal in Toronto's high-rise suburbs can be considered in terms of three frames, each of which appears unstable and in transition: neoliberalism as a mode of social regulation; urban political processes; and conceptualizations of suburbs. The ways in which these processes and concepts are in flux contribute to an understanding of current governance initiatives that are directed at Toronto's high-rise suburbs. I begin by considering the usefulness of applying the prefix "post" to neoliberalism, politics, and suburbs. Are we (or at least is Toronto) now post-neoliberal, post-political, and post-suburban?

Periodization of Neoliberalism

The global economic crisis of 2008 prompted some urban scholars to ask if what they were witnessing marked the end of neoliberalism as the dominant mode of social regulation. Could the crisis be considered such a spectacular failure of neoliberalism that the roughly thirty years run it enjoyed as a political and economic paradigm have come to an end? Larner, Lewis, and Le Heron, for example, writing in 2009, referred to "after neoliberalism" suggesting that, if not a sharp epochal transformation, at least a significant shift in the terrain of social regulation had occurred. Also, in 2009, Keil proposed an updating of Peck and Tickell's (2002) conceptualization of roll-back and roll-out neoliberalism, with what he labelled "roll-with-it neoliberalism" (Keil 2009). While cautioning against a declaration that the world had entered a post-crisis period of post-neoliberalism, he argued that significant changes were occurring in the neoliberal mode. The phases of aggressively "*dismantling* (the welfare state and Fordist compromise) and *rebuilding* (of less state and more market-oriented forms of capital accumulation)" (Keil 2009, 233, emphasis in original), often with a markedly revanchist inflection, had themselves generated new crises in regulation. New forms of regulation were required, forms that Keil (2009) called roll-with-it. While it was clear that roll-back and roll-out neoliberalization had generated crises and required rethinking policies and practices, Keil stressed that any new mode of regulation would still be "thoroughly bounded by the limits set through normalized neoliberal governmentalities" (238). Keil's thoughts reinforced those of David Harvey (2005) who had earlier noted that, while unstable overall, processes of neoliberalization had been quite successful in constructing a new notion of common sense. Neoliberalism had become hegemonic as a mode of discourse. It has pervasive effects on ways of thought to the point where it has become incorporated into the common-sense way many of us interpret, live in, and understand the world (Harvey 2005, 3).

Keil (2009, 239) suggested two possible types of roll-with-it neoliberalization: "more authoritarian, capital-oriented, market-serving policies and political constellations" or "more democratic, populist, reformist, ecological options." Or perhaps contradictory and contested approaches would evolve that incorporated elements of both types, though always within the constraints set by the naturalization and normalization of neoliberalism as a way of understanding urbanization and urbanism. The new approaches would see (as in the case study that is the subject of this chapter) new articulations of state, market and civil society, and the re-engagement of the state in realms from which it had previously rolled-back. The new approaches, however, would adhere to the primary obligation of the state to support continued capital accumulation. Keil seemed to anticipate and describe, not the demise of neoliberalism, but significant shifts in its constitution and operations in keeping with its historical and ongoing uneven evolution.

Post-political Governance

One of the operational shifts of evolving neoliberalism is to depoliticize urban governance; to render it post-political and post-democratic. In order to shelter contentious issues from widespread and time-consuming public debate, new policy initiatives are framed as win-win scenarios of consensus building by stakeholders. Their starting point is an assumption that consensus already exists among those stakeholders in the form of internalized neoliberal concepts of "common sense" (such as austerity); the purpose of any public consultation process is to massage the details of a largely pre-determined outcome. According to Wilson and Swyngedouw (2014, 4), "In post-politics, political contradictions are reduced to policy problems to be managed by experts and legitimated through participatory processes in which the scope of possible outcomes is narrowly defined in advance."

Post-suburbia

As a concept, post-suburbia is intended to capture a multitude of changes underway in urban regions around the world; changes in the relations between districts originally considered suburban and a single dominant historic urban centre. Also in flux are the residents of, patterns of built form of, and ways of living in once-suburban places. In many urban regions there are several centres of population density and intense economic activity rendering the simple binary of dominant centre and subsidiary suburb obsolete. Wu and Phelps (2011, 245) "see the primary value of the term [post-suburbia] in focusing attention on the multi-faceted, and multi-scalar, nature of transformations affecting metropolises." According to Phelps and Wu (2011, 4), post-suburbia should not be understood as representing a clean temporal break with suburbia. In this regard, "the state and its interventions represent a critical continuity between suburbia and post-suburbia." Phelps and Wu (2011, 5) note that "one key aspect around which the new post-suburban politics will coalesce concerns the 'retrofitting' of suburbia and the further urbanization of post-suburbia." They also note that post-suburbanization does not always entail growth but can also entail processes of decline.

In a conceptual article about, specifically, suburban governance, Ekers, Hamel, and Keil (2012, 406) note the relative absence of a suburban focus in most literature on governance. Their goal is to account "for the universalization of suburbanization, while maintaining a focus on the particular manifestations of this global process." They stress the importance of not approaching suburban governance through a lens of periodization as such historical accounts tend to privilege post–Second World War North American models setting them as a base point from which all other models can be traced. Instead the authors propose an approach based on the concept of three modalities of suburbanization – state-led, market-led, and private authoritarian action. In different locations across the globe, the three modes may have unfolded at different points in time reflecting different local histories and politics and "a changing articulation of state, private and self-led development patterns" (410).

How Useful Is the Post in Post-neoliberal, Post-political, and Post-suburban?

Rather than representing an epochal, universal, and coherent shift in each of those realms, "post" can be a useful descriptive, conceptual, and analytical tool in capturing broad shifts in modes of social regulation, politics, suburban form, processes of urbanization, and ways of living.

The particular nature and scale of those shifts, and the relationships among the actors and institutions involved, varies from place to place around the world. In the Toronto case, the concept of an in-between city captures one set of those changes. The Toronto Tower Renewal program offers valuable insight into the construction of post-political neoliberal governance in Toronto's post-suburban in-between city.

The Toronto Case

As noted in the introduction to this chapter, one of the city's responses to the problem of the inner suburb and its residential towers is the Tower Renewal program. It originated in a building science class taught at the Faculty of Architecture at the University of Toronto in 2000. The class examined the case of Toronto's postwar concrete apartment towers and determined that while they have a structural lifespan of 300 to 400 years, every fifty years or so, they will require a complete retrofit of all other aspects: exterior cladding, windows, roofs, mechanical systems, kitchens, and bathrooms. The buildings are now at the stage of needing their first total retrofit. Those ideas were further developed by a student in his master's thesis, picked up by a local architectural firm where he worked, and championed by the then mayor of Toronto, David Miller. In 2008, the city established a Tower Renewal office with a handful of staff reporting to the city manager.

The initial goal of the program was to create supply-side conditions that would entice building owners to rehabilitate their properties. It was argued that building retrofits would reduce energy consumption, improve the quality of everyday life for tenants, and enhance the exchange value of the building as an asset. Post-rehabilitation, the vast tower housing stock of more than 300,000 dwelling units would provide affordable, quality housing for another fifty years. Tower Renewal was presented as a win-win-win idea for building owners, tenants, and the city at large. Subsequently, the goals of the program have been expanded to think at the scale of tower neighbourhood revitalization, including improved amenities and social and community services.

The fundamental policy tool that the program hoped to initiate was for the city to establish a financial institution that would make low-interest loans to tower owners in order to fund improvements to their buildings. The city would use its good credit rating to borrow large amounts of capital at very favourable rates. In cases of non-repayment of loans, it would use its taxing powers to add the outstanding balance to the property tax bill. But, setting up this institution requires provincial approval which has not been forthcoming to date.

Tower Renewal was implemented at a time when suburban issues topped the local policy agenda – a "suburban moment" during which attention focused away from earlier preoccupations with fulfilling a perceived mandate to be culturally competitive (for example, investing heavily in major cultural facilities like a new ballet/opera house) and onto the inner suburbs and their residents. In part, this was driven by what appeared to be, in 2005, a rise in crime in the suburban tower neighbourhoods driven by the social exclusion of many of their residents. At that time, ten years into an aggressively neoliberal Common Sense Revolution launched by the provincial government, the "problem of the inner suburb" drew widespread concern. The inner suburbs came to be demonized as centres of crime, their social exclusion seen as threatening to the well-being of other Toronto residents, and the city's economic competitiveness.

A suite of new policies intended to address those concerns were generated. Transit City proposed the construction of several new light rail lines to connect a number of in-between city districts with each other and with the historic downtown. Priority Neighbourhoods identified thirteen neighbourhoods with intense social need and directed resources to them. Tower Renewal promised the possibility of the retrofit of a vast stock of high-rise rental housing. Together, the three program seemed to offer the potential to dramatically improve everyday life for a majority of Toronto residents, including the hundreds of thousands living in the inner suburbs.

Transit City, like Tower Renewal, was supported by then mayor David Miller, and was thus an easy target for attack by the right-wing populist who defeated him in the mayoral election in 2010. The new mayor, Rob Ford, made it clear that his first priority was to cut Transit City. Several years later, a few elements of the Transit City plan were under construction, but the dream of a well-linked city region with high-order transit throughout the inner suburbs had died. Priority Neighbourhoods did survive the defeat of David Miller and was later expanded to cover more neighbourhoods; although, like all area-specific programs, it has been criticized for its limited scope and its failure to address structural causes of social exclusion.

Tower Renewal also survived the Miller defeat. At the present time, several years into its implementation, we can measure its successes and failures and do so in three registers: discourse, policy, and the materiality of everyday life. In terms of discourse, Tower Renewal has been successful, at least among policy-makers, an informed newspaper-reading general public, and university students. It has helped in putting the question of the inner suburbs on the policy radar. And it has been successful in promoting itself as an ideal solution to housing problems in the inner suburbs. In terms of policy, it has had only limited success. Recently, it has initiated supply-side policy changes in terms of a loosening of zoning regulations to permit non-residential uses in suburban apartment neighbourhoods with the hope that this will spur new investment and retail infill. In terms of material improvements to everyday life of suburban tower residents, it has had extremely limited success.

The funding that the program has managed to cobble together to date has been minimal. Toronto Renovates provided a total of $7.5 million in federal and provincial funding for some upgrades. Eligible buildings had to be privately owned and have rents at the very low end of market. Owners could apply for up to $24,000 per unit. If rents remain low for a period of fifteen years, the loan is forgivable. These funds work out to a total of 312 units or approximately 0.01 per cent of the total high-rise rental stock. Another fund of $10 million was available for energy retrofits. What is actually needed, if the goal is to address the entire rental tower housing stock in a substantial manner, is funding in the range of several billion dollars. Assuming expenditures in the range of $25,000 per unit, about $7.5 billion would be required. Clearly in an age of austerity as common sense, the program will never access that much money.

On the housing policy horizon in Toronto in late 2016 and early 2017 are two initiatives that could have significant impact on the affordable housing stock in the city, though how much impact is difficult to predict. The provincial government announced in 2016 that municipalities will be allowed to implement inclusionary zoning, that is, to require private sector developers to include some affordable housing in their projects. However, if municipalities make such a request they will not be able to also require cash contributions (allowed under section 37 of the provincial Planning Act), which they have come to rely on in this age of austerity to fund such public improvements as park refurbishment, and improvements to community and recreation centres. Also of concern is the use of the term "affordable housing," which, as defined in Toronto's Official Plan, is not the same as "social housing" or "rent-geared-to-income housing." As defined, affordable housing targets households of modest income, but remains unaffordable to the very lowest income households. The very few affordable housing projects that have been recently completed (thanks to the tiny amount of funding still available from senior governments) also set income maximums that serve to reinforce the notion of subsidized housing being fundamentally residual in nature. In so doing, these recent projects reverse progressive housing policy first introduced in the 1973 National Housing Act (NHA) that facilitated social housing with a mix of incomes, including the very lowest and without income caps.

The second initiative is that of the federal government of Justin Trudeau, elected in the fall of 2015, which has promised substantial funding for social infrastructure, including housing. In the winter of 2017, the federal government committed to investing $11 billion in affordable housing. This figure, which at first glance appears quite impressive, is in fact disappointingly small when looked at closely. The money is to be spent over an eleven-year period and across the entire country. It will address all aspects of affordable housing: new construction, maintenance and repair of existing housing, rent supplements, homelessness, and housing on First Nations reserves. The details of a new national housing program have not been announced, but it is possible that it will also retreat from the deeply progressive nature of the 1973 NHA and include income caps and rely on public-private partnerships in its delivery.

Concluding Remarks

The tragedy of Toronto's progressive neoliberalism in the context of renewing high-rise suburban rental housing is twofold. First, because the Tower Renewal project requires the buy-in of private sector landlords and has only very limited resources to convince them to participate, it is certain that only a very small percentage of buildings will see substantial renewal. Private sector landlords are, by their very nature, disinclined to voluntarily engage in new forms of state regulation of their assets. Those landlords who understand the value of maintaining and improving the rental towers they own and operate will do so without the assistance (or as they might view it, the interference) of a local state agency. This is tragic for the half million residents of the buildings in question. Second, as a state project that considers rental high-rise housing to be a de facto stock of affordable housing in need of renewal and as a state strategy that enlists the city's assistance to do so, Tower Renewal diverts attention from the urgent need for new state-funded social housing construction. This is tragic for the 170,000+ households in Ontario that are waiting for state-subsidized non-profit housing (Ontario Nonprofit Housing Association 2016). Also of interest is the fact that Tower Renewal engages building owners rather than building tenants, perhaps with the intention of limiting expectations about the scope of the program.

The threads of progressivism woven through the Tower Renewal program (the promise of improving the housing of low-income renters and the apparent valorization of everyday life in the demonized inner suburbs) is overwhelmed by the essentially neoliberal foundation of the project. State action in and on the market must be cautious and self-limited, and in the specific realm of housing, state action cements the policy drift away from state provision of non-profit housing. The two are fused together in a way that makes it impossible to reject Tower Renewal precisely because, in the present-day context in Toronto, it does indeed have potential for some progressive outcomes, no matter how small and despite the program's neoliberal core.

In its tragedy lies the success of Tower Renewal: the state limits its direct participation in non-profit housing provision and dampens any calls for new state-funded non-profit housing programs; it buttresses the conception of the market as a good provider of housing; it creates a discourse in Toronto of progressive state action on housing for low income households and on suburban renewal that is contained within a neoliberal construction of austerity as common sense. Indeed, Tower Renewal demonstrates the power of discourse in an age of austerity-as-common-sense governance; it creates a perception of action in the policy arena while in fact allocating only a very small amount of resources to actual material accomplishments. It also establishes a potential precedent of the ways in which the distinction between private and public sectors and goods can be been muddied. Perhaps Tower Renewal's progressive neoliberalism will be transferred to other state/market projects in the Toronto urban region to the ultimate detriment of its citizens.

ACKNOWLEDGMENTS

This research was supported by the Social Sciences and Humanities Research Council of Canada through funding from the Major Collaborative Research

Initiative "Global Suburbanisms: Governance, Land and Infrastructure in the 21st Century (2010–2017)."

REFERENCES

Canada Mortgage and Housing Corporation. (2016). *Rental Market Report Greater Toronto Area*. https://eppdscrmssa01.blob.core.windows.net/cmhcprodcontainer/sf/project/cmhc/pubsandreports/esub/_all_esub_pdfs/64459_2016_a01.pdf?sv=2017-07-29&ss=b&srt=sco&sp=r&se=2019-05-09T06:10:51Z&st=2018-03-11T22:10:51Z&spr=https,http&sig=0Ketq0sPGtnokWOe66BpqguDljVgBRH9wLOCg8HfE3w%3D

Ekers, M., Hamel, P., and Keil, R. (2012). "Governing Suburbia: Modalities and Mechanisms of Suburban Governance." *Regional Studies* 46, no. 3, 405–22. https://doi.org/10.1080/00343404.2012.658036

Harvey, D. (2005). *A Brief History of Neoliberalism*. New York, NY: Oxford University Press.

Keil, R. (2009). "The Urban Politics of Roll-with-it Neoliberalization." *City: Analysis of Urbantrends, Culture, Theory, Policy, Action* 13, no. 2–3, 230–45. https://doi.org/10.1080/13604810902986848

Larner, W., Lewis, N. and Le Heron, R. (2009). "State Spaces of 'After Neoliberalism': Co-Constituting the New Zealand Designer Fashion Industry." In Keil, R., and Mahon, R. (eds), *Leviathan Undone? Towards a Political Economy of Scale*, 177–94. Vancouver: UBC Press.

Ontario Nonprofit Housing Association. (2016.) *2016 Waiting Lists Survey Report*. http://qc.onpha.on.ca/flipbooks/WaitingListReport/#4

Peck, J., and Tickell, A. (2002). "Neoliberalizing Space." In Brenner, N., and Theodore, N. (eds), *Spaces of Neoliberalism: Urban Restructuring in Western Europe and North America*. Blackwell, UK: Oxford and Boston.

Phelps, N.A., and Wu, F. (2011). "Chapter 1 Introduction: International Perspectives on Suburbanization: A Post-suburban World?" In Phelps, N., and Wu, F. (eds), *International Perspectives on Suburbanization: A Post-suburban World*? 1–11. New York, NY: Palgrave Macmillan.

Sieverts, T. (2003). *Cities without Cities: an Investigation of the Zwischenstadt*. London, UK: Spon.

Theodore, N., and Peck, J. (2011). "Framing Neoliberal Urbanism: Translating 'Commonsense' Urban Policy across the OECD Zone." *European and Regional Studies* 19, no. 1, 20–41. https://doi.org/10.1177/0969776411428500

Wilson, J., and Swyngedouw, E. (2014). "Seeds of Dystopia: Post-politics and the Return of the Political." In Wilson, J., and Swyngedouw E. (eds), *The*

Post-political and Its Discontents: Spaces of Depoliticisation, Spectres of Radical Politics, 1–22. Edinburgh, Scotland: Edinburgh University Press.
Wu, F., and Phelps, N.A. (2011). "Chapter 14 Conclusion: Post-suburban Worlds?" In. Phelps, N., and Wu, F. (eds), *Introduction: International Perspectives on Suburbanization: A Post-suburban World?* 245–57. New York, NY: Palgrave Macmillan.
Young, D., Wood, P.B., and Keil, R. (eds). (2011). *In-Between Infrastructure: Urban Connectivity in an Age of Vulnerability*. Kelowna, BC: Praxis (e)Press.

6 Redeveloping Montpellier's Suburban High-Rises: National Policy Meets Local Activism in the Debate over Public Space

ROZA TCHOUKALEYSKA

This chapter considers the production of public space in suburban high-rise neighbourhoods in Montpellier, France. Drawing on ethnographic work completed in the La Paillade neighbourhood of Montpellier in 2014 and 2016, I consider the nuanced negotiations over access to public space. Of particular interest are the overlapping and sometimes competing interests of neighbourhood, municipal, and national actors in the production of public space, and the different ways of imaging and producing high-rise spaces. Set on the north-western edge of the city of Montpellier, the La Paillade neighbourhood is at once representative of suburban, high-rise social housing neighbourhoods in France, and a site with a unique history and social dynamic. Through my focus on La Paillade, I am interested in the physical design and social architecture of high-rise public space, with a view to understanding how such sites are shaped, disrupted, and moulded by state redevelopment programs.

This approach is grounded in an understanding of public space as a site produced through social action: meeting neighbours in the street, interacting with fellow cafe patrons, and seeing family and friends in the market, in the park, and on the beach. In this sense, public space is a site of interaction – of engagement, conversation, bumping along, public spectacle, and imagination (Watson and Studdert 2006; Low 2010; Koch and Latham 2013). The "public" aspect of public space is created when a range of actors can meet, affirm each other's presence through overt or unspoken forms of recognition, and create a site where sociability can be the basis for inclusion in civic citizenship. Public space is a site of ritual and daily rhythm (Mayol 1998), and a place where collective memory is performed (Melé 2005) and a sense of community built through participation in a shared location (Low 2010). A busy sidewalk or plaza speaks to Jacobs' ([1961] 1992) argument that a sense of public safety is created when there are many "eyes on the street," while the flows and movements of people through such sites reveals much about the structure of daily life and the conglomerate of social, political, and economic functions embedded therein (Hayden 1997).

Public space is of course also a legal entity, and a sphere enacted through urban planning codes (Blomley 2007). It is a place where citizenship rights are tested (Inceoglu 2015), and where political agency is expressed (Kohn 2004). Yet viewing public space as a social site has proven to be particularly germane to understanding how a range of users "appropriate" (D. Mitchell 2003) a place, and the ways in which identity is formed and transmitted through such sites. Understanding public space as formed through social interactions also means that sites not formally zoned as "public" can function as such. This is particularly relevant to La Paillade, where the most intense forms of public interaction and engagement frequently occurred in liminal spaces or privately owned commercial sites: the outdoor and indoor markets, neighbourhood cafes and their outdoor terraces, shops, and outdoor sports clubs. Such sites encourage a range of users to meet and interact – especially the outdoor markets, held in a normally closed-off parking lot – and to do so without some of the underlying anxiety about policy action or state oversight that marks the use of streets, sidewalks, and local parks. While national high-rise renovation programs and municipal cultural planning initiatives have largely intervened in sites formally zoned as "public" – plazas, parks, sidewalks, streets – these spaces were not the most frequently used social sites in La Paillade. The result is the creation of parallel public spaces: an instance in which the sites undergoing renewal and renovation are not always recognized as the "public space" of the neighbourhood by residents, while those sites used for social interaction and engagement are frequently overlooked by state initiatives.

Montpellier in Context

Situated in south-central France, Montpellier borders the Mediterranean and has a population of 250,000 (or 450,000 for the greater Montpellier area). It is one of the fastest-growing urban regions in France, and has an economy built on tourism, viticulture, and the high-tech sector, along with tertiary education, medical services and research, and regional government offices (Morvan 2013). The city's vibrant growth is in part an outcome of events in the 1960s: the Algerian war of independence saw France repatriate citizens, with many settled in the southern parts of country, including Montpellier. The city also saw the arrival of political migrants from Spain and Portugal escaping the Franco regime and, later, workers from North Africa to support the 1970s economic boom in France. As a result, Montpellier's population nearly doubled between the early 1960s and late 1970s, and had tripled – compared to 1960s levels – by the end of the 1990s (Audric and Tasqué 2010).

In Montpellier, as in other parts of France, the development of suburban high-rise neighbourhoods was in part a response to these population shifts. The focus of this chapter is the La Paillade neighbourhood, one such suburban high-rise neighbourhood, which sits on the north-western edge of the city. Built on a former vineyard and bordering the Mosson River, the neighbourhood welcomed its first residents in 1967, and is now home to 21,000 people (SIG/INSEE 2013). Made up of high-rise and mid-rise concrete apartment blocks, with a selection of single-family homes in the surroundings, La Paillade is centred around a key public space: Le Grand Mail, a raised walkway that spans the length of the neighbourhood. On one end of Le Grand Mail is a municipal pool, a tennis club, and the city's celebrated football club, and at the opposite end of the 750-metre walkway sit the busy indoor and expansive outdoor markets (figure 6.1). A few steps from the market is a theatre and community arts centre, along with a pétanque court, schools, and social service facilities. While Le Grand Mail itself is underused, as will be outlined below, the neighbourhood sees steady pedestrian traffic, with residents and visitors converging for the markets, school-age students visiting the community centre, and the football stadium drawing a regional crowd on game days. The neighbourhood is also the terminus for two of Montpellier's tramlines, which also service the nearby Pierres Vives multimedia library, an expansive new building designed by Zaha Hadid.

While many residents refer to the neighbourhood as La Paillade, an approach I have also adopted, the area was formally renamed La Mosson in 2000. The renaming of the neighbourhood was a contentious issue, and represented an arguably misplaced effort to improve the image of the area through a rebranding exercise. At play was La Paillade's complex standing in the city: while many research participants and residents identify strongly with the community, the neighbourhood's challenging socio-economic status is often negatively

6.1 The outdoor market and the Halles de la Mosson. Photo by the author.

perceived in Montpellier. This perception is based on several indicators which differentiate La Paillade from the city centre, and from other new suburbs. For instance, while nearly 65 per cent of adults in Montpellier were employed in 2013 (INSEE 2013), over the last few years that figure has hovered around 35 per cent for La Paillade (SIG/INSEE 2010). The number of households classified as being at the poverty line sits at 57 per cent for La Paillade (SIG/INSEE 2012), compared to 26 per cent for Montpellier as a whole (INSEE 2013). As a social housing district, La Paillade also represented a high rate of households renting their primary residence (84 per cent; SIG/INSEE 2012).

Alongside these socio-economic details sits another characteristic of La Paillade, more difficult to define statistically: anecdotally, the neighbourhood is perceived as an immigration reception area, with a highly diverse population and a large North African community. The anecdotal nature of these views is due to the particularity of French laws, which ban the collection of statistical data based on ethnic, racial, or religious identity. The French motto of "*liberté, égalité, fraternité*" – and in particular the call for *égalité*, or equality – means that formally all citizens of the republic are equal, or, in the terms of the French republic, indistinguishable from each other in terms of status and identity (Weil 2005; Wieviorka 2005; Amiraux and Simon 2006). Maintaining that sense of *égalité* means that data which could be used to formally distinguish one group of citizens from another is not collected, the notion of multiculturalism is moot, and statistics on ethnic identity do not exist. The absence of such data has not, of course, stopped the politicized use of ethnic identifiers across the political and social spectrum. Nor has it halted the capacity of community organizations and individual actors in neighbourhoods such as La Paillade from publicly claiming complex identities (for instance French and North African, at the same time), and building social movements to challenged entrenched institutional racism (Wacquant 2008). The identification of La Paillade as a diverse neighbourhood, with an established North African community, is based on the perspectives of residents, the insistence of local NGOs, and popular media depictions and social discussion.

The complexity of public space described below works within these amorphous bounds: statistically distinguished through challenging socio-economic identifiers, La Paillade is anecdotally also associated with an ethnic identity that marks this space as "Other" within Montpellier. The tensions and spatial divides which exist between La Paillade and the rest of Montpellier – and the process through which this neighbourhood has been consistently identified as requiring strong state intervention – has as much to do with formal socio-economic data as with the equally potent, but formally hidden, ethnic categorizations. These dynamics are not unique to La Paillade or Montpellier. Rather, they are common across many French cities, where the high-rise suburbs have become synonymous with "immigration" – a term that often has little to do with citizenship status and is instead synonymous with "ethnically non-French" (Weil 2005) – and with a social, economic, and political disassociation with the city centre. The 2005 suburban riots in France emerged from these dynamics (Silverstein and Tetreault 2006), and as evidenced in the work of Dikeç (2007) on Paris and K. Mitchell (2011) on Marseille, among others (Bresson 2007; Wacquant 2008; Garbin and Millington 2012), continue to shape the tone of a range of state interventions. In this sense, La Paillade is both a place with its own history and identity, and one representative of broader tensions between national policy and local desires.

As with many suburban, high-rise social housing districts in France, over the last three decades La Paillade has been the object of regional and national urban renewal programs. Many such programs respond to the categorization of La Paillade as a ZUS (Zone Urbaine Sensible), a marker which suggests that this neighbourhood is spatially disadvantaged and in need of sustained social, economic, and structural intervention. Funded through a range of state and municipal sources, and carried out largely by local agencies, the renovations have aimed to reconstruct the outside of high-rise buildings, repair elevators and essential utilities, and provide more community amenities (Chédiac and Bernié-Boissard 2010). Alongside attention to the physical fabric of the neighbourhood, renewal programs have sought to alter the social patterns of La Paillade by intervening in the range of community programs offered, supporting arts initiatives, and providing funding for sports activities. While the extent of renovation and social programs is impressive, the reality of these projects often falls short of their intention. As Dikeç (2007) notes in relation to suburban high-rise developments in Paris, the process of designating funding and determining which community programs will be supported can serve to reaffirm negative stereotypes and stigmatizations associated with such neighbourhoods. Particularly in an instance where suburban high-rise communities have been associated with increasing rates of social and economic deprivation (K. Mitchell 2011) and a political rhetoric that labels their ethnic diversity as problematic (Amiraux and Simon 2006), continual state intervention into the physical and social fabric of such areas is frequently unwelcome (Bresson 2007). This is certainly true of La Paillade, where the seemingly never-ending renovations have been met with hesitation, or dismissed as tick-box politics.

Fenced in and Closed off

On a June afternoon, I was taken on a walking tour of the mid-rise apartment buildings to the east of the tramline in La Paillade by "A," a long-time resident who was born in the neighbourhood, went to school there, and after working and living in other locations, had chosen to come back to La Paillade as an adult. "A" viewed the myriad of neighbourhood redevelopment projects of the last decades with considerable pessimism. Rather than improve public space, the high-rise renewal process had closed down sites that were publicly accessible. Winding our way between the buildings, our tour took a walking-while-talking approach (Anderson 2004), an ethnographic and interview process where participants set the route and determine the sites and spaces they view as most relevant to their daily routines. During tours, "A" and other participants commented on how neighbourhood spaces were used, and identified key walkways and squares, spots for socializing or playing football, and the range of liminal spaces (interior courtyards, cafe terraces) which functioned as semi-public sites.

On this particular walk, "A" was keen to show me a specific kind of faltering space, one which used to be public but arguably no longer was: the interior building courtyards which were described as having been a desirable childhood play spot, and which were no longer accessible to the public. This closing down of public space was demonstrated as we walked around a series of five- and six-storey apartment buildings. Today, the buildings look rather unremarkable. They have an entrance facing the street, no stores or apartments on the ground floor, and balconies and apartment windows starting from the second and upper floors. The cluster of buildings is surrounded by low metal fences, with each block having its own electronic key entry, which gives access to the parking lots and surrounding areas. Bushes of greenery attempt to soften the metal fences, and the light pink and yellow colours of the buildings mimic the colour palette of buildings in Montpellier's historic city centre.

As we walked around the buildings and into the parking lot behind them, "A" explained that in the 1980s and 1990s the buildings looked different from today. The ground floor was open, and the buildings effectively sat on stilts: you could run underneath the building and into the interior courtyard. With these points of access now enclosed, the memories of playing games beneath the buildings were instead told with gestures, words, and small sketches on pieces of paper. While the buildings were in their original form in the 1990s, children could play hide and seek beneath them, zipping between buildings, sidewalks, and courtyards, and using all ground-level points of passage. The result was a distinct social architecture in La Paillade, where you met your neighbours often, and saw more of your friends; parents were out keeping up with their kids, and there was a sense that the neighbourhood was there for the taking. The distinction between public and private space was blurred, with large portions of each apartment building and the surrounding grounds viably open to use from the entire neighbourhood, with the interior courtyards of buildings being particularly enticing.

In the late 1990s, these interior spaces were closed off as a security measure (Chédiac and Bernié-Boissard 2010). This means that walls were built at ground level, enclosing the building so that the courtyards were no longer accessible to the public. Metal fences were also added around the buildings, and electronic key entry for the parking lots and surrounding green space was installed. The process effectively marked out the "private" from the "public," and in "A"'s view the neighbourhood shrank, with only sidewalks, streets, and plazas left as public space. As we did our walk through the neighbourhood, "A" frequently pointed out childhood haunts that we could no longer visit: we could not cross a parking lot because it required electronic key entry access; we could not open any of the building doors or metal fence entrances, because these too were key-entry only; we could not visit the courtyards, and instead drew out maps and diagrams of how they used to function. There was no way to walk underneath the buildings, as the stilts and entryways had long been built into a solid wall, and most apartment blocks had fencing or shrubs around then which made it difficult to get close to them. High-rise redevelopment in La Paillade has had the effect of enclosing the neighbourhood, with the impact of municipal policy – modelled on national policies – being, in "A"'s view, to fence in (or fence out) people from their everyday spaces.

Nowhere to Stand

This was a point picked up on by other research participants. During a mid-week walk, I was led on a walking tour of Le Grand Mail, the main pedestrian walkway of La Paillade, by two participants, "K" and "H" (both pseudonyms). Le Grand Mail is an elevated walkway and public space that links many high-rise buildings and spans over roads and streets. As noted earlier in the chapter, at one end of Le Grand Mail are the indoor and outdoor markets, and at the other you are within walking distance of the football stadium. On this particular day, we were walking along Le Grand Mail because I had asked to visit a site that "K" and "H" considered emblematic of public space redevelopment in La Paillade. The decision to visit Le Grand Mail was made quickly, with both participants in agreement on the importance of the site. Le Grand Mail, they argued, was emblematic because it had cost so much, yet failed so bitterly. The main renovation of this elevated walkway and public space had

6.2 Le Grand Mail, mid-afternoon in June 2014. Photo by the author.

been, seemingly, going on for decades. There had been a big renovation in the 1990s, and more recent renovations and cleaning projects; trees had been planted, a community garden was established in 2014, and the walkway's tarmac had been redone in sections over the years. A fountain had been installed in the centre of Le Grand Mail, though "K" joked that it looked more like a concrete road block sprinkled with colourful tiles than anything which might resemble a fountain. The space, "K" argued, was not used; and indeed on this day and several others when I had visited Le Grand Mail on my own or during a walking-while-talking tour, the space seemed sparsely used (figure 6.2).

As "H" bluntly said, Le Grand Mail is the sort of postmodern structure that looks good on paper but does not function in reality. It is poorly used, and sections of it are considered no-go zones by some residents because of worries over safety. Apart from market days and immediately after school, there are few people on Le Grand Mail, because there is seemingly little reason to be there. There are very few shops or cafes on the walkway, with many commercial buildings being boarded up and closed down. The tramway and roads are at street level, which means that they are effectively one floor below Le Grand Mail. Parking lots and the entrances to many buildings and other facilities are also at street level, which means that a few pedestrians in La Paillade have a reason to climb a staircase to access Le Grand Mail. Pointing out tarmac that was patched up, and spots on La Grand Mail that have been remodeled, "H" called the constant municipal attention to this space a "bizarre redevelopment," because it meant that considerable monetary funds were poured into a site that few residents use. The low usage was also due to the location of the national police on Le Grand Mail, meaning that if you congregated in groups of three or more – and particularly if you were young – you were seen as a gathering, and "K" noted that the police could take issue with that. The design of Le Grand Mail, a long flat corridor of concrete, meant that users felt visible, and both "H" and "K" noted that this was not a place where you would spend time with your friends.

As we continued to amble along Le Grand Mail, "K" went on to say that we were walking through this site because I had asked about it. Normally, pedestrians crossing La Paillade in a north-south direction would walk at ground level, following the tramway tracks and sidewalks. The renovations, "K" explained "were a bit of a strategy thing, you see, ok it's true that it made the place [Le Grand Mail] look a bit better. But it's not enough to put in giant flower pots, that doesn't change things because the buildings around here are still rubbish."

The many types of renovations on Le Grand Mail were pointed out: neighbourhood greening, renovations to the indoor markets, renovations to high-rise tower cladding, creation of sports facilities, creation of arts facilities, new parks and green spaces established, and economic and housing policies implemented. Many of these had been geared towards changing the visual landscape of La Paillade. Yet this was problematic because, as "K" put forward, you cannot change the social and economic problems faced by the neighbourhood by focusing so closely on redoing its exterior contours. Many of the renovation programs had changed the way the plazas and streets look, and yet none of the redevelopment programs had succeeded in making these places easier for residents to use, or more welcoming. Largely, the renovation measures had made La Paillade look neater and tidier.

The Splinter of Local Responses

The ongoing tensions between communities and security services in French social housing suburban areas, alongside the polarizing political rhetoric on identity and ethnicity in public space in France, has meant that high-rise public spaces are increasingly sites of tension (Dikeç 2007). In La Paillade, these broader trends have overlapped with ongoing renovation programs, as outlined above, and led to an aversion to using formal public spaces in the neighbourhood. Instead, many research participants indicated that they gravitated towards private and semi-private retail spaces to fulfil the sociability function normally associated with public space. Of these, the most important are the outdoor markets, followed by neighbourhood cafes and snack bars.

There are several markets in La Paillade. The Halles de la Mosson, the indoor market, runs five days a week with around twenty-five vendors, in a building directly behind a tramway stop. The expansive outdoor market, which surrounds the Halles de la Mosson and stretches out into neighbourhood parking lots with its 100 vendors, is open three days a week, and offers household goods, clothing, personal items, and speciality products. On Sunday mornings, the flea market takes over the football stadium parking lot, with upwards of 300 vendors from Montpellier and beyond who sell used items, antiques, clothing and shoes, mechanical parts, posters, and household decorations, among many other items. Finally, each Wednesday a flower, seed, and potted plant market settles on a vending space a few blocks from the football stadium and draws in a rural and urban clientele with a wide selection and competitive prices. Each of these markets takes place in privately owned space (parking lots) or semi-public space (the indoor market, in a municipally owned building that locks after each session).

In many respects, the markets function as public space: they bring together citizens and denizens of all ages and a range of backgrounds, genders, and economic positions. As many participants emphasized, the markets allow contact and informal conversation, a chance to see and be seen, and to recognize your neighbourhoods and fellow urbanites (or suburbanites). Speaking with market vendors, many emphasized that there was a pattern to how the space functioned: a rhythm of regular clients and also of friends and family passing by, along with neighbours and those coming from surrounding communities to do their shopping. One food vendor emphasized that the market was not just about shopping, but also about hanging out. You could stroll slowly and take your time, without being accused of loitering, which normally carries a fine and police intervention. It was possible to sit in a chair in the Halles de la Mosson, in the cafe or on one of the tables outside, and not be told to move on. The long walkways between stalls in the flea market and the outdoor market create dense pedestrian pathways, in sharp contrast to the relative silence of Le Grand Mail. The markets, a flea market seller noted, were impressive for bringing out people who would not normally go on promenades in the neighbourhood: women with young children, who took advantage of the market-as-event to enjoy an afternoon outdoors, particularly important in a climate where headscarves have become politicized and banned from many state spaces (Wiles 2007); and older residents, who frequently join vendors at their stalls and borrow seats from behind the market stand. When the market folds up for the day, these same places are quieter and relatively empty, and La Paillade loses a bit of its bustle and activity. The markets, and the social actions related to shopping and browsing, create an atmosphere that enables the sort of interaction that produces sociable public spaces.

The cafes of La Paillade are also important in this respect, and together with the market spaces they have become the focus of social and political activity in the neighbourhood. Many research participants identified cafes as their preferred location for meeting friends, and as the sites where they felt most comfortable discussing pressing local matters: the state of housing, access to social services, and questions around municipal representation, among many others. Taking part in these informal discussions often means sitting outside on a cafe terrace, ordering coffee, and having – as it was described by several participants – a frank debate with friends and with cafe-table neighbours you may not have met before about the problems faced by La Paillade, the impact of constant intervention and policing, and the meaning of being "Maghrébin" in today's France. The cafe meetings were often described as a serendipitous effort to create a space for safe political discussion: most participants knew that certain cafes served as venues where La Paillade residents would meet each other and confront the issues that touch their lives.

That these conversations took place in cafes is not accidental. As one participant explained, it was difficult to book a room in the Maison Pour Tous, the municipally run community space. While that space is open to many uses, it was not always made available for political discussions. Another participant explained that clientelism was rife in the neighbourhood, and that local associations supported through municipal funds were hesitant to open their doors to dialogue that is so often sternly critical of what the city is doing. The cafes of La Paillade are viewed as ideal for much the same reasons that the markets are good meeting places: they are owned and operated by neighbourhood residents who have a vested interest in seeing the neighbourhood enlivened. Cafés are also not as dependent on municipal politics for their function. Using Le Grand Mail as a meeting space or having any sort of pointed debate on that site was not viewed as a comfortable idea. This is in part because Le Grand Mail is underused, making any sort of gathering conspicuous, and in part because it is a contentious zone, more a municipal project than a neighbourhood public space. The most open and accommodating, the most public sites of La Paillade formed around commercial spaces, with cafes and outdoor markets being the best option for building a social site that encourages neighbourhood interaction, engagement, dialogue, and sociability.

Parallel Public Spaces

From this overview of public space use in La Paillade rises the notion of parallel public spaces. The sites that are formally zoned as "public space," and are the object of state renovation programs, are less frequently used as neighbourhood meeting places. The sites that are used by many residents are, instead, the cafes, markets, shopping areas, and other semi-public and private retail sites around La Paillade, which rarely benefit from state redevelopment funds or municipal attention and yet are integral to the social networks of the neighbourhood. It is particularly poignant that political debate and activism relies on cafes for their venues: spaces that are privately owned, yet are visible to pedestrians, prove to be the most accessible and stable places to meet, organize, and think through different approaches to supporting the neighbourhood.

Speaking to one research participant while seated in an outdoor cafe, I asked what they wanted to see in terms of neighbourhood public space. The response was: renovations, but ones that actually create local spaces, open spaces, and accessible and safe spaces. The mismatch between local experiences and national policy was stark, with one interview participant, "C," commenting that the sense of always being under renovation, but never having the community catered to, was disenfranchising. As this participant noted: "I've got the impression that in France we want these neighbourhoods to always stay what they are, difficult neighbourhoods, precarious neighbourhoods, and all that goes with it. At the level of the government, the municipality and local organizations, we are not their priority, even though we are tax-paying French citizens."

In an instance where public space is important to being recognized as a member of a wider urban public, and through this to forms of citizenship (D. Mitchell 2003), this mismatch of spaces – and parallel spaces – raises questions about broader forms of social and political engagement. In particular, it puts into question the disparate ways in which national high-rise renovation policy has engaged with the physicality of place – the concrete, walkways, and giant flowerpots of Le Grand Mail – while bypassing the much-needed social dimension of such sites.

REFERENCES

Amiraux, V., and Simon, P. (2006). "There Are No Minorities Here: Cultures of Scholarship and Public Debate on Immigrants and Integration in France." *International Journal of Comparative Sociology* 47, no. 3/4, 191–215. https://doi.org/10.1177/0020715206066164

Anderson, J. (2004). "Talking Whilst Walking: A Geographical Archaeology of Knowledge." *Area* 36, no. 3, 254–61. https://doi.org/10.1111/j.0004-0894.2004.00222.x

Audric, S., and Tasqué, O. (2010). "La population de Montpellier Agglomération a triplé au cours des cinquante dernière années." *INSEE* (Institut national de la statistique et des études économiques). https://www.insee.fr/fr/statistiques/1286050

Blomley, N. (2007). "How to Turn a Beggar into a Bus Stop: Law, Traffic, and the 'Function of the Place.'" *Urban Studies* 44, no. 9, 1697–712. https://doi.org/10.1080/00420980701427507

Bresson, M. (2007). "Peut-on parler d'un échec de la participation dans les quartiers 'sensible' en France? Réflexion sur la pluralité des attentes et les confusions autour de ce thème." *Pensée plurielle* 2, no. 15, 121–8. https://doi.org/10.3917/pp.015.0121

Chédiac, S., and Bernié-Boissard, C. (2010). "La résidentialisation." In Volle, J.P., Viala, L., Négrier, E., and Bernié-Boissard, C. (eds), *Montpellier: La ville inventée*, 110–25. Marseille, France: Edition Parenthèse/Gip Epau.

Dikeç, M. (2007). *Badlands of the Republic: Space, Politics, and Urban Policy*. Oxford, UK: Blackwell Publishing.

Garbin, D., and Millington, G. (2012). "Territorial Stigma and the Politics of Resistance in a Parisian *Banlieue*: *La Courneuve* and Beyond." *Urban Studies* 49, no. 10, 2067–83. https://doi.org/10.1177/0042098011422572
Hayden, D. (1997). *The Power of Place: Urban Landscape as Public History*. Cambridge, MA: MIT Press.
Inceoglu, I. (2015). "Encountering Difference and Radical Democratic Trajectory: An Analysis of Gezi Park as Public Space." *City* 19, no. 4, 534–44. https://doi.org/10.1080/13604813.2015.1051743
INSEE (Institut National de la Statistique et des Etudes Economiques). (2013). "Dossier Complet: Commun de Montpellier (34172)." https://www.insee.fr/fr/statistiques/2011101?geo=COM-34172#chiffre-cle-1.
Jacobs, J. (1961) 1992. *The Death and Life of Great American Cities*. New York, NY: Vintage Books.
Koch, R., and Latham, A. (2013). "On the Hard Work of Domesticating a Public Space." *Urban Studies* 50, no. 1, 6–21. https://doi.org/10.1177/0042098012447001
Kohn, M. (2004). *Brave New Neighbourhoods: The Privatization of Public Space*. New York, NY: Routledge.
Low, S. (2010). *On the Plaza: The Politics of Public Space and Culture*. Austin, TX: University of Texas Press.
Mayol, P. (1998). "The Neighbourhood." In De Certeau, M., Giard, L., and Mayol, P. (eds), Tomasik, T.J. (trans.), *The Practice of Everyday Life*, Volume 2: *Living and Cooking*, 7–13. Minneapolis: University of Minnesota Press.
Melé, P. (2005). "Conflits patrimoniaux et régulation urbaine." *ESO Travaux et Documents*, 23, 51–7. https://halshs.archives-ouvertes.fr/halshs-00005717/document
Mitchell, D. (2003). *The Right to the City: Social Justice and the Fight for Public Space*. New York, NY: Guilford Press.
Mitchell, K. (2011) "Marseille's Not for Burning: Comparative Networks and Integration and Exclusion in Two French Cities." *Annals of the Association of American Geographers* 101, no. 2, 404–23. https://doi.org/10.1080/00045608.2010.545290.
Morvan, S. (2013). "Montpellier peine à avaler sa croissance sans limites." *L'Express*, 26 March. http://www.lexpress.fr/region/montpellier-peine-a-avaler-sacroissance-sans-limites_1234706.html.
SIG (Système d'information géographique de la politique de la ville)/INSEE (Institut national de la statistique et des études économiques). (2010)."Emploi 2010." https://sig.ville.gouv.fr/Tableaux/QP034005/332.
– (2012)."Caractéristiques socio-démographiques des ménages 2012." https://sig.ville.gouv.fr/Tableaux/QP034005/337.

– (2013). "Quartier prioritaire: Mosson." https://sig.ville.gouv.fr/Territoire/QP034005.

Silverstein, P., and Tetreault, C. (2006). "Postcolonial Urban Apartheid. *Riots in France*, 11 June. http://riotsfrance.ssrc.org/Silverstein_Tetreault/.

Wacquant, L. (2008). *Urban Outcasts: A Comparative Sociology of Advanced Marginality*. Cambridge, UK: Polity Press.

Watson, S., and Studdert, D. (2006). *Markets as Sites for Social Interaction: Spaces of Diversity*. Bristol, UK: Polity Press.

Weil, P. (2005). *La République et sa diversité: Immigration, intégration, discrimination*. Paris, France: du Seuil.

Wieviorka, M. (2005). *La différence: Identités culturelles, enjeux, débats, et politiques*. Paris, France: L'Aube.

Wiles, E. (2007). "Headscarves, Human Rights, and Harmonious Multicultural Society: Implications of the French Ban for Interpretations of Equality." *Law and Society Review* 41, no. 3, 699–736. https://doi.org/10.1111/j.1540-5893.2007.00318.x

7 (De)Constructing Housing Estates: How Much More than a Housing Question?

STEFAN KIPFER

From a Marxist and socialist feminist perspective, research and action on "housing" provides openings to many aspects of totality. In his still-relevant comment in *The Housing Question*, Engels (1969) reminds us, however, that housing is never just about housing; it also mediates class formation and links up to broader dynamics of capitalist development. This line of argument has been taken up in various strands of neo-Marxist research that link housing questions to the problem of state intervention, collective consumption, and social movements (Castells 1978), gender and social reproduction (Mitchell, Marston, and Katz 2004), land-rent dynamics in the urban process (Harvey 1989; Smith 1996), "city trenches" in urban politics (Katznelson 1981), and state space (Brenner 2004). From these perspectives, public housing redevelopment and place-based policy appear as a combination of welfare state restructuring, reprivatized social reproduction, competitive state rescaling, and dispossessive gentrification. My research on housing estate redevelopment in Toronto and Paris largely corroborates these arguments. However, it puts a special emphasis on the (neo)colonial dynamics that shape urban policy and public housing redevelopment. As urban questions more generally, these dynamics have everyday implications and political ramifications that significantly exceed the realm of "housing" understood as a combination of built environments and spatial forms.

Rather than presenting new empirical findings, this chapter outlines major conclusions from a long-standing research project on a selection of housing estate redevelopment projects in Greater Paris and relates it to earlier research undertaken about public housing redevelopment and place-based policy in the City of Toronto.[1] All overseen by the *Agence nationale pour la rénovation urbaine* (ANRU), which was created in

2003, these Parisian projects are undertaken in a nationally coordinated fashion – no longer sporadically, as was the case in the France of the 1980s, or municipally, as has been the case in Toronto. The creation of ANRU consolidated the process by which demolition and renovation moved to the centre stage of what has been a two-generation-long place-based urban policy (*politique de la ville*), not a product of the last fifteen years, as in Toronto. As in previous incarnations, however, physical redevelopment and place-based targeting are tied to the socio-political purpose of mixing up resident populations. Officially, the ANRU's projects are all intended to valorize, mix, and securitize the target districts. While they are implemented more cautiously than in Toronto, through a typically gradual and selective approach to demolition and private-sector redevelopment that also anticipates resistance, these mixological goals are strikingly similar to those informing projects in Toronto, as are their contradictions.

The Paris and Toronto housing redevelopment projects are not isolated cases, of course. They articulate and constitute broader trends: the marketized deconstruction of social housing across advanced capitalism and territorially targeted state intervention in various parts of the world. Indeed, state spatial strategies in Paris are not simple derivatives of a unique French Republican policy "model" or "regime." While they developed in response to the spectre of the African-American ghetto (and its supposed British equivalents, Brixton and Toxteth), they are also an export model that serves as a negative foil (the danger of the banlieue) or a positive model for state intervention in cases that lack a comparable revolutionary urban history. The problem of the Paris banlieue is present as a symbolic reference point among policy-making circles not only in other European countries but also in North America, notably in Toronto itself, as Parastou Saberi (2017a, b) also discusses in her dissertation research. Indeed, the rise to the fore of France (alongside the US and the UK) as a reference point for Toronto planners is one rationale for our research. In this sense, redevelopment in Paris is not a case study but an entry point for relational comparison (Hart 2017).

"Housing" is of course a global problematic, and so is state-centred geographical targeting. If one can say, following Doreen Massey (1991), that Toronto and Paris are global places where people relate to each other and extra-local "flows and interconnections" in various, also antagonistic ways, studying these urban regions alone or together alerts to us dynamics that go much beyond Euro-America. My research corroborates Ananya Roy's (2005) insight: that the problem of informality, which is often seen as a paradigmatically "Southern" urban issue, is also a deeply Euro-American concern. This is not only because, today, the neoliberalization of labour markets, housing, and planning in the "Global North" generates or intensifies phenomena (homelessness, couchsurfing, precarious work, micro-entrepreneurship, corruption, organized crime) that one could call "informal," too. It is also because the very strategies that produced the public housing estates now under redevelopment did so, from the 1940s to the 1960s, through formal and informal imperial networks of institutions, state personnel, and expertise that considered "the slum" or "the shantytown" as a pathology and sociopolitical threat in (neo)colonies and metropoles alike. In France, the term *bidonville* was a colonial invention that was also mobilized in the hexagone to target shantytowns for eradication and find ways of easing their inhabitants (from North Africa and Southern Europe) slowly into "normal" public housing. This example speaks to Roy's (2005) point that urban research should bring into light an often-occluded fact: Northern and Southern urbanisms belong to a common, if fractured and deeply unequal, historical geography.

Theoretically and politically, my research is still centred on North Atlantic realities, however. It highlights both the promises and limitations of Marxist-influenced, socialist feminist, and left-Weberian housing research and their (otherwise crucial) emphasis on welfare state restructuring, social reproduction, competitive state rescaling, and dispossessive gentrification. These approaches do not always overlook the problem of the everyday to which the urban question refers. But even when they do not, they pay less attention to the (neo)colonial political dynamics that help shape urban policy and public housing redevelopment than is warranted. In order to focus on the implications (neo)colonial, multi-scalar urban policy has for urban and state theories, my starting point has been to draw on a complex of insights from Henri Lefebvre, Frantz Fanon, and Antonio Gramsci. In the last section of this chapter, I will push this complex further and call for additional materialist work on imperial and post-colonial dimensions of the state. The political urgency of such theorizing is difficult to overestimate, for the territorialization of social relations wrought by urban policy continues to disorganize both subaltern forces and oppositional politics today. In what is only a seeming paradox, they both further dynamics of neoliberalization and prepare the ground for right-populist projects.

What are the major lessons from my Paris-based research in cross-Atlantic perspective? I have organized my reflections into eight basic theses.

(1) The Meaning of Spatial and Socio-Political Peripherality and Centrality Is Not Self-Evident

The projects that I analysed in the Paris region vary even more than they do in Toronto, where they straddle the central city and the inner suburbs but extend neither to pre-war, market-dominated low-income housing districts nor to the outer, demographically dominant suburbs of the region. In Paris, the eighteen projects I analysed differ widely with respect to their position in land and labour markets, geographical distance from central Paris, transit connectivity, morphology, and political history. While many projects are focused on large-scale housing estates (*grands ensembles*) with a high proportion of social housing, a few zero in on pre-Fordist neighbourhoods dominated by working-class tenants in private rental housing, both in the City of Paris and adjacent municipalities. They both exceed and relate to some (but not all) of the dynamics of suburbanization discussed in this book (see the chapter by Güney, Keil, and Üçoğlu in this book). As a whole, the projects form a diverse socio-spatial patchwork that criss-crosses the urban region, a sort of a territorial Emmental cheese. All of this indicates that while renovation projects are peripheralized *socio-politically* from the perspective of state-spatial strategies, they are not all *locationally* peripheral within the urban region or within their respective municipal context. And what is more, daily realities in and across these social spaces cannot be read off mechanically from their peripherality, be the latter locational or socio-political.

Of course, state-bound and highly mediatized place-based conceptions of space – as well as dynamics of segregation – have left deep marks. This is even more the case in Paris than in Toronto, where place-based policy is for the most part a product of the last decade, not a result of almost four decades of explicit state intervention. These conceptions of space have homogenized targeted social spaces from the outside by *inventing* them as pathogenic spaces of social problems, whether termed "priority neighbourhoods" in Toronto, or "banlieues" or "quartiers" in France (Bertho 1997; Tissot 2007; Dikeç 2007). One can say that they have generated a visual common sense, a sort of territorialized "way of seeing" (Berger 2008) the social world as derivative of spatial form. However, the weight of this homogenization should not be taken as sufficient evidence to assume that everyday practices and aspirations in the targeted districts conform to the assumptions inherent in spatial targeting – or that these districts exhaust the realities of exploitation and domination. Following Lefebvre (Kipfer and Goonwardena 2013), the language of centre and periphery is neither a spatial metaphor only, nor is it a neat spatial expression of social relations. It alludes to the fact that territorial relations are *mediations* of social totality in general, and state-bound relations of domination specifically (see also the chapter by Kipfer and Dikeç in this book).

(2) Housing Estate Redevelopment Is a Housing Question

My subject of investigation dealt with housing understood in part as *habitat*: a combination of objects to shelter people in collectively produced built environments. The housing estates, large or modest, which constitute most of the projects I analysed were built between the 1950s and the 1970s as forms of collective consumption (to speak with Manuel Castells 1977) to rationalize social reproduction under Fordism. Partly a response to housing struggles, they were made possible by – and facilitated – new state capacities in relationship to land rent, building finance, and urban-regional development. In turn, the deconstruction of postwar housing estates, large and small, can be interpreted as a complex variation of state-managed gentrification (Hackworth and Smith 2001). While in Toronto redevelopment is typically total as well as strictly conditional upon private real-estate investment, state-managed gentrification in Paris is usually more limited to sections of housing estates and more variegated in economic terms. It proceeds through explicit as well as creeping forms of commodifying land, tenure, and housing supply, some of which is promoted by financially constrained housing authorities themselves (Tournon 2018). Where they are only punctual beachheads, renovation projects fragment and segment housing estates, thus opening opportunities for future rounds of real estate investment[2] such as those that are made possible by the *Grand Paris Express* regional transit system.

Moreover, while in Toronto redevelopment is a municipal affair (spurred on by the downloading of public housing finance from the Province of Ontario onto cities in the 1990s), the Paris cases are part of a national effort produced by a peculiar, in part "upward" form of rescaling of neoliberal state intervention. In both regions, redevelopment accentuates widely documented affordable housing shortages. The effects of ANRU projects (as well as the actions of Toronto Community Housing, the Toronto public housing authority), thus illustrate the need to reinvigorate collective consumption. Paradoxically, this need is more widely and publicly acknowledged in Paris than in Toronto, where the complete dependence of public housing redevelopment on private real estate is now raising the spectre of demolition without redevelopment for projects that are still at a distance from the gentrification frontier, notably in the postwar suburbs. In France, too, collective consumption runs up against the manifold private property and market biases of post-1977 housing policy (Lambert 2015). These biases are currently intensified under the Macron government and its plans to push housing authorities to residualize, marketize, and financialize their operations. Clearly, in this context, decommodifying land is one precondition to renewing collective consumption and resocializing housing.

(3) Housing Is Never Just about Housing

Following Lefebvre, housing can be grasped not only as *habitat* but also as *habiter*: thing-ified product and appropriating everyday practice, routine *and* aspiration, reality *and* possibility. In this view, housing opens to social totality and the contradictions that run through it. This insight allows us to see one of the great merits of Friedrich Engels's (1969) *The Housing Question*, which was to warn us that the research and practice of housing reform engages in false concreteness every time it treats housing matters in misleading isolation from the wider dynamics of capitalist development. At the time, Engels argued that the bourgeoisie just moves housing problems elsewhere because capitalism does not allow it to solve them. Submerged in Engels's account were the gendered dimensions of housing reform. These dimensions became central in the inter-war period, during the heated socialist and Bolshevik debates about ways of recasting planning and architecture to facilitate strategies of socializing social reproduction and partly collectivizing domestic space (Hayden 1981).

In the immediate aftermath of the postwar expansion (as well as social housing programs), it seemed that housing researchers with a focus on US-occupied Euro-America (but not elsewhere) had a point when they argued that social democracy *can* solve the bulk of housing problems within a capitalist context. However, from today's vantage point, from which the postwar period appears exceptional, Engels's argument is again particularly salient, even in Euro-America. Like the Toronto projects, ANRU-sponsored initiatives focus to a large extent on reshuffling the spatial relations between various housing forms (and increasing the market or market-like proportion of the housing supply). Prioritizing demolition over redevelopment, they cannot, and are not intended to, meet housing needs (even though various left-leaning municipalities in Paris remain committed to maintaining and even expanding social housing, in significant contrast to Toronto, where few argue today that the state should increase its role in the housing business, to use local language). On both sides of the Atlantic, there is no shortage of research that documents the extent and depth of "housing problems" today and their socially uneven impacts. Indeed, Toronto and Paris redevelopment programs, which contribute in particular ways to housing crises in these very expensive global-city regions, underscore how accumulation and dispossession under neoliberal capitalism have shrunk, reoriented, and fragmented social reproduction as a gendered element of state intervention (Mitchell et al. 2004).

(4) Demolition/Renovation Is (Also) a Political Project to Reorganize Core-Periphery Relations

Another merit of Engels's work – his *The Conditions of the Working Class* (1969) in England as well as *The Housing Question* (1970) – is to remind us that the social geographies in and beyond the points of waged production help form classes as lived realities. To speak in Lefebvre's (2003) own distinct terms, housing *(habitat* and *habiter)* is one medium through which relations of domination between socio-political centres and peripheries are organized territorially. As such, housing is an aspect of the urban. It mediates the relations of domination traversing the social order from macro- to micro-levels in ways that articulate relations of production and reproduction. Projects of building social housing on a larger scale under Fordism, for example, can be interpreted as a strategy to produce domesticated and respectable working classes. While in Toronto this took the form of social paternalism aimed at a small minority of the working class, in Paris postwar social housing was cross-class in overall composition and centred on the established segments of the Fordist working class in part to avoid the explosive proletarian concentrations that had defined Paris's revolutionary faubourgs, the working-class suburbs of the eighteenth and nineteenth century.

In practice, postwar housing yielded territorial compromises between unevenly spatialized social forces (Schmid 2003). In the Paris case, housing estates fortified the "red bastions" in Paris' suburban belt of Communist-controlled municipalities. Parallel to their Toronto counterparts, which constituted merely interstitial redistributive elements in the Fordist landscapes of urban renewal and mass suburbanization, the Paris projects helped sustain a territorial compromise between the "red belt" and the Gaullist central state, rooted socially in the central city and bourgeois suburbs. In turn, the new form of social mixing that is at the heart of current strategies of deconstructing housing estates wants to pre-empt organized opposition and rebuild paternalist forms of social control to manage the threat that is perceived to emanate from the increasing precarious fractions of the inhabitant population. In contrast to Toronto, in Paris this threat perception remains closely tied to radical, even revolutionary urban pasts and presents. More specifically, place-based mixing has helped reorganize political alliances. In the Toronto region, demolition has emboldened the right's hostility to social housing and solidified the hegemony of "the market" in progressive competitive coalitions (Abbruzzese 2017). In Paris, it furthered the deconstruction of the red belt and helped build a regional political constellation alternating between pink-green and liberal-right-populist alliances.

(5) The Political Project of Place-Based Social Mixing Has Strong Neocolonial Dimensions

The production of space is one crucial way in which racism materializes in the capitalist world. Yet neo-Marxist research on gentrification and state rescaling sometimes gives short shrift to the racialization of state and space. Loïc Wacquant's (2008) neo-Weberian research on the symbolic violence wrought by state-led territorial stigmatization (which has powerfully insisted on the depth of racism in American urban policy) also underestimates the realities of state racism in urban policy elsewhere (Kipfer 2012). In France and Canada, state racism has had neocolonial dimensions, including in housing. In both cases, redevelopment targets spaces largely or overwhelmingly inhabited by people of colour, many of whom, particularly in Paris, have family connections to former colonies. (Let me add that in Toronto, the most direct colonial aspect of the state concerns Indigenous peoples. This aspect is foundational there in the sense that public housing rests on Indigenous land. However, while public housing redevelopment in central Toronto contributes to a broader gentrifying dynamic that affects Indigenous people directly, the Indigenous question is not an explicit and primary determinant in the orientation and design of public housing redevelopment itself.)

Originally, regular postwar social housing in Toronto and Paris was built for varying segments of what we would now call the white working class (and, in France, also petit-bourgeois) populations. In France, for example, regular postwar social housing (which was a central component of postwar housing estate planning) was for the most part kept separate from immigrant housing (shantytowns, workers' hostels, transitional housing) with numerical thresholds, distinct housing programs, and specialized personnel, some of whom trained in Algeria. In Paris, as in Toronto, social housing became identified with non-white populations only since the 1970s, in the very period when the postwar expansion of mass standardized social housing came to an end (most recently, see Blanc-Chaléard 2016).In this context, the deconstruction of social housing and housing estates is based on the lingering assumption that non-white residential dominance is a threat: an explicit political threat in Paris and, in Toronto a more diffuse threat related to "guns" and "gangs" (see also Saberi 2017a, 2017b).

In Paris, planners sometimes see non-white resident concentrations as concrete obstacles to cleaning up the image of the banlieues, leveraging private investment, and restoring a "proper" social mix. In the quest to turn stigmatized estates as well as older pre-war immigrant neighbourhoods into "normal" neighbourhoods, current redevelopment projects are not only racialized but also reinvent state repertoires dating from the late colonial period. While in Canada the links to British colonial history are more mediated politically and sometimes indirect, current French urban strategies share a direct (if complex) link to late colonial state repertoires, and, indeed, histories of counter-colonial struggle. One can detect these links in attempts to limit the concentration of large (read: African) families in the attribution of social housing units and efforts to securitize projects with physical designs that facilitate the everyday deployment of mobile police units and the intervention of paramilitary forces during riots and rebellions. In both cases, public housing redevelopment allows us to understand the concrete socio-political and institutional, not-just-metaphorical weight of colonial pasts and (neo)colonial presents.

(6) Urban Strategies Are Riddled with Multiple Contradictions

To say that housing is both *habitat* and *habiter* means that spatial form and housing are never just top-down territorial projections of the social order. People's everyday practices are rarely a direct reflection of the goals that inform the production of the built environment. State-bound conceptions of space embedded in policy and plans are imperfect indicators of the outcomes of state action, not least because the latter is shot through with multiple contradictions and unintended effects. In Paris, the postwar grands ensembles failed as a project of controlled and socially mixed peripheralization: they were drawn into the *oppositional* core-periphery dialectic of 1968, hit by the fiscal and economic crisis of technocratic Gaullism, and eventually became, in the 1980s, basing points of a new cycle of mobilization against racism and police violence. In this late Fordist period, the "core"-"periphery" imaginaries that pitted old and new working-class suburbs against the bourgeois social spaces in the western part of central and suburban Paris were recast once more, at an expanded metropolitan scale. The Toronto story is much less grandiose in the sense that public housing projects, numerically less important and socially much more residual, never played a comparably significant political or symbolic role; there, too, however, residents began mobilizing against housing authorities only a few short years after the projects were built, and just before the end of large-scale public housing construction in the 1970s.

Today, in Paris and Toronto, the ongoing policy of deconstructing and reconstructing social housing (or housing estates more generally) is reinventing, not terminating, social segregation and lines of race and class. Because renovation includes reconstructing social housing on as well as off existing estates, it does not permanently displace all existing residents. Instead of being massive, displacement is selective and often fine-grained. As a result, renovation produces new forms of micro-segregation and segmented residential mobility.[3] These benefit some residents but not others, while also facilitating new forms of class-based or racialized political domination among various resident groups, whether old or new. In both Paris and Toronto, this is true for seemingly opposite reasons. To the extent that it is produced at all, physical proximity among social groups and class fractions cannot overcome "social problems" (such as economic precarity, bodily racialization, crises of social reproduction) and does not lead to social integration or political harmony. In turn, the goal of social and physical mixing is internally contradictory and confronts obstacles (private property, racism, and class hostility) that are hostile to mixing and tend to push redevelopment onto a path of re-segregation at various scales.

Some examples apply to both urban regions. First, racial profiling deployed against inhabitants moves with their bodies no matter where and in whose proximity they live. Second, the inevitable property demarcations in market housing (or partly marketized social housing) directly reduce opportunities for social interaction, particularly when they are reinforced by physical design borders such as fences or physical separations between buildings or building clusters. I add another example from France. Given that ANRU projects are planned with the help of a national agency but implemented by local housing agencies and municipalities, the way in which systemic mixophobia is tackled or accommodated varies according to local relations of force. For example, the (typically coded) imperative of ethnic mixing does not necessarily translate into strategies of ethnic dispersal. Depending on local political dynamics and the degree of resistance against rebuilding social housing in other neighbourhoods or towns, it can also be redirected into ethnic clientelism or controlled multiculturalism to manage existing spatial concentrations of inhabitants. The Paris case thus defies the idea according to which housing policy simply reflects French Republicanism (which, in its false universalism and seeming colour-blindness, is often seen as the opposite of Canadian multiculturalism).

(7) Territorialized State Intervention Can Politicize but Also Trap Opposition

Since the 1980s, in France, place-based urban policy and (sub)urban revolt have co-evolved in a mutually dependent cycle. The former is both a response to and a cause of the latter. For example, ANRU projects seem to have accentuated the intensity of the 2005 uprising in various places even as they were rationalized after the fact as responses to the latter. As in Toronto, they have also provoked less spectacular, often women-led opposition against (the modalities of) demolition and relocation. But those organizers (rooted in traditions of urban trade unionism or anti-racism) who tried to coordinate opposition across individual projects or even at the national scale had difficulties sustaining their efforts. In this specific sense, oppositional forces in Paris, which carry significantly more weight than in Toronto, have faced similar problems moving beyond the local scale of action as public housing tenants and activists in Hogtown, where opposition to redevelopment plans has not left the project scale. In this regard, redevelopment in both cases has been reasonably effective in binding opposition territorially while "fragmenting the subaltern" in and beyond the target districts – to use an expression by Anne Clerval (2012). These cases thus illustrate why Lefebvre thought that urban strategies can sustain hegemonic projects.

The difficulties of sustaining opposition to redevelopment point to more than the limits of neighbourhood action (*quartierisme*), as necessary as such action is. They also question the sustainability of political projects that, while opposed to neighbourhoodism, remain tied to the imaginary of *les quartiers* or *la banlieue* as self-evident bases of radicalism. In France in particular, these imaginaries remain potent and provide symbolic resources to mobilize residents under condition of segregation. To the significant extent that redevelopment recasts and modifies instead of challenging these conditions, it is eminently plausible to keep alive the idea that segregation provides "raw materials" for people to recognize their commonalities, build mutual aid, organize collectively, forge common interests, and thus appropriate existing spatial practices and imaginaries for emancipatory, even liberatory, purposes. In fact, place-specific forms of resistance based on the idea that segregation is a resource for radicalism rejoin older traditions, notably histories of anti-colonial, anti-racist, and working-class struggle.

However, redevelopment shows that relations of domination (and exploitation) can be organized in various spatial forms. The history of chattel slavery, for example, is full of cases where spatial proximity, not distance, was deemed essential to facilitate surveillance and exercise control. Today's logic of "mixing" populations along lines of class and race indicates that selective desegregation – or, rather, a combination of micro-segregation and relative spatial mobility within segregative conditions – can complement, if not replace, large-scale segregation as a means of spatializing social relations of rule. How? By scattering and re-concentrating existing residents (on various scales), putting them in competition with each other, and encouraging paternalistic relations between new and old inhabitants. For integral emancipatory projects, it is thus dangerous to be fixated on singularly territorialized notions of the subaltern: the "suburbs," the "ghetto," the "banlieue," or the "quartiers." The re-making and unmaking of these social spaces contributes to processes – resentment, interpersonal competition, micro-segregation along lines of class, gender, race, and housing tenure – that undermine everyday relations of solidarity and create conditions that can be exploited by the liberal-entrepreneurial or the far right. To counter the latter, it is imperative to rebuild sociopolitical ties (*refaire du collectif*) in territorially multifaceted ways, including through boundary-defying political organizing (see also Kipfer and Dikeç in this book).

(8) We Need to Revisit the Theories of the State

The construction and deconstruction of housing estates is a clear case of how state action is to be understood, in part, as a response to subaltern presence, even if this presence takes the form of a distant threat rather than substantive concessions or socio-political incorporation, as Nikos Poulantzas pointed out in his most Gramscian period (Poulantzas 2013; Sotiris 2014). Greatly aided by a rich, heterodox French literature on the colonial lineages that are recomposed in the Republic today, my Paris research confirms that this subaltern threat, and the way it is addressed institutionally, has never been only national in character. In the postwar era, responses to the twin threats of Communism and tricontinental liberation were organized in part through the late imperial networks of personnel, ideas, and expertise of France's not-quite-national state, even as that state was internalizing various North Atlantic aspects of *Pax Americana*. In turn, the redevelopment of housing estates recomposes neocolonial repertoires of action in part through transnational channels of media imaging, policy transfer, and inter- and transnational state intervention (in urban warfare, for example). While social housing and urban policy have been central to nation-state formation, the scales of urban strategy (housing included) have never been only national in character.

Theoretically, I have developed my approach to the state through a dialogue between Henri Lefebvre and Frantz Fanon, enriched by Antonio Gramsci. Through Lefebvre, one understands urban questions as a mediation of everyday life and the macro order, including the state, which articulates accumulation and domination in its institutional architecture. In this view, political domination is organized in part through relations between central and peripheral spaces at multiple – that is, never just local or national – scales. Urban policy, which helps in the organization of hierarchical territorial relations, can contribute to the organization of hegemony, which is a combination of coercion and consent and a convergence of material, ideological, and symbolic dimensions of life. What Lefebvre (and Gramsci) only broached but did not develop sufficiently, we can gather from Fanon. He insisted that to understand the (neo)colonial and imperial sides of the world, including the extended state, requires a reworking of historical materialism.

In some respects, my research thus returns to a problem that predates the postcolonial turn in social theory: the concern with the post-colonial state framed in the early 1970s by Hamza Alavi (1972). Alavi insisted that states be analysed not in a deductive fashion (derived from metropolitan state theory) but in relationship to the state-society relations and historic blocs characterizing a particular social formation. In places like Pakistan, this meant taking into account how states (and modalities of rule in general) are shaped by (neo)colonial relations. They can never be just national because they include within themselves relations between post-colonial and metropolitan states, which systematically blur the boundaries between internal and external forces. If one shifts focus from post-colonial to imperial situations, Alavi's point – which was taken up and recast by others since then, explicitly or otherwise (Amin-Khan 2012; Bayart 2006) – alerts us to the transnational linkages that can be formative in metropolitan states with imperial reach. The contingency of the national is thus not only a matter for former colonies; it also goes to the heart of state formation in the imperial heartlands (Elsenhans 1974; Cooper 2005). The national in imperial states results from post-independence nationalization, or ongoing settlement processes in settler colonies. In both cases, this nationalization of states remains incomplete because of a combination of imperial pasts and transnational presents, imperial or otherwise.

Today, the task of theorizing the relationship between the specifically (neo)colonial and other formative aspects of the state in the crisis-ridden imperial heartlands remains crucial. The fact that demands for linkages between expeditionary warfare, "domestic" anti-terrorism, and territorialized state intervention "at home" and "abroad" have multiplied and deepened once more in the last few years only underlines the urgency of this task. Following the Daesh attacks in Paris and Brussels in 2015 and 2016, some politicians in France demanded coordinated state action that could simultaneously target sites in Europe (Molenbeek, St Denis) and bomb Daesh strongholds in Syria and Iraq (Al-Raqqah, Mosul), while also securing borders in and around the European Union. These demands reinforced arguments towards transnational "anti-terrorism": counterinsurgency and states of emergency in the hexagone and Françafrique, the parts of Africa most closely tied to French neocolonialism (Rigouste 2016; Survie 2016). Contradictorily, such demands want to create new forms of transnational state action with one hand, while limiting or deconstructing them to strengthen national state capacities with the other. Conceptually, such demands can be captured under the rubric that has held together many contributions to state theory since the 1980s (the internationalization of the state), albeit only under certain conditions. One is that emphasis be put on tensions (between [para-]military and economic forms of state internationalization) and contradictions (between projects to intensify internationalization and projects to renationalize state action). The other is emphasizing relationships between forms of state internationalization that developed since the 1970s (which is the focus of most research on state internationalization) and those that predate that period.

The role of urbanism and urban strategy in the internationalization of the state remains to be explored. For the purpose of illustration and future work, let me identify two current strands of state theory that connect well with my research. First, the securitized and militarized aspects of place-based urban policy have contributed to "authoritarian statism," a tendency, which, as Poulantzas (2013, 203–47) once warned (see also Boukalas 2014), can create the conditions for conjunctural shifts to formally authoritarian states. However, the place of urban policy in authoritarian statism has garnered no sustained attention from researchers. Second, the multi-scalar, transnational aspects of territorialized state intervention in and around housing estates allow us to rejoin current debates on imperial state intervention. Among these, we can mention the works of Jerome Klassen (2014) and Claude Serfati (2017). Both pay detailed attention to the links between neoliberal, securitized, and militarist aspects of French and Canadian state intervention and their active role in US-dominated imperial dynamics, with additional reference to fledgling European imperial efforts (Forschungsgruppe Staatsprojekt Europa 2012; Wissel 2015) and North American settler colonialism (Coulthard 2014; Dunbar-Ortiz 2014). In these debates, too, the role of urban strategies (urbicide, military planning, "slum" redevelopment, securitized urban development, spatially extensive infrastructure planning) in imperial and (neo)colonial state intervention and state-society relations remains absent or undertheorized.

Conclusion

What are the political implications of these reflections? Students of housing policy may ask: What is a planner to do? They should consider rising "in defense of housing," understood as something other than real estate (Madden and Marcuse 2016). In both Paris and Toronto, where public housing redevelopment has not yet adopted the slash-and-burn method dear to the Chicago Housing Authority, there is much to defend in the way of social housing. Why? While social housing can *but need not* yield segregation, hierarchy, and bureaucracy, egalitarian conviviality cannot possibly come about by "the market." By creating affordable housing shortages and entrenching property lines, commodified housing and land-rent dynamics reinforce gendered and racialized urban segmentations and hierarchies. Both combine to undermine prospects for a culture of diversity-in-solidarity. Given the way in which urban policy, planning, and social housing programmes today have incorporated "security," the "market," and "entrepreneurialism" into their operations, the metaphorical space for planners and advocates to pursue strategies of expanding, decommodifying, and democratizing social housing is small, however. In turn, the deconstruction of social housing has made it even more difficult to keep together distinct, if sometimes converging, historical demands: those for generous and democratic collective consumption and those against segregation, neocolonial practices, and state racism as seen in practices like police violence and racial profiling.

Decommodifying housing and promoting liberation in and through segregated realities cannot be left to housing specialists. It is in fact inconceivable without transformative movements. But today, room for radical planning – that is, ways of articulating (para-)statal planning and transformative movements – is limited. A second question thus imposes itself. A hundred years after 1917, students of revolution may ask: What is to be done? In periods of radical change such as those of a hundred years ago, many specialists asked or were forced to consider how city-building and spatial planning could intensify revolutionary transformation. In periods that resemble the state of Paris and Toronto today, contradictory times fraught with danger and hope but without a sustained emancipatory upsurge, a broader perspective on housing forces one to search more deeply to unearth the conditions for serious innovations in city-building practice. Taking Lenin's idea of dual power beyond Lenin himself, building capacities for egalitarian and sustainable planning depends in part on the capacity of labouring peoples in various situations to mobilize, self-organize, and promote self-determined projects (Borra 2016; Aureli 2008). Multiple but potentially allied self-determined bases of operation, in housing and beyond, may also put in sharp relief the obstacles that stand in the way of egalitarian, hospitable, sustainable, internationalist lives.

ACKNOWLEDGMENTS

I would like to acknowledge the many helpful comments and questions I have received from the editors and reviewers of this volume as well as the members of the audience at Building and Rebuilding the Periphery (Istanbul, December 2015), Historical Materialism (London, November 2016), and the working group Justice, Espace, Discrimination Inégalités (JEDI) at the Université Paris-Est, Marne-la-Vallée (March 2018).

NOTES

1 For details and most relevant citations, see Kipfer (2016, 2013, 2012); Kipfer and Goonewardena (2014); Kipfer and Petrunia (2009).
2 For the latest evaluation research dealing with land rent and real estate, see Daniel Bourdon and Christophe Noye (2016).
3 In France, some evaluation researchers have taken the lack of mass displacement and the dearth of "typical" central city gentrifiers on re-developed housing estates to question the argument that housing estate development is a form of state-managed gentrification. However, in fact, they have assembled considerable evidence of just such forms of gentrification: property-related and class-specific dynamics of residential recomposition, selective on- or off-site displacement, and socially "upward" dynamics of re-segregation that favour some inhabitants, new or old, over others (see Lelévrier 2018).

REFERENCES

Abbruzzese, T. (2017). "'Build Toronto' (Not Social Housing): Neglecting the Social Housing Question in a Competitive City Region." In Keil, R., Hamel, P., Boudreau, J.-A., Kipfer, S. (eds), *Governing Cities through Regions: Canadian and European Perspectives*, 143–72. Waterloo, ON: Wilfrid Laurier University Press.

Alavi, H. (1972). "The State in Post-colonial Society." *New Left Review* 74, no. 1, 59–84.

Amin-Khan, T. (2012). *The Post-colonial State in the Era of Capitalist Globalization*. London, UK: Routledge.

Aureli, P.V. (2008). *The Project of Autonomy: Politics and Architecture within and against Capitalism*. New York, NY: Princeton Architectural Press.

Bayart, J.-F. (2006). *L'Etat en Afrique: La politique du ventre*. New edition. Paris, France: Fayard.

Berger, J. (2008). *Ways of Seeing*. London, UK: Penguin.

Bertho, A. (1997). *Banlieue, Banlieue, Banlieue*. Paris, France: La Dispute.

Blanc-Chaléard, M.-C. (2016). *En finir avec les bidonvilles: Immigration et politique de logement dans la France des Trente Glorieuses*. Paris, France: Publications de la Sorbonne.

Borra, B. (2016). "'Cooperation Rules the World. The Community Rules the Individual': On Hannes Mayer." In Aureli, V. (ed.), *The City as a Project*, 262–88. Berlin, Germany: Ruby.

Boukalas, C. (2014). "No Exceptions: Authoritarian Statism: Agamben, Poulantzas and Homeland Security." *Critical Studies on Terrorism* 7, no. 1, 112–30. https://doi.org/10.1080/17539153.2013.877667

Bourdon, D., and Noye, C. (2016). "L'impact des projets de rénovation urbaine: Mixité sociale et mobilité des ménages." *La revue fonçière*, 11, 13–16. http://www.revue-fonciere.com/RF11/RF11_Bourdon.pdf

Brenner, N. (2004). *New State Spaces*. Oxford, UK: Oxford University Press.

Castells, M. (1977). *The Urban Question*. Cambridge, MA: MIT Press.

– (1978). *City, Class, and Power*. London, UK: Macmillan.

Clerval, A. (2012). *Paris sans le peuple: La gentrification de la capitale*. Paris, France: La Découverte.

Cooper, C. (2005). *Colonialism in Question: Theory, Knowledge, History*. Berkeley, CA: University of California Press.

Coulthard, G. (2014). *Red Skin, White Masks: Rejecting the Colonial Politics of Recognition*. Minneapolis, MN: University of Minnesota Press.

Dikeç, M. (2007). *Badlands of the Republic: Space, Politics, and Urban Policy*. London, UK: Blackwell.

Dunbar-Ortiz, R. (2014). *An Indigenous History of the United States*. Boston, MA: Beacon.

Elsenhans, H. (1974). *Frankreichs Algerienkrieg 1954-1962: Entkolonisierungs-versuch einer kapitalistischen Metropole. Zum Zusammenbruch der Kolonialreiche*. Munich, Germany: Carl Hanser.

Engels, F. (1969). *The Condition of the Working Class in England*. London, UK: Panther.

– (1970). *The Housing Question*. Moscow, USSR: Progress Publishers.

Forschungsgruppe Staatsprojekt Europa (ed.). (2012). *Die EU in der Krise: zwischen autoritärem Etatismus und Europäischem Frühling*. Münster, Germany: Westfälisches Dampfboot.

Hackworth, J., and Smith, N. (2001). "The Changing State of Gentrification." *Tijdschrift voor Economische en Sociale Geografie* 92, no. 4, 464–77. http://dx.doi.org/10.1111/1467-9663.00172

Hart, G. (2017). "Relational Comparison Revisited: Marxist Postcolonial Geographies in Practice." *Progress in Human Geography* 42, no. 3, 371–94. https://doi.org/10.1177/0309132516681388

Harvey, D. (1989). *The Urban Experience*. Baltimore, MD: The Johns Hopkins University Press.

Hayden, D. (1981). "Madame Kollontai and Mrs. Consumer." In Hayden, D., *The Grand Domestic Revolution*, 281–9. Cambridge, MA: MIT Press.

Katznelson, I. (1981). *City Trenches*. Chicago, IL: University of Chicago Press.

Kipfer, S. (2012). "'Ghetto or not ghetto, telle n'est pas la seule question': Quelques remarques sur la 'race,' l'espace, et l'Etat à Paris." In Boggio Éwanjé-Épéé, F., and Magliani-Belkacem, S. (eds), *Race et capitalisme*, 125–46. Paris, France: Syllepse.

– (2013). "Urbanisation et racialisation: Déségrégation, émancipation, hégémonie." In Buclin, H., Daher, J., Georgiou, C., and Raboud, P. (eds), *Penser l'émancipation: Offensives capitalistes et résistances internationals*, 111–29. Paris, France: La Dispute.

– (2016). "Neocolonial Urbanism? *La Rénovation Urbaine* in Greater Paris." *Antipode* 48, no. 3, 603–25. http://dx.doi.org/10.1111/anti.12193

Kipfer, S., and Goonwardena, G. (2013). "Urban Marxism and the Post-colonial Question: Henri Lefebvre and 'Colonization.'" *Historical Materialism* 21, no. 2, 76–117. https://doi.org/10.1163/1569206X-12341297

– (2014). "Henri Lefebvre and 'Colonization': From Reinterpretation to Research." In Moravánsky, Á., Schmid, C., and Stanek, Ł. (eds), *Urban Revolution Now: Henri Lefebvre in Social Research and Architecture*, 93–109. Farnham, UK: Ashghate.

Kipfer, S., and Petrunia, J. (2009). "'Recolonization' and Public Housing: A Toronto Case Study." *Studies in Political Economy* 83, 111–39. https://doi.org/10.1080/19187033.2009.11675058

Klassen, J. (2014). *Joining Empire: The Political Economy of the New Canadian Foreign Policy*. Toronto, ON: University of Toronto Press.

Lambert, A. (2015). *Tous propriétaires: L'envers du décor pavillonnaire*. Paris, France: Seuil.

Lelévrier, C. (2018). "Rénovation urbanine et trajectoires résidentielles: Quelle justice sociale?" *Métropolitiques*, 12 March. https://www.metropolitiques.eu/Renovation-urbaine-et-trajectoires-residentielles-quelle-justice-sociale.html

Madden, D., and Marcuse, P. (2016). *In Defense of Housing*. London, UK: Verso.

Massey, D. (1991). "A Global Sense of Place." *Marxism Today* 38, 24–9.

Mitchell, K., Marston, S., and Katz, C. (2004). *Life's Work: Geographies of Social Reproduction*. Chichester, UK: Wiley.

Poulantzas, N. (2013). *State, Power, Socialism*. London, UK: Verso.

Rigouste, M. (2016). *Etat d'urgence et business de la sécurité*. Paris, France: Niet Editions.

Roy, A. (2005)."Urban Informality: Toward an Epistemology of Planning." *Journal of the American Planning Association* 71, no. 2, 147–58. https://doi.org/10.1080/01944360508976689

Saberi, P. (2017a). "The 'Paris Problem' in Toronto: The State, Space and the Political Fear of the 'Immigrant.'" PhD diss., York University.

– (2017b). "Toronto and the 'Paris Problem': Community Policing in 'Immigrant Neighbourhoods.'" *Race and Class* 59, no. 2, 49–69. https://doi.org/10.1177/0306396817717892

Schmid, C. (2003). "Raum und Regulation: Henri Lefebvre und der Regulationsansatz." In Brand, U., and Raza, W. (eds), *Fit für den Postfordismus? Theoretisch-politische Perspektiven des Regulationsansatzes*, 217–42. Münster, Germany: Westfälisches Dampfboot.

Serfati, C. (2017). *Le militaire: Une histoire française*. Paris, France: Amsterdam.

Smith, N. (1996). *The New Urban Frontier*. New York, NY: Routledge.

Sotiris, P. (2014). "Neither an Instrument nor a Fortress: Poulantzas' Theory of the State and His Dialogue with Gramsci." *Historical Materialism* 22, no. 2, 135–57. https://doi.org/10.1163/1569206x-12341355

Survie (2016). *Etat d'urgence, surveillance, et interventionnisme militaire en Afrique*. Paris, France: Survie, Ensemble contre la Françafrique.

Tissot, S. (2007). *L'Etat et les quartiers: Genèse d'une catégorie de l'action publique*. Paris, France: Seuil.

Tournon, M. (2018). "Du redressement économique au projet urbain: Gestion financière des HLM et génèse de la rénovation urbaine (1980–2000)." *Métropolitiques*, 19 March. https://www.metropolitiques.eu/Du-redressement-economique-au-projet-urbain-Gestion-financiere-des-HLM-et.html

Wacquant, L. (2008). *Urban Outcasts: A Comparative Sociology of Advanced Marginality*. Cambridge, UK: Polity.

Wissel, J. (2015). *Staatsprojekt Europa: Grundzüge einer materialistischen Theorie der Europäischen Union*. Münster, Germany: Westfälisches Dampfboot.

PART THREE

Spotlight on Istanbul

8 From Kayabaşı to Kayaşehir – A City Grows "Out in the Sticks"

ERBATUR ÇAVUŞOĞLU AND JULIA STRUTZ

The growth of Istanbul since the Second World War has followed the model of an oil stain: The constant densification of the already existing city, its informal extension, and the ex post facto legalization of its fringes (Tekeli 2013, 154). Until the Mass Housing Administration of Turkey (TOKİ) was equipped with extensive rights and resources in 2003, suburbia as a space planned by the state or the city was an exception. Kayaşehir is possibly the biggest and most important of these new satellite towns.

Located at one of the northern extremes of the city, Kayaşehir used to be called Kayabaşı, which literally means "on top of the rocks" and evokes a notion similar to "out in the sticks" or "in the middle of nowhere." With its announcement as a mass housing area in February 2006, it was renamed as Kayaşehir, the "city on the rocks." It is the continuation of the mass housing area Başakşehir, which emerged there since the mid-1990s under the auspices of then-mayor of Istanbul Recep Tayyip Erdoğan's plan to create a neighbourhood for his followers (Çavdar 2013). This characteristic continues in Kayaşehir: In the two general elections in 2015, Erdoğan's Justice and Development Party (AKP) gained 62.2 and 70.6 per cent of all Kayaşehir votes, significantly higher than the Istanbul average of 42.2 and 49.2 per cent.

Kayaşehir is close to the İkitelli industrial site that offers employment for 300,000. To its southeast, the huge Atışalanı military reservation was recently opened for urban redevelopment at the end of 2015. The western border of Kayaşehir is the Sazlıdere water reservoir, targeted for a new channel project to create another passage between the Black Sea and the Sea of Marmara. All these projects were exempted from local planning authorities as "reserve construction area for the New City," whose size was 244.750.000 square metres in April 2014 (Official Gazette 2014).

8.1 TOKİ Kayaşehir massive housing site as seen from a distance. Photo by Roger Keil.

The most dominant characteristic of the process of Kayaşehir's planning and construction, starting in 2006, is ambiguity (Roy 2009, 2015). The deliberate chaos and continued uncertainties created by the municipality and TOKİ in all stages of the production of Kayaşehir (expropriation, planning, project design, tenders, sale, housing acquisition, and the everyday life of the suburb), enables the latter to gain enormous profits that they redistribute to firms and individuals in their political networks, close to the AKP. It thus ensures political support both from investors and capital as well as from the "rural-urban poor" (Roy 2009, 82). The planning process of Kayaşehir shows that even state institutions can act transversally (Caldeira 2017) towards their own practice. It is maybe an extreme example of "suburban governance" (Eker, Hamel, and Keil 2012), in which the state has a strong and singular role in maximizing capital accumulation of private authoritarianism.

This chapter is based on two main sources. One source is several lawsuits and the accompanying official documents, plans, and plan revisions about the production of Kayaşehir. These lawsuits were opened after 2008 by owners of land in the area to hamper the expropriations

8.1 Plans made for Kayabaşı/Kayaşehir between 2007 and 2009

Date	Planned Area	Land Ownership	Number of Housing Units	Estimated Population
Until 2006	1,500 ha.	Private, agriculture, 439 outbuildings, 81 small industrial plants	959	~5,000
2007	1,500 ha.	38% private ownership, 32% public property, 30% object of litigation after expropriation	86,250	345,000
2008	1,300 ha.	42% private ownership, 58% public property	33,314	99,552
2009	Unknown	100% public property (TOKİ)	19,803 (?)	59,411

Source: PR (2007, 2008, 2009).

enforced by TOKİ. This litigation reproduces what the bureaucracy, experts, and formal planning procedures officially knew about Kayaşehir. The other source is a (strikingly high) number of Facebook groups and forums about Kayaşehir frequented by its inhabitants. The population seem to be confronted with a similar problem as we were during our research: The gated housing complexes are impossible to access for outsiders, the spaces between the fences are deserted, places to gather do not exist, and shopping areas are rare and rather empty. Online forums and groups often deliberately replace neighbourly contact, mutual help, information exchange, or advertisement of local products and services. In this landscape of ambiguity, statistical data, especially on the development of land prices, of course does not exist or is simply not reliable.

We are aware that long passages in this chapter abound with numbers, seemingly overwhelming details, and foreign names. Although we have tried our best to omit the nonessentials and to guide the reader through this mess, the feeling of incomprehension and the impulse to capitulate is constantly present. But the devil is in the details and meant to benumb anyone who tries to understand. Whether we want it or not, our text reflects the effects of ambiguity on everyone involved – including us.

Planned Mis-Planning

The selection of Kayabaşı as the site for Turkey's biggest satellite town was all but accidental. In 2006, the planning report prepared by TOKİ for their mass housing project counted 959 housing units, 439 outbuildings, and 82 light industrial plants, housing about 5,000 inhabitants on a project site of 1,500 hectares (PR 2007, 14; see table 8.1 in this chapter). Thus, agriculture was slowly replaced by light industry in Kayabaşı. The Republic of Turkey and TOKİ notably did not own any land in the area before 2006. Land was cheap here not only because it was peripheral land but also because the recently passed Master Plan for Istanbul, developed by an office of hundreds of specialists with high recognition after years of analysis and debate, restricted on principle any further growth of Istanbul to the north. It was plain sailing for TOKİ to convince owners to sell their plots, as obviously no private investor would be interested any time soon.

TOKİ declared the area a mass housing area and started expropriations in the early months of 2006. In 2007, TOKİ had already bought 32 per cent of the 1,500 hectares, and a second third of the land was matter of a lawsuit after owners had objected to TOKİ's expropriation (PR 2007, 19). At this point, expropriations were based on a plan foreseeing 86,250 housing units inhabited by a population of 345,000 (PR 2007, 37) in the area. A year later, TOKİ had already gathered two-thirds of the area's property under its authority. The report for the 2008 plan not only reduces the size of the project area by 200 hectares, it also harshly criticizes the 2007 plan (prepared by the same authority, TOKİ) for its excessive density. With less than half of the housing units (33,314) of the "original" plan, only about 100,000 inhabitants were projected (PR 2008, 12). The silent acceptance of the 2008 plan over the 2007 plan was because family sizes had dropped from an average of four to an average of three – ultimately amounting to 50,000 fewer people. As a consequence, much less infrastructure like roads, schools, and hospitals needed to be provided.

In 2009, TOKİ again deemed a new plan was necessary, as allegedly the connection roads for a planned third bridge crossing the Bosporus would affect the eastern part of the future Kayaşehir. While the heated debates about a possible third bridge continued in 2009, the Master Plan of 2009 did not include the bridge or its connection roads. Only in 2012 were both included in the plan, after Recep Tayyip Erdoğan had determined their location during a helicopter flight (TMMOB, 2012). Although from the perspective of the plan the bridge did not exist, it was given as a pretext in the 2009 plan to change the 2008 plan. The 2009 plan, however, failed to mention the expected size of the mass housing area and the borders of the project area. It further stated that in the end 100 per cent of the land of the mass housing area was allocated to TOKİ, and that 59,411 inhabitants will live in the future suburb (PR 2009, 22). If we invert TOKİ's calculations from 2008 and take three people as the average household size, TOKİ planned to produce 19,803 housing units. The current count on TOKİ's website, however, is 23,016 housing units (TOKİ 2017a). Publicly, TOKİ has also announced there will be 65,000 housing units once construction in the area is finished.

Three different plans in three years with three different names, three different project areas, numbers of housing units, expected inhabitants, and urban amenities. Still, only weeks after the 2009 plan revision the first housing units were sold. The first people started to live there in the spring of 2011.

Dispossession and Good City Planning

The ease and speed with which TOKİ managed to expropriate the owners of land in the area can be explained by low prices. Many seemingly were not aware of the increase in land rent to be expected, but even those who knew could not avert expropriation and displacement. This has something to do with the specifics of TOKİ's planning principles. Until 2006, the majority of plots were between 100 and 500 square metres, often gardens used for small-scale agriculture and housing. Arguing with good and contemporary planning principles, TOKİ determined a minimum size of 5,000 square metres for plots in the area. In addition, TOKİ has the authority to cut up to 60 per cent of the original plot size during expropriation in exchange for development rights and infrastructure. Consequently, an owner of 200 square metres, for instance, is left with a plot of 80 square metres after these legal cuts. The owner is then made a shareholder in a plot of 5,000 square metres somewhere "close by." Almost every owner of agricultural land tried to inhibit this dispossession, as witnessed by the enormous number of lawsuits they opened against TOKİ. But even if the owner successfully sued TOKİ, this would only have increased their share, not stopped the expropriation. Their only option was still to sell their share and leave the area, their agriculture, and their industry behind.

So, who are the people who moved to Kayaşehir and invested their little money in an area "out in the sticks," whose future was so unpredictable? At first glance, the housing provision system of TOKİ defines a very homogeneous social profile of future inhabitants, as it entitles some user groups to advantageous access to parts of its housing stock.

The initial inhabitants of Kayaşehir were thus supposedly mainly members of the lower middle class and middle class, working in the adjacent İkitelli industrial zone or as clerks or teachers. This is an effect of the size of the apartments on the one hand, and the requirements for application on the other. Flat sizes vary from 48 square metres (flat type DY), to 53 square metres (type B), 75 square metres (type B1), 92 square metres (type C), 110 square metres (type C1), and 122 square metres (type C2). Although the requirements (and the size of flats) vary from project to project, they usually require the applicant to prove that they have lived in Istanbul (sometimes also on the European side of Istanbul) for at least two years; that none of the family members, including children, owns another apartment; and that they have not bought property from TOKİ before and have not used a credit scheme managed by them. Sometimes requirements like age (e.g., a minimum of thirty or twenty-five years) or a maximum income per household (e.g., 3,700 TL) are added. For the low-income group, apartments are distributed to applicants by the drawing of lots, and here veterans and widows have a priority position, while a quota of 5 per cent exists for disabled persons and a quota of 25 per cent for retirees. At least officially, the housing type provided for low-income families cannot be resold or rented until the credit is paid (TOKİ 2017b). Kayaşehir provided the opportunity for many to become homeowners at rates that resemble monthly rents as a tenant. As TOKİ's slogan goes, they had the opportunity to become homeowners as if they were paying rent. Homeownership in Turkey is one of the only ways to access upward mobility and security for a lower middle-class family.

No Risk, No Fun: The Effects of Ambiguity

Seen from today, investment into Kayabaşı was more than profitable. If the projects are ever finalized, Kayaşehir will have direct access to the third bridge and the third airport, a hospital with an expected 80,000 visitors a day, and access to the metro network of the city. Most importantly, the wondrous election pledge of the 2011 elections is only a stone's throw away. During the 2011 elections, then-prime minister Recep Tayyip Erdoğan prepared the Turkish public over weeks for the disclosure of his "crazy project," a present to the populace (Bianet 2011). When Kanalistanbul – a second, artificial Bosporus canal – was finally announced, prices in Kayaşehir skyrocketed. Whether the airport will eventually be used as an airport, or whether the channel will be dug or not, is of as little importance to the price increases in Kayaşehir as is the quality of construction or the amount of social infrastructure in the neighbourhood. These sorts of promises often only take the form of rumours and are distributed in networks of trust. They are sufficient to encourage investors to buy and for land prices to explode. In a situation of constantly changing plans for the project area, and no or only insufficient public statements, the central question is thus: Who was informed about the future investments planned by the state in the area?

To convince a low-income family to invest their little money and run the risk of losing everything in this uncertainty could not be done with formal arguments, but only through their involvement in informal networks close to the government. Still, conditions for them to stay put are anything but favourable. The deliberate chaos of plans and construction also extends to the future inhabitants. Not only is it often not completely clear when to pay what kind of money to whom, the date when the owners will be able to move into their new flats is vague and often postponed for months. For the low-income family that attempts to "become homeowners as if they were paying rent," such delays cause serious trouble. To pay the first down payments on the credit each month while continuing to pay rent for their previous apartment is difficult. But it is impossible to do so for months longer than expected.[1] The failure to pay three monthly credit installments in a row results in losing the apartment. The family's connectedness to the right networks is crucial, as it can organize exceptions to this rule (Roy 2005).

Furthermore, the apartment buildings themselves take precedence over urban infrastructure in Kayaşehir, such as public transport, health care, schools, day care, and shops. Often opened years after the first people moved to Kayaşehir, their absence hits the poor hardest. Without owning a car, commuting to work is expensive and slow; women especially are basically trapped inside the apartment towers. Nearby employment opportunities in the service sector, which the inner city and informal neighbourhoods offer, do not exist in this location.

Another nuisance and serious financial burden is the monthly fees residents have to pay to the site management for cleaning, repairs, gardening, and security in the common areas. These extra fees can often amount to one-third of the monthly down payments. They are suddenly and arbitrarily increased – the expenditures and budgets of the site management are not controlled by the inhabitants. Residents who cannot pay their fees regularly risk losing their homes. The site managers often charge themselves with a "mission civilatrice" of educating

8.2 TOKİ high-rise apartment buildings in Kayaşehir, İstanbul. Photo by Roger Keil.

the urban poor about modern life in an apartment building. A constant source of conflict is, for instance, the hanging of laundry and carpets on balconies. As the public display of laundry suggests a low-income population with a rural background, site managers try to prevent residents from doing it. Low-income families often have to inhabit small apartments with large families and do not possess much furniture but carpets. As these serve as desk, bed, table, and sofa at the same time, being able to wash them regularly is a hygienic necessity. The poor are thus not only constantly under pressure to pay more than they can afford but are also mistreated and held responsible for all sorts of filth, noise, burglary, and other inappropriate behaviour.

Perhaps these mechanisms of exclusion are not necessary to force low-income owners to sooner or later sell their property. In the years 2010 to 2014, prices in Kayaşehir doubled every year. Real estate agents in Kayaşehir purported that an apartment that cost 60,000 TL in 2010 could be sold for 360,000 in 2015. The statistics of the web portal *sahibinden*, where owners sell their flats without real estate agents, shows that the price per square metre in Kayaşehir for buildings between five and ten years old (these are the TOKİ buildings) increased by 131 per cent from 1,524 TL in December 2013 to 2,813 TL in December 2016.

Even though the initial inhabitants had to struggle with a variety of problems and complained about the way they were treated, they would of course voluntarily and very gratefully resell their property as soon as its value had increased. Even if only for some time, becoming a homeowner in TOKİ allowed them to make a higher revenue than anything they would be able to make in a lifetime of work. The amount of real estate advertisement in Kayaşehir for apartments that are transferred to new owners with several years of credit still pending suggests that this is a not always legal, but that it is a very common practice. It is not very far-fetched to suggest that a majority of the flats have changed ownership several times since their construction, allowing probably several hundred thousand owners and sellers to profit from the increase in land rent each time.

The longer owners were able to stay put, the more money they gained. The larger, the longer, and the riskier the initial investment, the higher the revenue. Surely, low-income families are not among the main winners of the process.

The Real Profiteers

The real beneficiaries of the profits from Kayaşehir, however, are the private-sector entities close to TOKİ and the governing party. From the removal of rubble to the construction of luxury gated communities, down to the metres of heating and hot water in each flat – the companies involved have family ties to high-ranking AKP officials or have been collaborating and supporting the party in the last decades (Gürek 2008).

These networks become effective in the construction of housing in three phases. They create a system that guarantees revenue inside the network and persistent demand for housing. In the first phase, TOKİ makes a plan, gathers all property under its authority, and constructs mass housing on the plots with the highest density. High-density construction without additional taxation offers still reasonable revenue, although the flats are sold relatively cheap. But the cheap price of the flats is not the only guarantee for success. At the same time, TOKİ and the

8.3 Kayacity Residence, an ongoing luxury housing project, which is being constructed by the private company Celaloğlu, in Kayaşehir, İstanbul. Photo by Roger Keil.

municipality announce all sorts of mega-projects to earmark the area for expected increases in land rent.

It is not easy to get reliable information about TOKİ's subcontractors. Many of them do not even have websites, and their management boards remain featureless to the outsider. One of the more well-known subcontractors is, for instance, Özkar İnşaat, owned by the Özcan brothers. Niyazi Özcan was an AKP Member of Parliament for the city of Kayseri in the years 2002 to 2007. TOKİ commissioned Özkar İnşaat to construct 1,814 housing units in the fifth district in Kayaşehir (TOKİ 2017a). Helped by construction firms like Özkar İnşaat, TOKİ claims to have completed 23,016 housing units up to the end of 2017.

Özkar İnşaat is also a shareholder in the upscale condominium development Bulvar İstanbul in Kayaşehir. In what we propose to call the second phase, TOKİ and the state-owned real estate investment company Emlak Konut form partnerships with construction firms to perform what is called "revenue-sharing projects." Each project has its own specificities, but the general lines of these revenue-sharing projects are as follows: TOKİ contributes the plan, land, and infrastructure; Emlak Konut provides credit schemes; and, in this case, Özkar İnşaat and another firm, Özülke, take on the construction work, the advertisement, and the sale of properties. After the sale, each shareholder gets a percentage of the revenue agreed upon beforehand. To reduce the risk for the private sector, TOKİ has not only built 23,016 housing units and the accompanying infrastructure for mass housing but has also allowed the same firms to share revenue in projects for the upper middle classes. Prices for flats in Bulvar İstanbul range between 780,000 and 1,804,000 TL.

As of late 2017, there are nine revenue-sharing projects in the project area planned by TOKİ. Four of the nine involve a partnership with the construction companies Makro İnşaat and Akyapı. For at least two of the projects completed by this partnership, plots that were intended to house urban infrastructure in the 2009 plan were subsequently opened for housing development. The gated town Seyranşehir is built on a plot designated for a commercial area. The development Bahçetepe offers its inhabitants allotment gardens on a plot that was formerly planned as recreational area – a rather creative interpretation of the plan.

Repeatedly, the same firms surface in big construction projects, and they also create their own demand. At least some of the first inhabitants in Kayaşehir came from urban regeneration projects. In exchange for their former shelter, legal or informal, in the zone designated for a regeneration project, they were offered the opportunity to buy into TOKİ housing at better conditions than market price. Both the Özkar İnşaat/Özülke and Makro İnşaat/Akyapı partnerships have been contracted in revenue-sharing projects for the urban redevelopment projects in Ayazma. Here, the municipality and TOKİ took on the eviction of 1,730 recently migrated, mainly Kurdish families from their informal housing (Pérouse 2006; see Kuyucu 2014 for the ambiguities involved in this process). The families that could prove ownership were moved to the TOKİ estate Bezirganbahçe (Bartu-Candan and Kulluoğlu 2008). After years of organization and resistance, some of the families that were once tenants in Ayazma gained the right to buy in Kayaşehir (Kurşuncu 2011). Özkar İnşaat/Özülke and Makro İnşaat/Akyapı won tenders for the new construction in Ayazma.

After the finalization of TOKİ's mass housing in the eastern part of the area (phase 1), and the condominiums in revenue-sharing projects between TOKİ and the private sector (phase 2), the private sector can now join the venture in the third phase. These fully private developments will be concentrated in the area closest to the planned channel and in the interstices between mega-projects and mass housing. The 2009 plan foresees the lowest densities for the private developments. Investors acquiring land trust that all the promised mega-projects and state investments will further drive up the prices in the area and that Kayaşehir will continue to transform from a suburban somewhere, out in the sticks, to a new centrality with hundreds of thousands of inhabitants. They probably calculate that they will be able to convince the local authorities and TOKİ to slightly increase the allowed densities on their plot retroactively. In any case, risk for the private sector is diminished to a minimum.

Ambiguity and Suburbia

Admittedly, though not all winners benefit equally, the deliberate chaos in Kayaşehir created a win-win situation for all actors involved. Low-income families have doubled their initial spending, investors and constructors have made extensive profits, and the AKP has guaranteed political support. The number of disappointed can be as easily ignored as the damage done to ecosystems, forests, and water reservoirs.

Still, it begs the question why the municipality and TOKİ bother to draw plans and write plan reports at all. If plans are not meant to make the future predictable, to create trust among investors, and to make transparent how profits are distributed, why make the effort? We believe that it has something to do with legitimacy – to fulfil the formalities, at least on the outside. Lawsuits against plans and projects are (or once were) the only way to legitimately question or slow down a project. As a nice side-effect, specious plans keep oppositional forces like the chambers of architects or city planners occupied in this maelstrom of formalities and court cases, which they may win without being able to stop the project. In the end, someone who does not trust the success of the project without a reliable plan is therefore someone who does not blindly trust the governing party. To have someone like that as a neighbour and to share revenue with such people is not desirable in the end.

From the perspective of formal, codified procedures, what has happened in Kayaşehir is criminal cronyism. From the construction firm who wins the corrupted tender to the owner of a subsidized flat who sells it with debt, all are now accomplices in corruption who need to fear prosecution should their network ever be dismantled. They won't spare pains to make the network and the head of the network – the AKP and Recep Tayyip Erdoğan – continue in their powerful positions – another nice side-effect.

Ambiguity purposefully replaces formal types of knowledge and control like plans, plan reports, press releases, law, and regulations. Instead, actors resort to informal forms of knowledge, rumours, and promises that are distributed in networks of mutual acquaintance, mutual trust, and eventually also religious belief. The power of the state is converted into personal power, especially instrumental for the strong leadership position of Recep Tayyip Erdoğan. Many of his followers liken him to a caliph whose announcements are to be obeyed and interpreted as holy words. Recep Tayyip Erdoğan has continuously been very vocal in support of mega-projects and the further development of Istanbul.

For decades Istanbul's peripheries have been borders and edges that were continuously overrun by the expansion of the city. The development, densification, and redensification of these peripheries have created much revenue throughout the modern history of the city. No other political actor, though, has understood how to use the differences in land rent at the peripheries for profit as deliberately as the AKP. The party and its leader have managed to channel the revenues so that "HE" appears as the redistributor. We believe that much of the political success of the AKP can be explained by their spatial politics. The AKP is in many senses suburban: As a political movement it came from the rural-traditional-religious fringes of society; as political figures they started their careers in the informal, self-built, and self-developed peripheries. Excluded from politics, economy, and the dream of the good life in middle-class suburbia, they now take brutal revenge. From the centre of politics and economy, the AKP creates its own suburban dream, where the pious Muslim lives an orderly life in moral integrity. Kayaşehir, the city out in the sticks, will soon be all but suburban. It is the imagined centre of the new Istanbul and the new Turkey.

ACKNOWLEDGMENTS

We would like to thank our friend Professor Neil Korostoff for proofreading our article, and for his gentle remarks and improvements of our work.

NOTES

1 This urgency becomes tangible in a comment posted by user şehri on the page www.TOKİkayabasikonutlari.net on 25 February 2012. We prefer to keep it in the original, as it conveys much more, and the text is basically not translatable. To summarize it briefly, a new inhabitant of Kayaşehir complains bitterly about TOKİ. His family had already packed up everything and gave away their stove in June 2011. But they still can't move into their new flat, winter has passed, and they had to buy a new stove. He cannot pay the rent on their flat because he needs to pay the credit to TOKİ. If he doesn't pay TOKİ, they will lose their new apartment, and they are already four months in delay with the rent. He threatens to go on TV if they cannot finally move to their flat in March.

... biz vatandaşları madur bırakıyosunuz yazdan beri eşyalar kolide bekliyo yazın teslim edilcekdi haziranda dendi gitdik evdeki sobayıda başkasına verdik kış geldi geciyo hala evler teslim edilcek birde soba aldık üzerine yeniden mart ayı dendi mart ayıda girdi cok mart ayları gecer gibi geliyo şimdide kalmışsınız yol zorunu ya kardeşim bu binaları dikerken yaparken bir mutaipde cıkıp yol düşünmedimi mühendiz yokmuydu nasıl izin verdiler o evleri dikmenize mühendizsizce mutahip yol düşünmeden kimi kandırıyonuz yeter artık oyalamayın insanları bu iş leyleyin vaklamasıyla gecer sizde öyle getiriyosunuz madem TOKİ yaparken söz verdi hiç bir vatandaş madur olmuycak kardeşim ben madurum kiramı ödeyemiyorum TOKİnin ev taksitine yetişdircem diye kirayı ödesem TOKİyi ödeyemiyorum TOKİ ödemezsem TOKİ evi alcak elimden kirayı ödemezsem ev sahibi evden atıyo 4 ay kira borcum var hiç bir ev sahibi 4 ay beklemez madurum TOKİ ödesin kiramı yada her ay düşsün ev taksitinden inanın cevremdeki insanlar dalga geciyo daha evleriniz teslim edilmedimi diye yokdsa kandırıldınızmı diyorlar bizde bilmiyoruzki işin aksini nedir ben söylüyorum saten bu konuyuda TOKİden ev alan arkadaşlar var bekliyoz teslimantı mart geldi martda edilmezse cıkalım kanala haberlere işin aslını ögrenelim kandırıldıkmı yoksa oyalanıyoruzmu inanın artık para kalmadı elimizde teslim edilse teslim almaya para ödeyemiycez sayın baş bakanım tayip ordogandan rıcam bir acıklama yapsın sayın başbakanım tayip ordoganım inan bir umutla girdik bize sunmuş oldun evlere yazık günah bizlere inan gelin bakın evlerimize ac susuz kalıyoz yetişdiremiyoz aldımız maaş üç beş kuruş buda taksit kira ancak ödeyoz bireken kiralarımız oldu artık çözüm bulun 11 bölge ve diyer bölgeler teslim edin artık ben böyle giderse almış oldum daire mecburen geri devretmek zorunda kalcam ...

See http://www.tokikayabasikonutlari.net/kayasehir-baglanti-yollari-sorunu/#comment-13138.

REFERENCES

Bartu-Candan, A., and Kulluoğlu, B. (2008). "Emerging Spaces of Neoliberalism: A Gated Town and a Public Housing Project in Istanbul." *New Perspectives on Turkey* 39, 5–46. https://doi.org/10.1017/s0896634600005057

Bianet (2011). Erdoğan Çılgın Projesini Açıkladı. 27 April. http://bianet.org/kurdi/siyaset/129580-erdogan-cilgin-projesini-acikladi

Caldeira, T. (2017). "Peripheral Urbanization: Autoconstruction, Transversal Logics, and Politics in Cities of the Global South." *Environment and Planning D: Society and Space* 35, no. 1, 3–20. https://doi.org/10.1177/0263775816658479

Çavdar, A. (2013). "The Loss of Modesty: The Adventure of Muslim Family from Mahalle to Gated Community." PhD diss., Europa-Universität Viadrina Frankfurt (Oder).

Ekers, M., Hamel, P., and Keil, R. (2012). "Governing Suburbia: Modalities and Mechanisms of Suburban Governance." *Regional Studies* 46, no. 3, 405–22. https://doi.org/10.1080/00343404.2012.658036

Gürek, H. (2008). *AKP'nin Müteahhitleri*. İstanbul, Turkey: Güncel Yayıncılık.

Kurşuncu, H. (2011). "Ağaoğlu'nun Ayazması işte budur!" *Ayazma Mağdurları*, 10 February. https://ayazmamagdurlari.wordpress.com/2011/02/10/agaoglu%e2%80%99nun-ayazmasi-iste-budur/#more-366

Kuyucu, T. (2014). "Law, Property, and Ambiguity: The Uses and Abuses of Legal Ambiguity in Remaking Istanbul's Informal Settlements." *International Journal of Urban and Regional Research* 38, no. 2, 609–27. https://doi.org/10.1111/1468-2427.12026

Pérouse, J.-F. (2006). "Ayazma (Istanbul): Une zone sans nom, entre stigmatisations communes et divisions internes." In Arnaud, J.-L. (ed.), *L'Urbain dans le monde musulman de Méditerranée*, 155–74. Paris, France: Maisonneuve and Larose.

PR. (2007). *Küçükçekmece İlçesi Kayabaşı Toplu Konut Alanı. 1/5000 ölçekli Nazım İmar Planı ve 1/1000 ölçekli Uygulama İmar Planı Raporu*. Municipal government planning document, February 2007.

PR. (2008). *Kayabaşı Kentsel Gelişme Alanı. 1/1000 ölçekli Revizyon Uygulama İmar Planı. Açıklama Raporu*. Municipal government planning document, November 2007.

PR. (2009). *İstanbul İli Başakşehir ilçesi. Kayabaşı Toplu Konut Alanı Doğu Bölgesi. 1/5000 ölçekli Revizyon Nazım İmar Planı Raporu*. Municipal government planning document, March 2009.

Official Gazette of the Republic of Turkey. (2014). "Resmi Gazete 2014/6028." 30 April. http://www.resmigazete.gov.tr/eskiler/2014/04/20140430-5-1.pdf

Roy, A. (2005). "Urban Informality: Toward an Epistemology of Planning." *Journal of the American Planning Association* 71, no. 2, 147–58. https://doi.org/10.1080/01944360508976689

Roy, A. (2009). "Why India Cannot Plan Its Cities: Informality, Insurgence, and the Idiom of Urbanization." *Planning Theory* 8, no. 1, 76–88. https://doi.org/10.1177/1473095208099299

Roy, A. (2015). "Governing the Postcolonial Suburbs." In Keil, R., and Hamel, P. (eds), *Suburban Governance: A Global View*, 337–47. Toronto, ON: University of Toronto Press.

Tekeli, İ. (2013). *İstanbul'un Planlanmasının ve Gelişmesinin Öyküsü*. Istanbul, Turkey: Tarih Vakfı Yurt Yayınları.

TMMOB. (2012). "İstanbul 3. Köprü ve Bağlantı Yolları Hakkında." *TMMOB Mimarlar Odası, İstanbul Büyükkent Şubesi*, 14 November. http://arsiv.mimarist.org/basin-aciklamalari/3295-3-kopru-ve-baglanti-yollari-hakkinda.html

TOKİ. (2017a). "Projects by Provinces." http://www.toki.gov.tr/illere-gore-projeler

TOKİ. (2017b). "Conditions for Application." http://www.toki.gov.tr/basvuru-sartlari

9 Building Northern Istanbul: Mega-Projects, Speculation, and New Suburbs

K. MURAT GÜNEY

Global suburbanization, defined as "the combination of non-central population and economic growth with urban spatial expansion" (Ekers, Hamel, and Keil 2012, 422), is a definitive phenomenon of the production of space in the early twenty-first century. Whereas contemporary growth in metropolitan areas mainly materializes through suburban expansion, suburbanization occurs in considerably diverse forms throughout different parts of the world (Hamel and Keil 2015). Istanbul, a true global city that bridges West and East, Europe and Asia, and the developed and developing worlds, provides a unique example of this diversity in massive suburban development. Three main factors make Istanbul a special case in the study of global suburbanisms: (1) the unprecedented speed of population growth, which makes Istanbul the biggest and fastest-growing city in Europe, with a population of 14.8 million in 2016; (2) the authoritarian governance of mass housing development through the distinctive institutional body, the Housing Development Administration of Turkey, better known by its acronym "TOKİ," a centralized public institution that acts in practice as a private enterprise; and (3) the rapid yet speculative valuation of housing prices, which have almost tripled in just seven years following 2010 (TCMB 2017), and made Istanbul rank among the top three cities of the global house price index (Knight Frank 2015, 2016).

(1) Unprecedented Population Growth and Suburban Expansion through Massive Informal Housing Construction

Since 1980, Istanbul's population has tripled from under 5 million to 14.8 million, mostly due to high rates of migration from villages, towns, and cities in Anatolia (the Asian part of Turkey) to the largest city of Turkey, which represents about 18 per cent of the national population and more than 27 per cent of the national economy (TUIK 2016). Istanbul has always been a hub and a place that attracts new migrants, the major actors fuelling demand for suburban expansion. Historically, Istanbul has presented particular spatial patterns and social characteristics of massive suburban growth that range from "gecekondus" (shantytowns) to state-built, high-density public mass-housing projects.

During the 1980s and 1990s, high rates of migration resulted in the formation and growth of massive shantytown settlements in the peripheries of Istanbul, much like the self-built informal housing that could be found in and around urban peripheries almost everywhere in the Global South. Turkey was and is to some degree a developing capitalist state. As such it historically took a passive role, as Sonia Hirt (2007) argues in her discussion of the modes of state involvement in suburbanization more generally, largely because of lack of resources. In the particular case of Turkey, in the 1980s and 1990s the Turkish state deliberately overlooked informal housing construction as it lacked the required institutional and financial resources to establish formal public mass housing, despite the need for new working populations to be employed in Istanbul's rapidly growing industries.

(2a) TOKİ and the Formalization of the Mass-Housing Development

With the coming to power of the neoliberal and Islamist Justice and Development Party (the AKP, in its Turkish acronym) in 2002, the passive role of the state was replaced by the formalization of mass housing development, which was quite actively led by TOKİ. Turkey began to follow a particular path of highly centralized, authoritarian, state-led urban development, according to which the construction of informal housing was stopped and completely prevented by the AKP government's newly introduced legal and executive regulations (Işık and Pınarcıoğlu 2008).

In contrast to neoliberalism's premise that the government should be small and not interfere in the market, in neoliberal Turkey one can observe a gradually increasing role and intervention of the authoritarian government in urban planning through legislative and institutional power. TOKİ, which was actually founded in 1984 as a government institution yet remained passive until the end of 1990s due to a lack of funding and legal power to facilitate mass housing projects, was empowered by the AKP government to be the major actor in urban transformation and growth. With the reactivation of TOKİ by the AKP through a series of significant amendments to the "mass housing law" in 2004 (Law number 2985, first introduced in 1984), TOKİ was granted extraordinary rights to facilitate housing and urban development in line with the government's "Emergency Action Plan for Housing and Urban Development" (introduced in 2003). With the same legal amendments TOKİ was converted to an institution directly governed by the Office of the Prime Minister, while ownership of 64.5 million square metres of land was passed immediately to TOKİ (TOKİ 2017). In the following years, decisions about land acquisitions, valuation, and marketing were made in the Office of the Prime Minister in the capital, Ankara, while both public opinion and the opinions of the affected local municipalities were disregarded.

In 2012 a series of new laws were introduced by the AKP that tremendously increased the power and capacity of TOKİ. The "Law for the Regeneration of Areas Under Disaster Risk" (Law 6306, passed in 2012, known as the "disaster law") grants the government permission to expropriate land and housing based on the loosely defined claim of "protecting residents against earthquakes and other natural disasters." Relying on this controversial legal regulation, TOKI began to acquire land through government-led displacement and dispossession and facilitated new public mass-housing development, as well as infrastructural mega-projects, on this acquired land in the peripheries of Istanbul.

(2b) Production of New Suburbs through State-Led Mega-Projects in Northern Istanbul

Despite fierce criticism by urban planners and environmentalists, the northern forested area of Istanbul's European side was selected by the AKP government as the location of a "New City," where the government of Turkey hopes to attract about 1 million new residents (AKP, 2017). The production of the new northern suburbs is being facilitated through the introduction of state-led infrastructural mega-projects such as the so-called "Canal Istanbul," a proposed waterway on the European side of Istanbul parallel to the Bosporus; a third bridge over the Bosporus near the Black Sea that was completed in 2016 and cost over $1 billion USD; and a third airport in Istanbul, supposedly the largest one in the world and the most expensive ongoing infrastructure project in Turkish history to date (with a proposed budget of $30 billion

USD). The proposed construction projects in this new northern suburb of Istanbul include marinas on the Black Sea shore, a financial centre, and luxury housing projects on the banks of the new canal. The scale of these ongoing and planned mega-projects is enormous. The third airport is being built on 77 square kilometres of land, which is about four times larger than the Atlanta Airport in Georgia, USA, which has a size of 19 square kilometres and is currently the busiest airport in the world. Istanbul's third airport will be even larger than the whole of Manhattan Island in New York City, which is 59 square kilometres. When completed, the new Istanbul airport is planned to reach an annual capacity of 200 million passengers, more than double the capacity of the Atlanta Airport.

According to the projections of the AKP government, the third airport and bridge projects will generate tremendous employment and job opportunities for the incoming residents of the northern suburbs. In just a few years, the population that lives either in TOKİ's mass social housings in Kayaşehir or in other nearby luxury housing sites close to the ongoing mega-projects has already exceeded 100,000 people. However, some critical economists argue that the enormous size and scale of the third airport project is unnecessary, because the capacity of 200 million passengers will never be reached due to the declining number of tourists and overall economic growth rates (Gürsel and Toru-Delibaşı 2013). Thus, critical economists claim that mega-projects can never generate the job and employment opportunities proposed by the government. Moreover, environmentalists protest the cutting down of over 3 million trees for the construction of the arguably unnecessary third airport and third bridge. However, despite criticism and protests, the AKP government continues to introduce the ongoing construction mega-projects as a visible and observable show of state-led economic development and power. In one of the most dramatic examples of such shows of national pride and economic power, in 2017 the contractor of the third airport construction project celebrated the 564th anniversary of the conquest of Istanbul by the Ottomans in 1453 with a parade of 1,453 trucks in the construction area! Whether critics are right or wrong about the unnecessity of the ongoing mega-projects, a massive convoy of white-yellow trucks that carry construction materials to and from the mega-projects zone for twenty-four hours uninterruptedly along the routes connecting central Istanbul to the recently deforested northern areas is suggested as the sign of a well-working economy.

9.1 Trucks carrying construction materials to and from Istanbul's third airport construction area. Photo by Roger Keil.

While facilitating the AKP government's mega-infrastructural projects, TOKİ provides a singular example of a centralized public institution acting as an entrepreneurial company. The workings of TOKİ reveal the complex relations between state and capital in Turkey. TOKİ, as the authoritarian body of suburban governance, plays a major role in land acquisition for these mega-projects. While facilitating both mass housing development and infrastructural mega-projects, TOKİ contracted out construction work to selected private construction companies that were close allies of the government. In return for their loyalty to the AKP government, these private construction companies made enormous profits while carrying out public projects. The controversy over TOKİ's role as a public institution is not limited to its contracting out public projects to profit-generating private companies. TOKİ and the government also directly involved themselves in land speculation by withholding declaration of the exact locations of mega-projects until the last moment.

Here, one must note that none of these mega-projects were included in the Master Plan of the AKP-led Istanbul municipality from 2009, the major urban planning document for the region. On the contrary, in the 2009 Master Plan any attempts at suburban growth towards the north of Istanbul's European side, which was the only remaining forest area within the limits of the metropolitan municipality, was banned. In the words of the 2009 Master Plan of Istanbul, "As the life support systems of the city are under a process of destruction, the ecological values at the northern axis of the city running along the Black Sea coasts, primarily the water basins and forest areas of Istanbul, are closed to risky economic endeavours" (translated into English by the activist group Northern Forests Defence; Kuzey Ormanları Savunması 2015).

The construction of the mega-projects in northern Istanbul, however, were announced directly by then-prime minister Erdogan in 2010, without consulting the municipality or considering the 2009 Master Plan. Moreover, two years elapsed between the first announcement of the third airport project in 2010 and the official declaration of its exact location in 2012. The location of the "Canal Istanbul" project was also disclosed years after its first announcement. In the meantime, the aforementioned mass housing and disaster laws were imposed arbitrarily to expropriate land around the construction site of the third bridge and airport as well as the canal, and to transfer that land to TOKİ for sale. During that process, TOKİ sold cheap land in the peripheries of the city to government-allied companies that used the newly acquired land between the new third bridge, the construction site for the third airport, and the proposed new canal to develop luxury housing and condo projects.

TOKİ's direct involvement in for-profit luxury housing projects makes the role and position of this supposedly non-profit public institution more controversial. Indeed, the mass housing law (number 2985, amended in 2004), clearly states that "TOKİ is a non-profit government administration," and its main purpose is to satisfy housing needs of "low- and middle-income families who are not able to apply for bank loans due to income and savings levels that are inadequate to meet mortgage payments." These same statements are also repeated in the "Administrative Position" section of TOKİ's official website (TOKİ 2017).

Despite this clear legal framework, TOKİ, through its subsidiary institution Emlak Konut, also a public institution, is directly involved in for-profit luxury housing and condo construction around the ongoing mega-project area. While the Kayaşehir mass housing project (discussed by Cavuşoğlu and Strutz in detail in the previous chapter

9.2 Ongoing construction of a massive TOKİ housing complex in Başakşehir, a suburban district that has experienced rapid housing price appreciation thanks to the introduction of the mega-projects in northern Istanbul. Photo by Roger Keil.

of this book), located in close proximity to the mega-project area, is TOKİ's biggest and major ongoing social housing project that targets low- and middle-income families, the nearby Bahçetepe, MISStanbul Evleri, Avangart Istanbul, and Başakşehir Evleri are only a few examples of ongoing Emlak Konut luxury housing projects that benefited from the land speculation and housing price increase followed by the construction of the mega-projects in northern Istanbul. Currently Emlak Konut is one of the most profitable public institutions in Turkey. In order to justify the ongoing luxury housing projects, TOKİ claims that its for-profit projects facilitated by Emlak Konut aim to subsidize the ongoing mass social housing projects that target low- and middle-income groups. However, it is not possible to check the correctness of TOKİ's claim and learn how profits earned through the luxury housing projects are really used, because the budgets of TOKİ and Emlak Konut are not transparent (Sönmez 2011). Therefore, TOKİ is widely criticized for acting as a for-profit entrepreneurial institution (Sayek-Böke 2015), in clear violation of its foundational law (number 2985), which defines TOKI as a non-profit public institution and frames its obligations as facilitating housing development for disadvantaged segments of society.

As the above-described examples regarding the role of the TOKİ – whose functions in practice overreach its legal framework – clearly demonstrate, the government of Turkey, with the help of TOKİ's authoritarian interventions, is attempting to revitalize economic growth by generating capital and value through a speculative housing market fostered by new infrastructural mega-projects in the peripheries of Istanbul. Currently about 30 per cent of Turkey's total economic output derives from construction, real-estate, and related sectors (ParaAnaliz 2017). Construction, housing and mega-infrastructure projects are the key drivers of economic growth as well as a central mechanism of rent generation.

The question concerning the effects of massive sub/urban growth is complex and many-sided. Today in Turkey, the AKP government and private construction companies support housing-sector-led economic growth at seemingly any environmental or humanitarian collateral cost. In an opposing stance, urban planners, the Union of Chambers of Turkish Architects and Engineers (TMMOB), and environmentalist organizations such as Kuzey Ormanları Savunması (Northern Forests Defence) oppose suburban expansion towards north of Istanbul through infrastructural mega-projects on the grounds that they damage the environment. Moreover, as Istanbul experienced gentrification in the core through urban renewal projects, vulnerable populations who lived in informal squatter housings in and near the city centre were displaced and resettled in the newly built suburban mass housing projects in the peripheries of the city. Critics argue that such suburban growth disempowered and dispossessed already disadvantaged populations, who lost their homes in or close to the city centre and their access to the vast variety of educational, healthcare, and recreational facilities available in core locations (Kuyucu and Ünsal 2010; Ünsal 2013). In brief, critical urbanists, NGOs and activists state that uneven suburban development has triggered severe social and environmental consequences, including mass displacement of local neighbourhoods, reduced life and ecosystem quality, and environmental degradation.

(2c) Resistance to the Authoritarian Governance of Urban Development

The authoritarian impositions of the AKP government in urban planning and development were one of the major triggers for the nationwide anti-government protests in Turkey during June 2013, which began with the claiming and occupation of Gezi Park by hundreds of thousands of Istanbul residents, as they protected one of the few remaining green spaces in central Istanbul from being converted into a shopping mall. The Gezi Park uprising, which continued throughout the summer of 2013, was a turning point in opening up new questions, particularly about the varying and contested perceptions of urban development. The educated upper-middle and middle classes of Istanbul society supported the Gezi revolt. The supporters of the uprising were successful in bringing questions concerning the humanitarian and environmental harm inflicted by government-led development projects, particularly construction projects including shopping malls, mass-housing projects, transportation, and infrastructural mega-projects to the forefront of public opinion. Yet, at the same time, the more financially deprived poor segments of society, which compose a significant portion of Turkey's population, took sides with the AKP government and continued to support those government-led development and construction programs, including the ongoing massive housing and infrastructure mega-projects in Istanbul, despite the human and environmental damage that these projects have inflicted. Housing and urban development, which bolster home and land prices and provide homeownership opportunities for the low and lower-middle income groups, continued to be an important source of aspiration for the impoverished segments of the society and have been largely seen as an opportunity for upward mobility. Such a significant rupture in society in terms of diverging groups' conflicting perceptions of urban development, a rupture unearthed by the Gezi Park uprising, demands further investigation.

Conventionally, in Turkey, research focuses mainly on the dramatic negative effects of such rapid sub/urban development on vulnerable populations (Bartu-Candan and Özbay 2014; Erder 1996; Kuyucu 2014). Urban expansion facilitated by the government-led massive housing and infrastructural construction in the suburbs of Istanbul is identified a "state-led property transfer" by Turkish critical urban scholars (Kuyucu and Ünsal 2010), and fundamentally criticized because it raises the burden of debt for low-income households through the financialization of the housing market, subordinating poor segments of society to the mechanisms of financial discipline (Karaçimen 2013, 2015). Moreover, critics emphasize that the low-quality high-rise social housing projects constructed in the suburbs of Istanbul by TOKİ for the new migrants and populations affected by the urban renewal projects have become areas where problems related to weak infrastructure, poverty, and social conflict are concentrated (Erman 2016). Scholars also criticize the environmental degradation caused by such projects (Akbulut and Bartu-Candan 2014; see also Mürat Üçoğlu's chapter in this book).

While there are robust economic and ethnographic studies on vulnerable populations that point to the negative effects of the ongoing suburban mass housing development and infrastructural mega-projects, the diverging views of homeowners and new buyers in the newly built suburban housing projects have not been studied in similar detail. The support by certain segments of society for sub/urban development is mostly disregarded in oppositional-activist reports and academic studies. Therefore, it is equally important to analyse this pressing and yet underdeveloped area and identify not only disadvantaged groups but also the beneficiaries of suburban mass housing development, in order to better understand the reasons underlying public support for urban development and construction-led economic growth.

The AKP government, construction companies, and financial institutions appreciate infrastructural mega-projects as a major driver of modernization, economic development, and growth, and introduce mass social housing projects in the peripheries as an opportunity for low-income households, formerly settled in informal self-built housing, to be formal homeowners, enjoying a better quality of life and inclusion in the formal mechanisms of the market (ParaAnaliz 2017). TOKİ, central and local governments, and private construction companies are not alone in supporting mega-projects and housing-sector-led economic growth, which bolsters home and land prices in both the old urban centres and newly constructed suburban regions. Indeed, such economic growth benefits not only big capital owners and the pro-government business sector, but also individual land- and homeowners.

According to the Turkish Statistical Institute, 61 per cent of Istanbul's households were homeowners in 2011 (TUIK 2011). To be sure, major housing and construction companies, big capital owners, and foreign real estate investors, who possess most of the real estate in Istanbul, are the major beneficiaries of the rapid urban and suburban growth. Yet despite the highly unequal distribution of the wealth generated through construction-led economic growth, one cannot disregard the fact that homeowners possessing at least one apartment, who compose the majority of Istanbul's population, including upper-middle class households who bought an apartment in the luxury housing projects of Emlak Konut near the mega-projects as well as the low- and middle-income families who recently bought a home in the TOKİ mass housing projects, also benefit more or less from the rapid increase in house prices.

(3a) Financialization of Housing through Formalization: Speculative Housing Price Increase

In Turkey, the demolition of shantytowns and squatters and the formalization of mass housing was a process that was experienced simultaneously with the financialization of housing as a commodity. In that period, while housing-finance opportunities significantly expanded, the amount of bank and TOKİ loans and household debt rates increased dramatically. The financial weight and contribution of the housing and real estate sectors to economic output and capital accumulation rose notably. As a result, despite diminishing global and national economic growth rates, Turkey in general and Istanbul in particular continue to enjoy construction-sector-led growth accompanied by a rapid yet speculative valuation of housing prices.

Housing prices in Istanbul almost tripled between 2010 and 2017 (TCMB 2017). Today, Istanbul is considered one of the top European and global cities for real estate investment, with one of the highest returns (PricewaterhouseCoopers and Urban Land Institute 2013).

Home prices in Istanbul increased at a modest rate between 2003 and 2007, and then fell dramatically during the 2008–9 economic crisis. The current trend of rapid growth in house prices started in the years following the global economic crisis, particularly after 2010. The Central Bank of Turkey (TCMB) began to publish the House Price Index for Turkey and for various cities including Istanbul starting in the same year. Today the Central Bank's House Price Index is accepted as the major source of information on price changes in the housing market in Turkey. According to the Central Bank's Index, on average home prices in Istanbul have increased about 284 per cent as of January 2017 compared to the prices in 2010 (TCMB 2017). This rate is significantly above the already high national house price increase of 228 per cent during the same period.

So then, what are the possible reasons for such rapid increases in house prices in Turkey and particularly in Istanbul?

Here, one can name three major post-economic crisis efforts and policies by the AKP government that played a significant role in the swift valuation of real estate in both the old urban centre and the newly

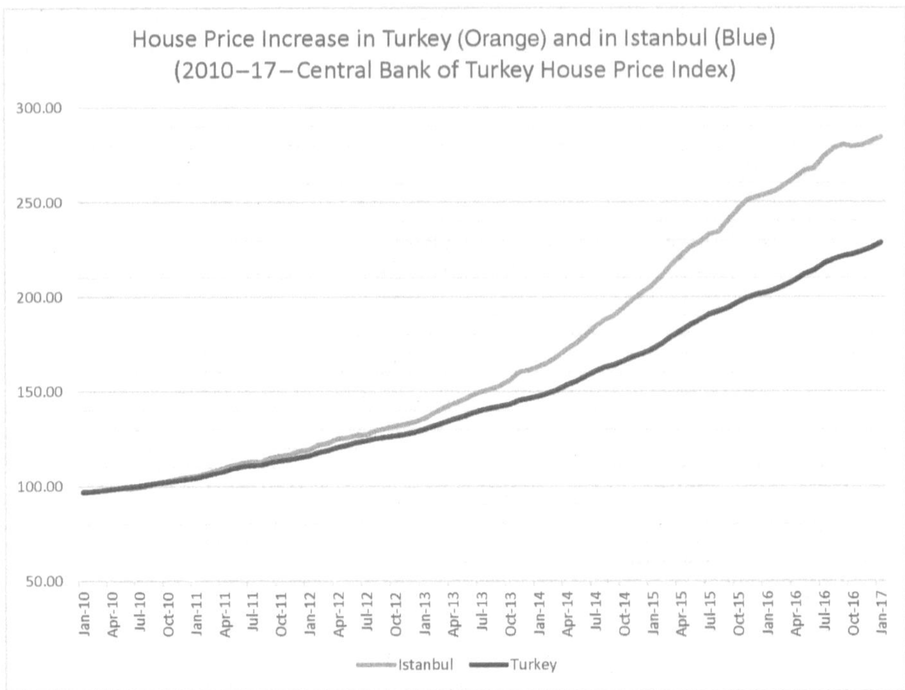

9.3 House price increase in Turkey and Istanbul. Based on public data provided by the Central Bank of Turkey. Chart prepared by the author.

constructed suburban regions of Istanbul. The first policy was in the field of finance. The government made housing financing accessible for larger segments of the society by holding down the historically high interest and mortgage rates in Turkey. This policy helped to boost domestic demand. The second policy was legal. The AKP government introduced several legal privileges for foreign capital to attract foreign investors to buy housing in Turkey. This policy helped to foster international demand. The third policy was the introduction of the previously described infrastructural mega-projects, particularly in and around Istanbul. These projects aimed to transform land into new social as well as luxury suburban housing development areas and create demand for investors, new migrants, and new home-buyers. All these efforts indicate a significant change in the AKP's economic and urban policies starting in 2009 and 2010.

(3b) Financialization of the Housing Market through Lowering Interest Rates

Following the global economic crisis, the very high interest rates seen in Turkey during the pre-crisis period between 2002 and 2009 under the AKP government sank to single digits after 2010, to as low as 8 per cent. More importantly the real interest rates (rates after inflation), which were more than 10 per cent before the economic crisis, fell to near zero after 2010. This dramatic fall in interest rates had two significant effects on the housing market in Turkey.

First, it helped to secure the availability of low-interest housing loans that encourage more and more middle-income households to buy new housing. This policy corresponds with the financialization of housing through the expansion of the mortgage market, which makes housing-finance opportunities available for larger segments of the population. In return, this policy converts home-buyers into long-term debtors and subordinates them to the mechanisms of financial discipline.

Second, with the stabilization of near-zero real interest rates, Turkish investors, who since the 1990s had traditionally invested in government bonds and savings accounts that provided high real returns, were forced to change their investment habits and search for new profitable ways such as real estate investment in order to continue to increase their wealth. Thus, the economic policies of the AKP government to maintain low interest rates in the years following the global economic crisis was a turning point for the Turkish housing market, which benefited from the rise in demand from both middle-income new home buyers and high-income investors.

(3c) Opening Housing Market to Foreign Investors

Another AKP government policy to support the housing market in the post-crisis period aimed at attracting foreign investors through legal privileges such as providing residence permits for foreigners who bought property in Turkey. Moreover, as part of such a policy, in May 2012 the AKP government introduced a new regulation popularly called "the reciprocity law" (in Turkish, "mütekabiliyet yasasi"), which indeed cancelled the reciprocity principle and paved the way for selling homes and land to citizens of countries where Turkish citizens cannot buy real estate. Before 2012, Turkey had a strict reciprocity law, which restricted many foreign nationals from buying property in Turkey unless in return a Turkish national was allowed to buy a home in the respective country. Therefore, until 2012 only nationals of several Western European countries could buy real estate in Turkey. However, the new law implemented by the AKP government lifted a large amount of these restrictions and opened the Turkish property market to nationals of about 130 different countries, including Arabs from the oil-rich Gulf countries as well as Russians.

After the introduction of the reciprocity law, foreign direct investment in Turkey in the form of land and housing purchases surpassed $3 billion USD for the first time in 2013. According to the United Nations Conference on Trade and Development's report (UNCTAD), in 2014 "real estate acquisitions in Turkey increased for the third consecutive year and at a faster rate (29%), reaching $4 billion and accounting for 25% of total Foreign Direct Investments flows to Turkey" (UNCTAD 2015, n.p.). To be sure, this is part of a global trend of rising investment in urban land and financialization of real estate (Sassen 2014). As Saskia Sassen (2016, n.p.) emphasizes, "In mid-2014 and 2015 alone, more than one trillion dollars was invested in real estate in just 100 cities across North America, Europe and Asia ... Urban land – not just buildings, but also undeveloped lots – is considered a good investment at a time when financial markets are shaky." At the cost of the displacement of local populations and dramatic transformation of historical neighborhoods, global cities are competing to attract foreign investors who are interested in buying real estate, and Istanbul is clearly one of the frontrunners in that race.

(3d) Boosting Housing Market through the Introduction of Mega-Projects

Turkey ranked first in the world in terms of house price increase in the year 2015 (Knight Frank 2016). And according to *The Economist*'s house price index, which measures the cumulative house price increase since 2012, Turkey comes second only after Hong Kong (The Economist 2016). Here one must note that these calculations reflect the average house price increase in all of Turkey. Istanbul, where the house price increase significantly surpasses Turkey's average, would lead by far in both of these indexes.

And not surprisingly, within Istanbul the highest increase in housing prices was observed in suburban neighbourhoods near Istanbul's ongoing mega-project areas. While in several neighbourhoods of Catalca province house and land prices increased about 76 per cent in only one year following the construction of the third bridge over the Bosporus, suburban neighbourhoods of Eyup district that are close to the new airport construction site have seen property increase in value by about 30 per cent in the same period (sahibinden.com 2015). All these data show that mega-projects have an overall positive effect on the tremendous increase in housing prices in Istanbul, and this trend is even more pronounced in the peripheral neighbourhoods in direct proximity to the mega-project sites. Mega-projects may have also a more or less positive effect on home prices in other parts of Istanbul. As one can observe, the trend of house price growth in Istanbul since 2010 coincides with the first official announcement of the exact location of the third bridge over the Bosporus, which was followed by the official announcements of the third airport and Canal Istanbul mega-projects in 2011 and 2012, respectively.

Conclusion: The Unequal Distribution of Urban Rent

As mentioned earlier in this chapter, from big real-estate investors to the owners of a single apartment in Istanbul, residents who are already homeowners benefit from such rapid housing and land price appreciation. I suggest that this observation helps to better understand the underlying reasons for the support of certain segments of society for the government-led suburban growth, massive housing development, and mega-projects in Istanbul and other parts of Turkey.

However, although rapid sub/urban development creates enormous economic growth and capital accumulation, the distribution of this added value is highly unequal. Construction-led rapid yet uneven development differentially affects populations according to their respective access to ownership rights and income levels. The unequal distribution of the benefits and harms of such development includes capital accumulation on the one hand and rising indebtedness on the other.

The AKP government's particular legal and economic regulations that I have discussed throughout this chapter caused a dramatically unequal distribution of urban rent. While government policies facilitate the accumulation of real estate in the hands of few already-rich foreign and domestic investors, homeownership rates for low-income

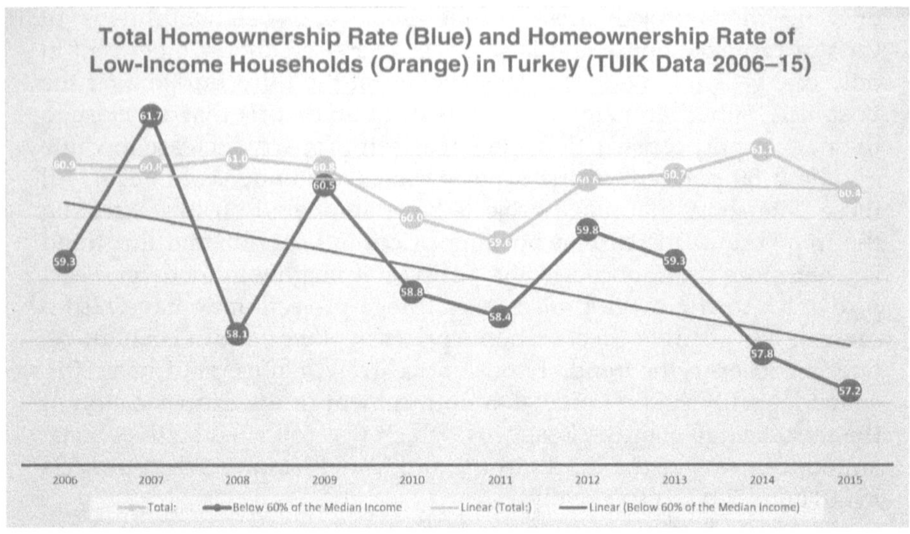

9.4 Homeownership rate and homeownership rate for low-income households. Chart prepared by the author.

families, TOKİ's target group, have not increased, while the debts of the same group have increased gradually in the last decade. Despite the massive scope of TOKİ's suburban social housing projects, which helped a certain segment of middle- and lower-income families to be new homeowners, these ongoing social housing projects were far from being enough to satisfy the affordable housing need of the lowest income families, who remained tenants, indebted and financially vulnerable. Instead of mobilizing all of its resources to create affordable social housing opportunities for the lowest-income families, TOKİ through its subsidiary institution Emlak Konut continues to focus on the production of luxury housing projects that target upper- and upper-middle-class domestic and foreign investors. As a result, homeownership rates for low-income groups have been decreasing in the last ten years (Coşkun 2016, 2017).

As seen in the graph in figure 9.4, according to the Turkish Statistical Institute's "Income Distribution and Living Conditions" research data (TUIK 2016), between the years 2006 and 2015 the homeownership rate in Turkey was about 61 per cent, and this rate did not change much during the ten-year period covered in the statistics. What is more striking, however, is the decline in homeownership rates of low-income groups, whose income is 60 per cent or lower than the median income in Turkey. As the official data of the TUIK demonstrate, the homeownership rate of low-income groups in Turkey decreased from 59.3 per cent in 2006 to 57.2 per cent in 2015, as part of a trend of declining homeownership rates for the poor classes particularly observed since 2012.

That is to say, despite the significant construction-led economic growth, mega-infrastructural investments, and massive housing development in Turkey, buying a home and being a homeowner become gradually harder for low-income families, while the rich and real estate investors who own more than one apartment increased the number of their properties and have been the major beneficiaries of the construction-based economy (Sayek-Böke 2015).

The wealth accumulation in Turkey that was mostly created through the construction, infrastructure, and real estate sector-led economic growth fostered by the AKP government in the last decade caused a speculative house price increase, and this speculative increase in home prices dramatically decreased the possibility of being a homeowner for low-income groups.

I want to conclude my chapter by emphasizing the need to expose the big picture and the overall effects of the economic and legal regulations related to housing policy in Turkey. As demonstrated in this chapter, despite the initial and supposed aim of TOKİ and the AKP government's "Emergency Action Plan for Housing and Urban Development" to increase homeownership rates for disadvantageous groups, the construction-led economic growth model fostered by TOKİ's and the government's authoritarian urban planning strategies resulted in the accumulation of real estate in the hands of the richest segments of the society, while homeownership rates for low-income groups are decreasing. Therefore, I tried to attract attention to the problematic trend regarding the homeownership rates of low-income groups in an economy that since 2010 has particularly grown through housing price speculation. This may be a starting point for a more comprehensive investigation and discussion of the long-term effects of the construction-led economic growth model, and on how to rethink mass housing policy and regulations in order to eliminate obstacles facing members of the poor classes who seek to become owners of their own homes.

REFERENCES

Akbulut, B., and Bartu-Candan, A. (2014). "Bir-İki Ağacın Ötesinde: İstanbul'a Politik Ekoloji Çerçevesinden Bakmak" [Beyond a Few Trees: Approaching Istanbul through the Framework of Political Ecology]. In Bartu-Candan, A., and Özbay C. (eds), *Yeni İstanbul Çalışmaları* [New Studies on Istanbul], 282–302. Istanbul, Turkey: Metis.

AKP. (2017). *2023 Political Visions*. https://www.akparti.org.tr/english/akparti/2023-political-vision

Bartu-Candan, A., and Özbay, C. (eds). (2014). *Yeni İstanbul Çalışmaları* [New Studies on Istanbul]. Istanbul, Turkey: Metis

Coşkun, Y. (2016). "Türkiye Konut Piyasasında Talep Eğilimleri ve Bilgi Bakışımsızlığına Yönelik Politika Önerileri" [Housing Demand Trends and Policy Suggestions for Asymmetric Information Problem in Turkish Housing]. *Bankacılar Dergisi* 96, 122–43. https://ssrn.com/abstract=2744639

– (2017). "Konut Sahipliğinin Artmaması Sizi de Düşündürüyor mu?" [Do You Worry, Too, Because the Homeownership Rate Is Not Increasing?]. *GMTR*, 27 March. http://gayrimenkulturkiye.com/2017/03/27/konut-sahipliginin-artmamasi-sizi-de-dusundurmuyor-mu/

Economist. (2016). "The Economist House-Price Index." Infographic. *The Economist*. https://infographics.economist.com/2016/globalhpi_20160331/

Ekers M., Hamel, P., and Keil, R. (2012). "Governing Suburbia: Modalities and Mechanisms of Suburban Governance." *Regional Studies* 46, no. 3, 405–22. https://doi.org/10.1080/00343404.2012.658036

Erder, S. (1996). *İstanbul'a bir semt kondu: Ümraniye* [A Neighborhood Has Landed in İstanbul: Ümraniye). Istanbul, Turkey: İletişim Yayınları.

Erman, T. (2016). "Distopik Bir Yaşam Çevresine Dönüşen Bezirganbahçe TOKİ Sitesi, Ayazma-Tepeüstü Kentsel Dönüşüm Projesi, Küçükçekmece" [Ayazma-Tepeüstü Urban Transformation Project in Küçükçekmece District: Bezirganbahçe TOKİ Site That Became a Dystopian Living Environment]. In Koçak, D., and Koçak, O. (eds), *İstanbul Kimin Şehri: Kültür, Tasarım, Seyirlik ve Sermaye*, 169–94. Istanbul, Turkey: Metis.

Gürsel, S., and Toru-Delibaşı, T. (2013). "Mega Havalimanın Kaderi Büyümeye Bağlı" [The Destiny of the Mega Airport Depends on the Growth Rates]. *BETAM*, 28 June. http://betam.bahcesehir.edu.tr/2013/06/mega-havalimaninin-kaderi-buyumeye-bagli/

Hamel, P., and Keil, R. (eds). (2015). *Suburban Governance: A Global View*. Toronto, ON: University of Toronto Press.

Hirt, S. (2007). "Suburbanizing Sofia: Characteristics of Post-socialist Peri-Urban Change." *Urban Geography* 28, no. 8, 755–80. https://doi.org/10.2747/0272-3638.28.8.755

Işık, O., and Pınarcıoğlu, M. (2008). "Not Only Helpless but Also Hopeless: Changing Dynamics of Urban Poverty in Turkey, the Case of Sultanbeyli, Istanbul." *European Planning Studies* 16, no. 10, 1353–70. https://doi.org/10.1080/09654310802420060

Karaçimen, E. (2013). "Political Economy of Consumer Debt in Developing Countries: Evidence from Turkey." PhD diss. University of London.

– (2015). *Türkiye'de Finansallaşma: Borç Kıskacında Emek* [Financialization in Turkey: Labour under the Pressure of Debt]. Istanbul, Turkey: Sav Yayınları.

Knight Frank. (2015). *Global Residential Cities Index Q4 2015*. http://content.knightfrank.com/research/1026/documents/en/q4-2015-3653.pdf

– (2016.) *Global Residential Cities Index Q1 2016*. https://kfcontent.blob.core.windows.net/research/1026/documents/en/q1-2016-3904.pdf

Kuyucu, T. (2014). "Law, Property, and Ambiguity: The Uses and Abuses of Legal Ambiguity in Remaking Istanbul's Informal Settlements." *International Journal of Urban and Regional Research* 38, no. 2, 609–27. https://doi.org/10.1111/1468-2427.12026

Kuyucu, T., and Ünsal, Ö. (2010). "Urban Transformation as State-Led Property Transfer: An Analysis of Two Cases of Urban Renewal in Istanbul." *Urban Studies* 47, no. 7, 1479–99. https://doi.org/10.1177/0042098009353629

Kuzey Ormanları Savunması (Northern Forests Defence). (2015). *Campaign: Stop the Killer Projects! Be the Breath of Istanbul!* 3 July. http://www.kuzeyormanlari.org/2015/07/04/campaign-stop-the-killer-projects-be-the-breath-of-istanbul/

ParaAnaliz. (2017). *TOKI'den 7 yılda 425 bin konut* [TOKI produced 425 thousand housing units in the last 7 years]. http://www.paraanaliz.com/2017/ekonomi/tokiden-7-yilda-425-bin-konut-8874/

PricewaterhouseCoopers and Urban Land Institute. (2013). *Emerging Trends in Real Estate: Europe 2013*. https://europe.uli.org/emerging-trends-in-real-estate-europe-2013/

Sahibinden.com. (2015). *Sahibindex Emlak Endeksi* [Sahibindex House Price Index] https://www.sahibinden.com/emlak-endeksi

Sassen, S. (2014). *Expulsions: Brutality and Complexity in the Global Economy*. Cambridge, MA: Harvard University Press

– (2016). "Investment in Urban Land Is on the Rise – We Need to Know Who Owns Our Cities." *Global Thought, Columbia University*, 6 August. http://cgt.columbia.edu/news/sassen-investment-urban-land-rise-need-know-owns-cities/

Sayek-Böke, S. (2015). "TOKI Ranta Çalışmış" [TOKI Worked to Increase Rent]. *Hürriyet: Ekonomi*, 27 April. http://www.hurriyet.com.tr/selin-sayek-boke-toki-ranta-calismis-28853859

Sönmez, M. (2011). *TOKI'de Aslan Payı Kimlerin*? [Who Gets the Lion Share from the TOKİ?]. Blog post, 1 June. http://mustafasonmez.net/?p=684
TCMB (Central Bank of Turkey). (2017). *House Price Index for Turkey (THPI) 2010-20175 (Data Set)*. http://www.tcmb.gov.tr/wps/wcm/connect/TCMB+TR/TCMB+TR/Main+Menu/Istatistikler/Reel+Sektor+Istatistikleri/Konut+Fiyat+Endeksi/
TOKİ. (2017). "Background." http://www.toki.gov.tr/en/background.html
TUIK (Turkish Statistical Institute). (2011)."Nüfus ve Konut Araştırması 2011"[Research on Population and Housing 2011]. http://www.tuik.gov.tr/PreHaberBultenleri.do?id=15843
– (2016)."Gelir Dağılımı ve Yaşam Koşulları Araştırması"[Research on Income Distribution and Living Conditions]. http://www.tuik.gov.tr/PreTablo.do?alt_id=1011
Ünsal, Ö. (2013). "Inner-City Regeneration and the Politics of Resistance: A Comparative Analysis of Sulukule and Tarlabaşı." PhD diss., City University, London, England.
UNTAC (United Nations Conference on Trade and Development). (2015). *World Investment Report 2015: Reforming International Investment Governance*. http://unctad.org/en/PublicationsLibrary/wir2015_en.pdf

10 Massive Housing and Nature's Limits? The Urban Political Ecology of Istanbul's Periphery

MURAT ÜÇOĞLU

Introduction

This chapter claims that the housing market dynamics in Istanbul that lead to rent-seeking have a severe impact on the natural environment in the Bosporus region. Environmental decay has become a serious problem in Istanbul. Beginning in the 1990s, Istanbul's ecological diversity and environmental richness started to decline. Increasing population and the discourse of creating a global mega-city led to the expansion of urban areas, and this expansion caused the elimination of many ecologically diverse areas or forests.[1] From 2003 onwards, environmental decay gained momentum as the neoliberal and developmentalist policies of the Justice and Development Party (AKP) regime started to carry out a massive construction mobilization throughout Turkey, but particularly in Istanbul. Istanbul stands at the core of all the AKP's developmentalist strategies. By the mid-1990s, after Recep Tayyip Erdoğan's election as the mayor of Istanbul, the city started to undergo a new expansion process together with the projects of Emlak Bankası (Real Estate Bank – today's Emlak Konut), KIPTAŞ (Istanbul Residence Development Plan Industry and Trade Inc.), and TOKİ (Mass Housing Administration of Turkey). These developmentalist strategies have been relying on a new discourse of homeownership and a new financial architecture that have engendered massive indebtedness among homeowners. As a result, the commodification and the financialization of housing have become a severe process that affects all aspects of society. There are of course winners and losers in this process; however, the developmentalist strategy of the AKP has resulted in bringing the inhabitants of Istanbul into the pool of housing market dynamics and mortgage debt.

The very success of this developmentalist strategy has been causing unprecedented environmental problems in Istanbul, through the impact of the construction process and the often sub- and exurban mass housing that is placed in environmentally sensitive areas, causing stress on the metabolisms of the region. The city's expansion has turned into a massive process that is trying to absorb the northern greenbelt, water resources, farming zones, and major forests. The environmental damage is clouded by a representation of so-called economic growth. Remarkably, then, the environmental destruction itself has not caused a major uproar, as it appears that the majority of dwellers in the new projects do not want to deal with these problems. Rather, they are trying to maximize the potential profit that they might get in the deepening game of the Istanbul housing market. This chapter aims at explaining the connection between the new construction-based capitalism and the changing urban political ecology in Istanbul's periphery. The expansion of Istanbul to its periphery has been forming a massive suburbia as the major landscape of the city's peripheral life. This new massive suburbia constitutes a new political economy. This chapter attempts to investigate this new political economy and how this expansion also has consequences for urban political ecology in the region. Taking my guidance from newer work on the urban political ecology of the periphery (see, for example, Keil and Macdonald 2016), I view societal relationships with nature as sets of complex human and non-human interactions. In this chapter, I use the terms "environment" and "ecology" as synonyms to accentuate the importance of ecological problems. Where I use the terms "new housing projects" or "further housing projects," they refer to the suburban developments.

The Changing Urban Political Ecology of Istanbul's Periphery

Istanbul has been experiencing serious environmental problems, leading to the deterioration of environmental biodiversity and an upsetting of the region's ecological balance since the late 1990s. However, the process became even more serious after 2003 with the new regime of state-led economic developmentalism. In the previous decades, the city was somehow able to protect its environmental resources despite its growing population. Indicators show that Istanbul has been facing serious environmental erosion since the 1990s. The policies of urbanization that began in the 1990s produced a significant decrease in productive agricultural lands and created more impervious areas

(Özacar, 2013). The Marmara earthquake in 1999 led to the acceleration of urban expansion to agricultural areas or environmentally protected areas, since new housing projects were presented as safe and enduring.

The expansion of the city and the growing population have caused an increase in automobile dependency. In the massive suburbia of Istanbul, auto usage has incredibly proliferated. As a result, air pollution is now a serious problem in the city. Istanbul, as a mega-city, still hosts many industrial production sites in its suburbs. The expansion of the city therefore also has the dimension of the enlargement of industrial land. As long as many economic investments are made in Istanbul, due to its "reserve army of workers," Istanbul could still be regarded as an industrial city as well as a financial and tourist hub. In fact, with its industrial and suburban development, the city's air quality has dramatically declined. Traffic jams are the central factor for NOx, NMVOC, and CO emissions. Istanbul is now one of the leading cities in the world for air pollution (TMMOB, 2017). The other major factors of air pollution are industrial production in the city and shipping pollution over the Bosporus (Incecik and Im 2012, 110).

Apart from air pollution, there is another problem that also has an impact on air quality: deforestation, which also causes the decline of biodiversity. According to the general scientific approach, deforestation causes loss of green spaces and changes in the quality of air and soil, and it limits the production of organic food. Istanbul has been losing its thousand-year-old forests and its green areas which used to host an ecologically diverse life and served as the habitat of many species. In order to boost the demand for new housing projects, the government facilitates the establishment of mega-projects in the periphery of Istanbul. Since 2013, the third bridge over the Bosporus, in the north of Istanbul, the third airport, again in the north part of the city, and a northern highway have been constructed. The construction of the third bridge on the Bosporus and the third airport resulted in the destruction of 2.7 million trees in the northern greenbelt of Istanbul (Ocak and Sönmez 2014). This number is subject to increase since many new housing projects near the northern forests have been built or will be built, and these areas also used to have several rich water resources and environmentally needed wetlands. Most of them have been destroyed by the mega-projects and further housing projects. Istanbul's clean water is now provided from other cities; thus the city is consuming the resources of other regions.

10.1 The production of suburban space with massive housing projects and gated communities in Halkali, Istanbul. Photo by Roger Keil.

Deforestation is still going on due to the mega-projects and new housing projects. Up to 4 million trees will be removed when the ongoing projects come to an end. As Konrad Gürtler (2016) writes,

> This calculation consists of approximately 2,5 million trees on the area of the third airport and 1.5 million trees in connection with the third Bosphorus bridge. For the third airport, the total project area is 7.650 ha, of which more than 80% is forested land and another 9% are lakes and ponds. For the bridge, 740 ha of oak, conifer and other woodland species will be lost just within the 60,5-meter-wide construction corridor, while reports acknowledge that a corridor between 500 m and 5 km has to be taken into account in order to assess deforestation and other ecological damages. (n.p.)

Deforestation also causes the decline of biodiversity; many species have already left the Istanbul area, and Istanbul is losing its role as a transit point for many birds. These are the major ongoing developments in terms of the environmental erosion in Istanbul, and indeed, this process continues to severely damage the environment.

In addition to these factors, Istanbul is under a serious threat from climate change. A recent study indicates that Istanbul will be one of the most affected cities in terms of the economic costs of climate change in the coming decades (Abadie, Murieta, and Galarraga 2016).

Housing Market and Rent-Seeking as a Justification of Environmental Decay?

Located between the developing and developed worlds, Turkey has been following up policies of developmentalist growth with intense state-led investments in infrastructure and construction since the shift to neoliberal political economy in the 1980s, in order to compete with the West. The influence of the private sector and corporate capitalists has become more significant in the course of economic distribution in the last three decades. Since 1980, the private sector has become more interconnected to state investments and vice versa (Pamuk 2014, 8).

The housing market is one of the major spheres of the commodification of use values, and the commodification of social use values has been part of widespread neoliberalization throughout the world. Housing constitutes the major part of social use value because, as Madden and Marcuse (2016) point out, housing is not only a place of accommodation but also the intersection of social, political, and ontological existence. Therefore, the commodification of housing is not a mere issue of economic progress; it is a social and ontological process that has been ongoing, in different geographies, within a violent system since the rise to dominance of the neoliberalization process. The commodification of housing is also linked to ecological concerns because it inevitably involves the production of new spaces that were formerly known as forests or protected ecological areas. The housing market has been regarded as part of the second circuit of capital by neo-Marxist scholars; however, it is now one of the primary ways of making profit within a newly emerged digitalized financial architecture. This financial architecture inevitably needs to articulate as many people as possible to accelerate the accumulation of wealth.

Since the late 1990s, the housing market has become a significant investment mechanism in Istanbul. However, the important boost in housing construction started after 2003, with a new political structure that made the Ministry of Urban and Environmental Affairs, TOKİ (Mass Housing Administration), and KİPTAŞ

(Istanbul Residence Development Plan Industry and Trade – an organization of Istanbul Metropolitan Municipality working with investors in the housing market), the basic engines of the commodification of housing. Since construction became a profitable sector, many corporations that used to do business in different sectors (e.g., textile, tourism, and mining) switched to the construction and real-estate sectors. TOKİ and KİPTAŞ started to act as the major tools of the AKP regime and its ministries. As part of this practice, TOKİ started to build many mass housing projects, not only for people in need of social aid but also for the middle- and upper-middle classes. Ironically, social housing for people in need of social aid has been low on TOKİ's agenda. The buyers of real estate commodities provided by TOKİ are not the people in need of housing (Perouse 2013, 81). Usually, middle classes and young couples are eager to buy these new houses. The process works with a new financial mechanism that operates within a regime of debt creation. Many young professionals enthusiastically took on a mortgage credit agreement and started to dedicate their lives to paying those debts (Üçoğlu 2016). The boosting of construction has been accompanied by both a financial system and a developmentalist discourse. With a new, aggressive, developmentalist discourse, the state in Turkey has been transformed into a cluster of partnerships with the private sector; thus, having close ties with governmental bodies has become the major means of capital accumulation. The developmentalist discourse is reflected in the construction and real estate sectors, and a system of crony capitalism has prevailed since the boom of the housing market industry. This developmentalist discourse and state-led construction regime have tremendous effects on the environment.

The trauma caused by the 1999 earthquake changed many things in the perception of Turkish citizens, and housing emerged as the most important commodity to be acquired. Istanbul, in the process of complete urbanization, has been playing a crucial inter-referencing role, and within this role the city is continuing to expand its periphery. In a broad sense, Turkey has chosen the path of pursuing an economy of development through speculative instruments since 2002, and this choice gave rise to investments in the domains of land-rent speculation and real estate/construction. These sectors have become heavily linked to the allocation of governmental resources, and a new kind of political economy has prevailed in the country. *The Economist* (2014, 2016) published an index called "The Crony-Capitalism Index," referring to a specific type of wealth accumulation based on "crony sectors" reliant on adherence to governments. The index covers twenty-three countries, with a perspective that sets democratic governance against "bribery" governance. The basic assumption asserted by the index is that where crony wealth is increasing, economic growth decreases, and the current productive system becomes unsustainable because the allocation of governmental resources into the hands of a few plutocrats or crony capitalists damages competition as well as creativity. Turkey was in the fourteenth rank in 2014, but an updated index in 2016 had Turkey ranked as eighth. The country was seen as one of the major destinations of crony wealth accumulation. The strongest crony sectors in Turkey are real estate, infrastructure, pipelines, and mining. The expansion of urbanization in Istanbul in the last fifteen years is entirely related to this new political economy. In a sense, in Turkey growth is continuing despite the increase in crony-wealth accumulation. However, the problem lies with the quality of this growth. The growth is maintained by providing debt-creation resources to people and by re/producing space; however, this growth does not increase the prosperity of the whole society. Nor is it increasing the quality of urban life and social services.

A massive process of suburbanization operating with a developmentalist ideology has been surrounding Istanbul essentially in the last fifteen years with a state-driven construction and property market policy. As a result, the reproduction of the periphery with new mega-projects as well as gated communities (see the chapters by Güney as well as Çavusoğlu and Strutz in this book) goes hand in hand with the erosion of the environment, in this case essentially that of the northern forests of Istanbul. In order to legitimate the erosion of the city's ecology, two strategies have been used since 2006. On the one hand, a traditional discourse linked to the status of homeownership is promoted and, on the other hand, the threat of another potential earthquake in Istanbul is used as a reason to displace people from their neighbourhoods. In addition to these elements, a discourse of developmentalism (incessantly exalting "growth") has become widespread, with a common narrative of "catching up with the West" by constructing numerous projects, including infrastructure investments and mega-projects such as airports, highways, shopping malls, and luxurious gated communities. These new projects are designed as part of re/producing space in the periphery of Istanbul. As Benjamin Friedman (2005) points out in *The Moral Consequences of Economic Growth*, economic growth that cannot achieve improvement of the living standards of people can easily turn into an uncontrollable failure. In other words, growth must improve living standards by having certain constraints to prevent damages that might be caused due to developmentalist practices that consider only quantitative growth. Friedman says:

> Environmental concerns in particular have expanded from their initial focus on the air and water to encompass noise pollution, urban congestion, and such fundamental issues as the depletion of non-renewable resources and the extinction of species ... Ever larger segments of our society accept that it is not just economically foolish but morally wrong for one generation to use up a disproportionate share of the world's forests, or coal, or oil reserves, or to deplete the ozone or alter the earth's climate by filling the atmosphere with greenhouse gases. (13)

Therefore, a developmentalist ideology that describes growth only within the framework of quantitative expansion has no intention of protecting or improving living standards that are literally linked to environmental concerns: protection of ecological balance, providing a clean environment for all species, and reducing the damages of technological advancement. Hence, this ideology of growth can easily turn into an uncontrollable failure when it is combined with the rent-seeking mechanisms of crony capitalism. The financial wealth that is created by the state and public banking in Turkey is distributed into the hands of a few corporations in the real estate/financial cluster through governmental practices. The state, in this case, acts as an enterprise that facilitates commodification and speculation. Land-rent speculation is one of the major fields of crony wealth in Turkey, and the ministries of the AKP regime and its tool, TOKİ, are the major institutions that manage this developmentalist ideology.

Therefore, to better understand the current changes in the urban political ecology of Istanbul, one needs to see them as embedded in the political economy of housing, construction, and real estate (including infrastructural mega-projects), and the ancillary discourses that drive this expansion. While it may be puzzling at first to see such widespread support for the government's developmentalist strategy in face of the ongoing damages inflicted by that policy to the environmental sustainability of Istanbul, tables 10.1 and 10.2 present a few indicators that can help to explain how the ongoing process masks environmental concerns. These indicators may help us to answer the crucial question: "How is consent for eroding the ecology of Istanbul created?"

The first indicator in table 10.1 shows the number of housing sales in Turkey from 2013 to 2016. The indicator in table 10.2 shows housing sales in Istanbul for the same period.

10.1 Housing sales in Turkey from 2013 to 2016

Year	Total number	New house	Second hand	Mortgaged	Other sales
2016	1,341,453	631,686	709,767	449,508	891,945
2015	1,289,320	598,667	690,653	434,388	854,932
2014	1,165,381	541,554	623,827	389,689	775,692
2013	1,157,190	529,129	628,061	460,112	697,078
Total	4,953,344	2,301,036	2,652,308	1,733,697	3,219,647

Source: Turkish Statistical Institute (http://www.turkstat.gov.tr/Start.do).

10.2 Housing sales numbers for Istanbul, 2013 to 2016

Year	Total number	Mortgaged	Other sales
2016	232,428	87,350	145,078
2015	239,767	93,564	146,203
2014	225,454	87,757	137,697
2013	234,789	106,977	127,812
Total	653,333	273,013	380,320

Source: Turkish Statistical Institute (http://www.turkstat.gov.tr/Start.do).

In addition to these statistics, according to data from the Banks Association of Turkey, the total number of households that were still paying mortgage debt was 2,071,467 as of September 2016. In the meantime, "other sales" are usually in the form of direct sales carried out by crony capitalists (to sell the houses for a higher price in the future) or by foreign investors for potential investment projects. It should also be noted that the majority of the population in Istanbul are already owners of real estate (around 60 per cent, but this number has been constant, not increasing, in the last years). Finally, there were only 18,189 housing sales to foreign citizens in Turkey in 2016. Among them, Iraqis had first rank with 3,036 houses.[2] Most of the demand is created domestically, through a discourse of growth and conservative homeownership. The statistics basically indicate that there is a high demand for housing in Istanbul, and this demand comes from the local people – unlike the general assumption that claims most of the new houses are sold to foreign investors.

The system of crony-wealth accumulation goes hand in hand with the individualization and financialization of the neoliberal subject. This refers to the aspect of risk management, because the neoliberal subject must constantly take risks in order to secure himself in the age of precarious work and workfare (Dardot and Laval 2013; Leitner et al. 2007; Soederberg 2011). In the case of the housing market, taking financial risk for potential land-rent speculation is now the most crucial discourse for the new middle classes, particularly for young professionals dwelling in Istanbul. The very success of this internalized neoliberalism (Keil 2009) relies on its capacity to combine conservative discourse, which works to the advantage of rent-seekers, and the neoliberal entrepreneurial self, which acts with the incentives of rent maximization. In the case of Turkey, and particularly in Istanbul, the combination of these two social realities turns into a disciplinary mechanism that works to spread the apolitical code of indebtedness.

The statistics in table 10.1 and table 10.2 show that there is a very high demand in Istanbul for housing. The demand is generated by the new consumption habits of Istanbul's new middle classes. The new middle classes (essentially people working in financial sectors, culture industries, and service sectors) have very similar consumption habits regardless of their social status and ideological formation. Whether they be conservative or secular, they come together in a unanimity of consumption habits even though they have spatially segregated themselves (Kurtuluş 2012, 11). In fact, many of them have a similar kind of monthly income, and they are mainly conservative in religious terms; also, almost 70 per cent of them are already homeowners (Üner and Güngördü 2016). Housing is now the commodity of the primary capital circuit that determines consumption habits of Istanbul's new middle-classes. These habits are promoted by a discourse of homeownership as a security mechanism, not only for an idealized family life but also for further risky investments (which constitutes the majority of the demand constituted by the middle classes). Hence, the majority of housing projects are not constructed for those who really need it. Indeed, the idea is to maintain the state-led developmentalist construction sectors by facilitating financial investment tools and by articulating people into this new consumption game by engendering an

10.2 "Watergarden" luxury housing project and shopping mall and in Ataşehir, Istanbul. Photo by Murat Üçoğlu.

enticing real-estate bubble. One evidence of this new mobilization is the rate of homeownership. Although many construction projects are being constructed, the rate of homeownership remains stagnant in Istanbul (see Güney's chapter in this book). To wit, the projects are not designed to make people homeowners, but to maintain the increasing rate of land rent.

Investment in housing has become the most profitable financial apparatus in Istanbul. Investing in savings accounts, government bonds, or the stock exchange are not as profitable as purchasing housing in Istanbul. Risk-taking has become a common game in Istanbul, and speculation as a mode of production has become the rule. The gambling can be seen via the numbers of housing sales and of indebted households; the numbers show that active demand is still high, and that people have enough belief in the potential profitability of housing that they get mortgage credits. A recent study published in April 2016 indicates that the annual average housing price increase rate in Istanbul is 18.73 per cent (Dunya 2016). Also, in the peripheral districts of Istanbul, where there is massive suburban development, the average increment in housing prices is more than 20 per cent (Yeniemlak 2016).

Another aspect of the political economy of this new risk society in Istanbul is the concentration of wealth. The housing projects are usually luxurious gated communities, condominium gated communities, and lower-class TOKİ-style housing. The crony companies building those projects are now forming a new type of capitalist elite in the country, and this new crony regime is increasing both the concentration of wealth in the hands of a few corporate capitalists and the social inequality in the society. The developmentalist policy depends on land-rent speculation and, by its very nature, has been generating its winners and losers. The neoliberalization of urban process has made the city a tool of a growth machine, and wealth accumulation through real estate – essentially in the periphery – is allocated among those who are adherents of the ruling AKP, which distributes property-driven wealth among its supporters (Sönmez 2012). It basically bypasses all the democratic and social levels of the decision-making process, especially at the municipal level, by using networks of both private and public nepotism (social capital). As a consequence, this process has changed the economic profile of the country, which indicates that wealth accumulation is concentrated in the hands of few people and the rest have been subordinated to a mechanism of indebtedness.

Crony capitalism based on the construction sector in Istanbul leads to the socialization as well as the individualization of rent seeking, since everyone in society becomes their own enterprise, ready to make an investment in the housing market to get the highest revenue as soon as possible. This may sound like a simple generalization; however, the indicators show that demand is still high and the indebtedness rate based on mortgage credit is also increasing. The process of crony-wealth accumulation damages economic growth, because the rate of unemployment is dramatically increasing (12.17 per cent as of December 2016, and this number does not include those who have lost hope and given up searching for a job). It also hurts fair competition and thereby reduces the belief in democratic and moral values, and reduces consciousness of protecting the ecology. This crony-capitalist system must constantly build new projects and destroy the environment to maintain its existence, and the number of new houses will continue to increase while the credit system continues to spread indebtedness, due to the fact that the country has lost its ability to invest in technological and scientific research. Real estate speculation is now the only major tool boosting the economy. The urban economy is now subject to deep economic inequality, and the distribution of wealth has been totally segregated since 2001 (Pamuk 2014, 314). The active demand causes an increase in prices, which leads people to think of making an investment in housing or of participating in the game by transforming themselves into a risk-bearer. The process is also relevant for people who seek housing only for its use value. When an opportunity for profit becomes clear for those people, they will act as a potential investor. Displaced people or people who are seeking housing only for its use value also form special networks of social capital, and they become ready to risk whatever savings they possess to become part of this housing market game (see Çavusoğlu and Strutz's chapter in this book). As a result, dwellers of Istanbul do not really reflect their concerns around environmental issues. The new consumption habits dominate the social reality, and the housing market dynamics mask the ecological erosion in Istanbul. There have been certain assumptions in the last years indicating that the Gezi Park uprising in 2013 was an environmentalist revolt, and that people showed their ecological concerns during that movement. However, this does not reflect the reality for two reasons: first, the Gezi movement was a middle-class movement to defend a lifestyle (Akgün et al. 2014), and few of the Gezi movement participants protested the mega-projects and housing projects that destroyed the ecological balance of Istanbul.

10.3 Luxurious suburban housing site on the Bahçeşehir-İstanbul Highway. Photo by Murat Üçoğlu.

Instead, they are a driving force of this new political economy.[3] Even though certain initiatives are still trying to protect the ecology of Istanbul, their number is extremely low, and they have not been able to create social awareness, as the majority of the population ignores these courageous struggles. This means that ecological concerns can easily be disregarded for the sake of land-rent speculation, which creates a lack of affordability.

Conclusion

The very success of the recent neoliberal developmentalist ideology implemented by the AKP regime in Turkey relies on the combination of rent-seeking and rent-maximization. A few crony capitalists are dominating the economy with the help of governmental policies that favour the developmentalist regime, which disregards social equality and environmental concerns. Even though the construction sector has been constituting 6–8 per cent of Turkey's GDP in the last decade, its effect with other sectors and its link with the financial sector (hence household debt) makes it the core of economic growth discourse. This discourse of economic developmentalism can become a very dangerous technique that could lead to economic and ecological failure. The developmentalist discourse may expand the economy in quantitative terms, but if it is not supported with a mechanism of social and environmental protection policy, it could easily turn into a failure. Turkey's transformation based on this conservative developmentalist discourse has taken the real estate/financial cluster as its core, and Istanbul constitutes the centre of this rent-seeking cluster. Massive suburban development is presented as an inevitable fact to meet the demand for housing. Essentially, in Istanbul the population is still growing, and so more housing projects are advocated with the assumption that it will fix the problem of shelter, and everyone will be able to find their dream housing. Necessary financial tools provided by the banks and backed by TOKİ are serving this very purpose.

However, the reality shows us that the situation is the other way around. Unlike the claims of TOKİ and politicians, Istanbul is experiencing a housing crisis, and this crisis causes the lack of affordability. In fact, the housing crisis is now more than a simple housing crisis – it is now an urban crisis. New housing projects are built to maintain the developmentalist system of crony capitalism; however, these projects push people to become more indebted, and they also decrease affordability due to the increasing demand coming from the new middle classes, even though they dedicate their lives to paying their debts. Deliberately or unintentionally, these new classes contribute to the changing political ecology of Istanbul and to the challenges posed to the city's sustainability by creating consent for the formation of massive suburbia. Istanbul will witness more problems pertaining to environmental damage. The social awareness of people has been eliminated by the risky game of the housing market, and the rent-maximization motivation of dwellers is tied to the rent-seeking of a few corporations. The result is the deterioration of societal relationships with nature, and the extensive use of resources. Hence, this new political economy based on the dynamics of the housing market is used by policy-makers to legitimate their actions that threaten the ecological biodiversity and environmental sustainability of Istanbul. Massive suburban development of Istanbul has several dimensions, but ecological problems will be the most controversial ones in the coming years.

NOTES

1 One might think that Istanbul's population growth is unexpected, but on consulting the urban history of the city as well as the role of the Turkish government's economic decisions in Istanbul's management, one can observe that the increase in population is somehow a policy planned in order to bring about a global mega-city. It is possible to say that the growth has been very fast, and that this is why it was not easy to organize (Belge 2012, 327). However, the migration to Istanbul involves a deliberate strategy – providing an industrial reserve army for the growing new industries of the city and establishing a global-scale mega-city.

2 There is an assumption among some Turkish people, according to whom the houses are being constructed for rich people from the Gulf countries and Russia. However, the indicators genuinely show that mostly local people are buying the units. Therefore, the discourse of attracting foreign capital alone cannot explain the tremendous rise in housing prices.

3 The assumption about the Gezi movement that claims it was a revolt against the AKP's developmentalist regime destroying the environment is a pseudo-scientific approach that has no evidence or research-based proof. During the time of the Gezi movement several research projects were conducted. For instance, one of them was conducted by Akgün et al. (2014), and it became a very important scientific resource for understanding that movement. The study involved broad research in which they conducted several interviews with the participants of Gezi movement. In their report titled *What Happened in Gezi Park?* one can see that the majority of participants were in the movement to defend their lifestyle, contrary to the assumption of environmentalist concerns. The participants were usually middle-class people with a university degree, or university students. Indeed, it would also be incorrect to say that the developmentalist regime is maintained by the lower classes who vote for the AKP. This view disregards the urban studies literature produced in the recent years in Turkey. In the case of Istanbul, it is obvious that this new urban regime is maintained by including as many middle-class people as possible, regardless of their lifestyles (Atayurt 2009; Kurtuluş 2012; Rutz and Balkan 2010; Üner and Güngördü 2016; Balkan and Öncü 2014).

REFERENCES

Abadie, L., Murieta, E.S., and Galarraga, I. (2016). "Climate Risk Assessment under Uncertainty: An Application to Main European Coastal Cities." *Frontiers in Marine Science* 3. https://doi.org/10.3389/fmars.2016.00265

Akgün, M., Cop, B., Emre, Y., and Yesevi, Ç.G. (2014). *What Happened in Gezi Park?* Istanbul, Turkey: İKÜ Press

Atayurt, U. (2009). "Tarlabaşı'nda Rantsal Dönüşüm" [Rent-based Gentrification in Tarlabaşı]. *Bianet*, 13 May. http://www.bianet.org/bianet/siyaset/114481-tarlabasinda-rantsal-kusatma

Balkan, E., and Öncü, A. (2014). *İslami Orta Sınıfın Yeniden Üretimi, Neoliberalizm, İslamcı Sermayenin Yükselişi ve AKP* [The Reproduction of the Islamic Middle Class, Neoliberalism, Rise of Islamic Capital, and AKP]. Istanbul, Turkey: Yordam

Belge, M. (2012). "İstanbul: Geçmiş ve Gelecek" [Istanbul: Past and Future]. In Bilgili, A.E. (ed.), *Istanbul: Şehir ve Kültür*, 319–42. Istanbul, Turkey: Profil Publishing.

Dardot, P., and Laval, C. (2013). *The New Way of the World: On Neoliberal Society*. London, UK: Verso.

Dünya. (2016). "Konut Fiyatları Bir Yılda Yüzde 15 Arttı" [Housing Prices Increased by 15 Percent in a Year]. 28 June. https://www.dunya.com/finans/haberler/konut-fiyatlari-bir-yilda-yuzde-15-artti-haberi-321278

Economist. (2014). "Our Crony Capitalism Index: Planet Plutocrat." 15 March. http://www.economist.com/news/international/21599041-countries-where-politically-connected-businessmen-are-most-likely-prosper-planet

– (2016). "Our Crony Capitalism Index: The Party Winds Down." 7 May. https://www.economist.com/international/2016/05/07/the-party-winds-down

Friedman, B.M. (2005). *The Moral Consequences of Economic Growth*. New York, NY: Vintage Books.

Gürtler, K. (2016). "Trees versus Concrete: Deforestation in the Bosphorus Region and Civil Society Responses." *Heinrich Böll Stiftung Türkei*, 15 April. https://tr.boell.org/de/2016/04/15/trees-versus-concrete-deforestation-north-bosphorus-region-and-civil-society-responses

Incecik, S., and Im, U. (2012). "Air Pollution in Megacities: A Case Study of Istanbul." In Khare, M. (ed.), *Air Pollution: Monitoring, Modelling, and Health*, 77–116. London, UK: IntechOpen. https://www.intechopen.com/books/air-pollution-monitoring-modelling-and-health

Keil, R. (2009). "The Urban Politics of Roll-with-it Neoliberalization." *City: Analysis of Urban Trends, Culture, Theory, Policy, Action* 2, no. 3, 230–45. https://doi.org/10.1080/13604810902986848

Keil, R. and Macdonald, S. (2016). "Rethinking Urban Political Ecology from the Outside in: Greenbelts and Boundaries in the Post-suburban City." *Local Environment* 21, no. 12, 1516–33. https://doi.org/10.1080/13549839.2016.1145642

Kurtuluş, H. (2012). "Orta Sınıfların Ortak Paydası Neoliberal Değerler" [The Common Sense of Middle Classes: Neoliberal Values]. *Express* 127, 10–13.

Leitner, H., Sheppard, E.S., Sziarto, K., and Maringanti, A. (2007). "Contesting Urban Futures: Decentering Neoliberalism." In Leitner, H., Peck., J., and Sheppard, E. (eds), *Contesting Neoliberalism*, 1–25. New York, NY: The Guilford Press.

Madden, D., and Marcuse, P. (2016). *In Defense of Housing*. London, UK: Verso.

Ocek, S., and Sönmez, Y. (2014). "An Aerial Tour of Istanbul's Disappearing Forests." *Hürriyet Daily News*, 2 September. http://www.hurriyetdailynews.com/an-aerial-tour-of-istanbuls-disappearing-forests.aspx?pageID=238&nID=71201&NewsCatID=340

Özacar, B.G. (2013). "Impacts of Urbanization on Flood and Soil Erosion Hazards in Istanbul, Turkey." PhD diss., University of Arizona.

Pamuk, Ş. (2014). *Türkiye'nin 200 Yıllık İktisadi Süreveni*. Istanbul, Turkey: Türkiye İş Bankası Yayınları.

Perouse, J.F. (2013). "Kentsel Dönüşüm Uygulamalarında Belirleyici Bir Rol Üstlenen Toplu Konut İdaresi'nin (TOKİ) Belirsiz Kimliği Üzerinde Birkaç Saptama" [A Few Findings on the Ambiguous Role of TOKİ, Which Has the Key Determinant Role in Urban Transformation Projects]. In Çavdar, A., and Tan, P. (eds), *İstanbul: Müstesna Şehrin İstisna Hali*, 81–96. Istanbul, Turkey: Sel Yayıncılık.

Roy, A. (2016) "Anti-Eviction Movement in Chicago." Paper presented at the American Association of Geographers Annual Meeting, San Francisco, CA, 1 April.

Rutz, H.J., and Balkan, E.M. (2010). *Reproducing Class: Education, Neoliberalism, and the Rise of the New Middle Class in Istanbul*. New York, NY: Berghahn Books.

Soederberg, S. (2011). "Cannibalistic Capitalism: The Paradoxes of Neoliberal Pension Securitization." *Socialist Register* 47, 224–41.

Sönmez, M. (2012). "Turkey's Second Privatization Agency: TOKİ." *Reflections Turkey*, May. http://www.reflectionsturkey.com/?p=489

TMMOB. (2017). "Dünya Çevre Günü Raporu" [Report for World Environment Day]. 2 June. http://www.cmo.org.tr/resimler/ekler/5de03236ea7c4b2_ek.pdf?tipi=72&turu=X&sube=0

Üçoğlu, M. (2016). "Istanbul's Suburban Dream Is Fueled by Debt." *Citylab*, 25 February. https://www.citylab.com/equity/2016/02/istanbuls-suburban-dream-is-fueled-by-debt/470550/

Üner, M.M., and Güngördü, A. (2016). "The New Middle Class in Turkey: A Qualitative Study in a Dynamic Economy." *International Business Review* 25, no. 3, 668–78. https://doi.org/10.1016/j.ibusrev.2015.11.002

Yeniemlak. (2016). "Istanbul'da Konut Metrekare Fiyatları: Hangi İlçede Ne Kadar oldu?" [Housing Prices by Square Metre in Istanbul: In What Districts Have Prices Changed?]. *Yeniemlak.com*, 10 March. https://www.yeniemlak.com/istanbulda-konut-metrekare-fiyatlari-hangi-ilcede-ne-kadar-oldu-787-istanbul-emlak-haberleri.

PART FOUR

The Suburban Century

11 Morocco's "Pirate Suburbs" from Punishment to Controlled Integration: Neoliberalizing the Regulation of Casablanca's "Chechnya"

WAFAE BELARBI AND MAX ROUSSEAU

Introduction: Taking Back Control over Morocco's "Pirate Suburbs"

Maghrebin[1] suburbanization is characterized by the juxtaposition of strong social disparities (Ben Othman Bacha and Legros 2015). Morocco follows this path, the rapid urbanization of its agricultural lands currently leading to the creation of fragmented spaces (Florin and Semmoud 2014). In large metropolitan areas (Tangier, Kenitra, Rabat-Salé-Témara, Casablanca, Agadir, Fez), the creation of new suburbs currently takes two main forms. The first one consists of the major equipment and property development operations, carried out jointly by public authorities and major promotion groups supported by national and international investments, and intended to accommodate tourists, wealthy classes, and international companies, among others. There is an undeniable alliance between the state and the market in the creation of luxury (sub)urban spaces in Morocco (Barthel and Planel 2010; Bogaert 2011; Rousseau, Amarouche, and Salik 2017). Yet, besides the creation of these high-visibility globalized spaces, the Moroccan suburbs also continue to be structured around informal urbanization and the construction of low-cost housing intended to shelter the many who are excluded from the globalizing major cities (Navez-Bouchanine 2001). Nonetheless, if the former type of suburbs – planned by the state and the private sector – is relatively well-covered by urban research at present, the latter – suburbs we choose here to qualify as "pirate," based on Simone's (2006) work on the African city – is much less explored. However, as we'll get to see in this article, one of today's main objectives for the Moroccan state is to regain control of the autotomizing pirate suburbs.

The suburbs of Casablanca, the main metropolis in North Africa, constitute a particularly enlightening example of the evolution of the regulation of a territory which escapes the division of space that has traditionally been exercised by the Ministry of the Interior. As the economic capital of Morocco, Casablanca grew from 25,000 inhabitants in the early twentieth century to nearly 1 million in 1960, and today exceeds 4 million inhabitants. In this chapter we will enquire into the neoliberal experiments on new forms of social and political regulation that we identify in the "non-reglementary" – or illegal[2] – peripheries of Casablanca. Such experiments may be construed as an attempt by the Moroccan state to adapt itself to suburbanization, a process threatening the traditional control exercised by the authoritarian state according to a marked urban-rural cleavage. In order to do this, we will first present the emergence of the "pirate suburb" by re-encapsulating it into the neoliberalization of the Moroccan state's regulation of space during the last three decades (Peck and Tickell 2002; Bogaert 2014), a phase of "roll-back neoliberalism" characterized by a "punitive" abandonment of the poor and of the informal suburbs – which then move into "piracy" – followed by a "roll-out neoliberalism" phase characterized by the creation of new political instruments in order to reintegrate these spaces and their inhabitants into the metropolitan society. Then, in a second step, we will use Lahraouiyine, one of Casablanca's poorest illegal suburbs (figure 11.1), as a case study.[3]

"Roll-Back" Neoliberalism and the Emergence of Suburban "Grey Areas"

In order to understand the structuring of the pirate suburbs, it is useful to reintegrate them into the analytical grid proposed by Peck and Tickell (2002). Although initially created to analyse the neoliberalization of spatial policies in the UK, it has then been successfully adapted to the Moroccan case by Bogaert (2011). For Bogaert, it is possible to distinguish a first phase of "roll-back" neoliberalism in Morocco, in which the developmental functions of the Moroccan state were dismantled following the World Bank's structural adjustment program starting in the 1980s.

The consequences of this early phase of neoliberalization on social inequalities immediately triggered a series of revolts and riots in the poor neighbourhoods of Moroccan cities, which explains why roll-back

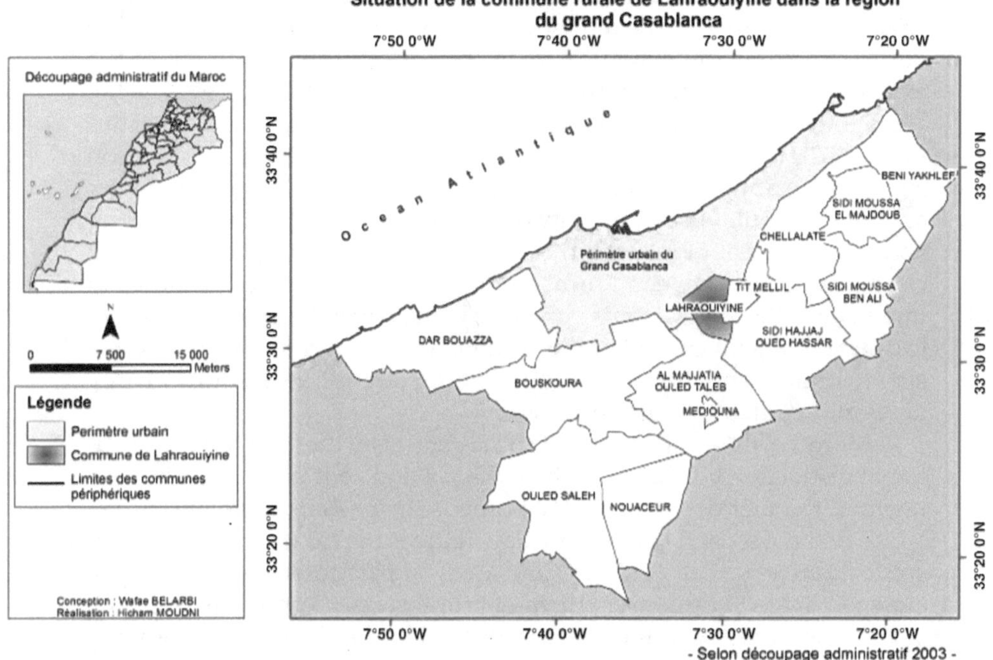

11.1 Localization of Lahraouiyine in the metropolitan area of Casablanca. Map by Wafae Belarbi.

neoliberalism also materialized into a particularly punitive stance towards urban Morocco. The traditional authoritarian state powers were therefore mobilized to accompany the structural reform of public policies and prevent their economic, social, and spatial effects from destabilizing the social order. Morocco's roll-back neoliberalism is therefore characterized by a state focus on building urban safety through city planning. In Casablanca, a "laboratory" city for the control of the poor neighbourhoods, it reacts, for instance, by placing the city's planning agency under the tutorship of the Ministry of the Interior (Rachik 2002). Public expenditures become more concentrated in the city centre, in which the slums become eradicated and large boulevards and structuring facilities are constructed. As a consequence, the illegal suburbs are cut off from the rest of the major metropolis, by the coastal highway in Casablanca or, in the case of Rabat, by a greenbelt introduced in the late 1980s. However, if these measures make it possible to limit the process of unregulated suburbanization, the state's control of the illegal peripheries is also limited by the reduction of public resources under the effect of the structural adjustment plan. In the end, the abandonment of the existing illegal suburbs by the Moroccan public authorities evokes the "benign neglect" of the Black and poor districts of the American cities in the 1970s (Aalbers 2014; Béal, Fol, and Rousseau 2016). This situation explains the appearance of "grey areas," newly (sub)urbanized, dense, and (over)populated by poor city dwellers originating from rural flight, and which do not fit into the traditional control of the state based on a clear distinction between "rural" and "urban" landscapes. In order to understand the experiments currently being carried out by the Moroccan state to regain control of these suburban territories, it is first necessary to understand the structuring of these semi-urban, semi-rural grey areas into veritable pirate suburbs.

In Morocco, the organization of social control by the state lies in a traditional distinction between two hierarchical and centralized supervisory regimes, the first one (whose key figure is the pasha) controlling urban space, and the second one (whose key figure is the caïd) controlling rural space (Rousseau et al. 2017). In rural areas, in particular, the surveillance aims to reinforce the conservatism of "rural society" under the supervision of the Ministry of the Interior, relayed by local landlords co-opted by the government (Leveau 1976). Yet, the clear urban-rural distinction, on which the stability of the authoritarian regime governing colonial and then post-colonial Morocco has depended, started to erode by the end of the 1970s, under the double effect of the acceleration of urban sprawl and of the state's policy of abandonment of the illegal suburbs. Rural flight and demographic redeployment within major cities, coupled with land speculation as well as the resulting rapid rise in real estate prices, led to the continued expansion of the poor suburbs. These are created and structured around a localized social and political regulation that blurs the traditional rural-urban distinction and therefore threatens to rival the close control wielded by the Ministry of the Interior (Rousseau et al. 2017). Neither urban nor rural, created outside of the control of urban planning, abandoned by the public and private sectors and governed by informal economic activities, the illegal, poor urban peripheries appear first of all as spaces of experimentation. We have chosen to analyse them by using Simone's (2006) concept of "pirate cities," a concept initially intended to describe the sub-Saharan African cities where the formal system of governance collapsed, leaving the gap to informal and collective access to urban services and resources.

Neoliberalizing the Regulation of the "Pirate Suburbs"

Referring to piracy to describe the illegal suburbs of the Moroccan metropolises does not necessarily imply that these spaces are not regulated, as they develop rapidly under roll-back neoliberalism. They are regulated, but according to norms that institutionalize their illegal character. In doing so, they escape the narrow control traditionally exercised by the Makhzen (a term which designs the political elite, but also the repressive manner in which power has been traditionally exercised in Morocco). On the one hand, the state withdraws from these suburban territories, after cutting them off from the rest of the metropolis. On the other hand, the differentiated regulation of this new suburban space is further reinforced by the representations of the urban elites who perceive these illegal spaces as signs of a "ruralization" of the metropolis, their inhabitants being depicted as unable to integrate into a large city whose modernization is threatened by their mere presence (Zaki 2005).

Abandoned by the state, the illegal suburbanization of Moroccan metropolises of the 1980s and 1990s is accompanied by the rapid rise of a clientelistic regulation of these territories (Iraki 2003; Abouhani 2012). Resulting from the convergent interests of the former rural landowners at the periphery of the major cities and of the suburban newcomers, this clientelistic regulation is clearly opposed not only to legal planning but also to the formal economy – suburban residents often occupy jobs in informal activities controlled by so-called notables. These notables, who benefit from a familial process of accumulation of wealth and status resulting from their long-time establishment in (formerly) rural territories, extract an important economic and political profit from the urbanization of their suburbanizing land estates and are now able to model the urban sprawl of the major Moroccan metropolises thanks to their links with local councillors. As Abouhani (2012) writes: "Notables are powerful agents of spatial fragmentation. They play a very important role in stretching the urban fabric and expanding peripheral urbanization ... The development process of these neighborhoods is always the same: the landowners parcel out agricultural land without planning or authorization, organize buyers into pressure groups, get elected to the municipal council, bring in various infrastructure equipment and block all draft plans for the development and restructuring of the neighborhood" (70–1).

By allowing the construction of illegal housing estates that are accessible to formerly rural populations deprived of resources, enabling them to integrate into the official market, the notables secure their unfailing support. The notables are thus able to mobilize the local population to pressure the central administration in order to obtain more equipment, which in turn enhances the value of their land. The rapid transition between rural and suburban areas is thereby accompanied by an institutional vacuum temporarily occupied by the notables. The pirate suburbs thus appear more and more clearly as the place of "new" local societies autonomous from the centralized control of the Makhzen. Moreover, they come more frequently under contestation: through the example of Tangier, Navez-Bouchanine (2003) shows how the corruption of local representatives of the Ministry of the Interior, who permit the resettled populations to settle in the illegal areas in exchange for a bribe ("bakchich"), is more and more publicly considered a cause of the densification and consequent degradation of living conditions in these areas (floods, problems with garbage collection, saturation of urban networks, etc.). Similarly, a rise in power of a religious fundamentalism opposed to the Makhzen can also be observed in the pirate suburbs. Moroccan researchers have generally interpreted that as a response to the periphery's abandonment by the state.

However, two channels allow the illegal suburbs to have their voice heard nationally. The first is local elections, which often see the election of suburban notables (for example, illegal developers frequently become council presidents in urban peripheries) before they use their local mandate as a stepping stone towards national politics. The second channel, most noteworthy today, is social mobilization, by which the pirate suburbs manage to influence local and national policies (Belarbi 2015). Local associations, which proliferated in the pirate suburbs in the 1990s and 2000s, aim to improve living conditions through the improvement of the urban environment, the construction of collective amenities, and the strengthening of security (Ameur and Filali-Belhaj 1997; Escallier 2001). These associations become more and more important as the power of the notables decentralizes while urban sprawl continues.

Starting from the end of the 1990s, a new state strategy emerged to confront the structuration of the pirate suburbs, as a result of several factors: the desire to attract international investment in the name of economic growth by opening new territories to urbanization; the will to reintegrate the illegal urban peripheries into metropolitan society and the formal economy in the name of national competitiveness – a main concern for the new regime; and finally, the Casablanca attacks,[4] which highlighted the rise of Islamism in the pirate suburbs. It is possible to consider that these combined factors have opened a new area of "roll-out" neoliberalism in Morocco, in which the state reinvests in the pirate suburbs in an attempt to take back control over localized regulation. In order to achieve this, the selected solutions consist on the one hand of more official integration of the inhabitants' demands, notably through programs of social project management (Le Tellier 2009; Bogaert 2011), and on the other hand the integration of the illegal into the market. Indeed, the slum dwellers are relocated into state-subsidized housing that they have to acquire, which requires the development of strategies of integration into the market economy (training to understand water and electricity invoices, micro-credit programs, etc.). However, such a "return" of the state in the pirate suburbs does not happen without clashes. It highlights the inhabitants of these neighbourhoods' intense mistrust of public actors, local elected officials, and local representatives of the Ministry of the Interior (Navez-Bouchanine 2003).

The Evolution of the Regulation of Casablanca's "Chechnya": From Punitive Abandonment to Controlled Integration

Lahraouiyine, a suburb of slums located on the outskirts of Casablanca, appears particularly revealing of this political evolution aimed at embedding the pirate suburbs, long abandoned by the public authorities, back into the metropolitan society and economy (Belarbi 2011). Indeed, since the beginning of the twentieth century, Casablanca has witnessed the principal concentration of large slums or "bidonvilles" in the country, including Carrières Centrales and Ben M'sik, which housed the nationalist parties at the end of the protectorate. For several decades, these slums have been the object of differentiated treatment by the authorities: although tolerated, they were kept in a status of "permanent precariousness"; that is, the government refused to legalize this type of occupation of space and did not provide them with urban services (Zaki 2005).

If the illegal periphery of Casablanca has long appeared to be channelled, controlled, and managed by public authorities, the acceleration of the construction of slums in the 1990s was experienced as a "hard knock" by the institutional urban actors in Casablanca: it revealed the limits of their strategy and threatened the status quo that had aimed to maintain a population already economically and spatially marginalized in a state of social and political marginality. The commune of Lahraouiyine constitutes one of the poorest and most stigmatized areas emerging out of the most recent wave of illegal suburbanization of Casablanca. It

11.2 Lahraouiyine. Photo by Max Rousseau.

is characterized by the explosion of its population (the number of inhabitants is increasing at an average rate of 10 per cent per year, growing from 16,000 inhabitants in 1992 to 75,000 in 2014); by a resident population that does not originate from rural flight, but from displacement out of the metropolitan core; and, finally, by its strategic location on both city and regional scales. The rapid increase of the population of Lahraouiyine is a direct consequence of the entrepreneurial redevelopment of the centre of Casablanca (Berry-Chikhaoui 2007). Former residents who are excluded from the metropolitan core are retreating into a suburb dedicated to self-construction. As one inhabitant explains:

> I lived with my family in a house I rented in the Oasis neighbourhood at 2000 dirhams per month in Casablanca ... In 1974, I bought a land of 100 *mètres carrés* here. It was agricultural land. After a health problem, I was expelled from (the company that employed me) and I found myself without resources. In 1996, in the time of Chichan (Chechnya), I heard that everyone was building in Lahraouiyine. So, I said to my wife, the poor Fassiya (from the city of Fez) who is not used to Tbahdil (humiliation), we will go to Lahraouiyine, build our house and flee the renting. (Interview with a resident in Douar Baiz, 2007)

Roll-Back Neoliberalism: Punishing "Chechnya"

At first, the rapid development of Lahraouiyine was perceived as a threat by the state, which reacted with demolitions to contain its expansion that was considered as uncontrollable and happening in a strategic zone (on both sides of the metropolitan bypass of Casablanca). But these demolitions were soon seen as selective; moreover, they were implemented by the local authorities and local representatives of the Ministry of the Interior, who the local population see as corrupt. Following the demolition of the house of a pregnant woman in 1996, growing resentment eventually led to an uprising, which prevented the police and the demolition company from entering the neighbourhood. Despite the subsequent siege of Lahraouiyine by the police, informal construction of new housing continued in the pirate suburb thanks to the organization of its inhabitants. As one of them said, "When the soldiers and policemen came to prevent us from building, we'd go down the hill and attack them with slingshots. While the battles were going downhill, people were building their houses up the hill – do you see Douar El Mediouni, well that's how it was built" (Interview with a resident).

This insurrectionary episode has left traces. Since the "riots," Lahraouiyine is seen as having an image tarnished by *al-fawda* (social disorder) and appears as a spectacular manifestation of suburban misery, tainting the fringes of a metropolis that the central government, by encouraging the development of numerous mega-projects, tries to promote as a showcase of the globalization of Morocco. Commonly called Chichan (Chechnya), Lahraouiyine has since become a laboratory of social and political control of the pirate suburbs. After the riot, the commune was surrounded by the police and several dozen people were imprisoned. Furthermore, all new constructions were banned, and blocks were dug north of the commune in order to prohibit the transport of construction materials. However, the pirate suburbs did not stop developing, as the inhabitants succeeded in circumventing the embargo on building materials:

> The trucks brought back the blocks from all corners of the country. When a truck arrived at Lahraouiyine, we had to give him 1,500 dirhams just for the cost of transport. The truck driver had to give 500 dirhams to watch the machine. At the time of his arrival, people rushed to look for the materials with carts. We were well organized ... When the Makhzen encircled Lahraouiyine on the north side, digging deep blocks to prohibit the passage of the gear, the materials were sent from the south, via the Berrchid

> road, even from Marrakech. Yajour's trucks came to Lahraouiyine from all over the country. At that time a water tank cost 120 dirhams. There were people selling used or polluted water for construction. A bag of cement cost 100 dirhams. (Interview with a resident)

Given the government's inability to prevent the construction of new slums, two additional measures were taken in order to punish the insurgent suburb during roll-back neoliberalism: the administrative embargo and the administrative division. The introduction of an administrative embargo (*al-hissar al-idari*) implied that newcomers could not have a residence certificate, a national identity card, or even a birth registration. Only people native to the area had the right to get administrative papers, justifying their identity on Lahraouiyine's soil. This embargo, which resulted in the denial of citizenship to the new inhabitants, lasted until 2002:

> There was al-hissar al-idari (administrative embargo), starting from the riots until 2002. I live in this place, but I have no address here, I have an address in Sidi Othmane. But the others, they have neither an address, nor anything else, and this goes on until today. To receive a letter, they have to go to the post office, or they look for someone to give them an address. Here, we are considered as "apaches." The newcomers, Chichaniyine (the Chechens), could not have al-ikama (the certificate of residence). Even the dead do not have the right to have a burial permit here. You have to pay in order to enroll your children in school. Everything is traded here, 200, 300, 400 dirhams per child. People continue to build belfawda (anarchically). The people who live here are wronged: no post office, no address, no social fund for the people working in factories, no listing in the civil register. Even to get married you have to give money in order to get the papers. (Interview with a resident in Douar Baiz, 2007)

According to the inhabitants, it was only after the removal of the former Interior Minister Driss El Basri in 1999 that granting applications for administrative papers began to be processed more flexibly. Nevertheless, the criteria, which were random and informal because they were linked to seniority identification, eventually led to the reinforcement of the clientelistic system that already existed in Lahraouiyine. This therefore appeared as a move from a clientelistic regulation based on access to a parcel of land to another form of clientelism based on access to an administrative identity.

The second significant institutional response concerns the succession of administrative divisions that Lahraouiyine has experienced since the riots of 1996. Their size and frequency reflect the desire to control the territory by destabilizing its organization. In the span of a decade, the illegal suburb has gone from a simple *douar* (a small unincorporated group of homes) belonging to the commune of Tit Mellil to a rural municipality sheltering approximately 9,000 to 10,000 households in 2009, which is a part of the province of Mediouna. In 2009, Lahraouiyine was once again fragmented, but its status was switched from a rural to an urban commune.

The creation of the Lahraouiyine caïdat in 1997 is another security measure that was implemented after the riot. An administrative division to facilitate the administrative supervision of the rural communes, the caïdat constitutes a prolongation of local authority. Its primary role is to maintain order and public security. Assisted by several agents, the caïd is the main interlocutor with the population. Moreover, he assists the municipal authorities with the task of providing technical assistance in the performance of their administrative and technical tasks associated with basic equipment. He also acts as a facilitator to the service providers of the commune. In this way, he controls the actions of the urban services. Moreover, he is accountable for the application of a series of legislative texts related to associations, public gatherings, trade unions, and the press. These measures have the same objective: reinforcing territorial control of Lahraouiyine, restraining it politically, and weakening its representative authorities and segmenting them, as well as reasserting the control of the state security apparatus on the pirate suburb.

Such a punitive regulation of Casablanca's "Chechnya" gave birth to a decline in the influence of local notables, which in return contributed to shape the "pirate suburb." In Lahraouiyine, it was indeed the local elected representatives who took over the administration of illegality. Strongly marginalized by the decision-making bodies of Casablanca's metropolitan area, the notables of Lahraouiyine, although they had access to local elected positions, had no close ties with the governors, let alone the wali (the regional administrative head). Disenfranchised because they were mostly illiterate or had only a primary level of education, they did not participate in any major decisions affecting the territories of their commune and remained outside the formal territorial governance system. Faced with this situation, they have developed opposition strategies through the facilitation of illegal occupation of the land and mobilization of inhabitants and media coverage of their demands, but also by strong control over land use and its turn towards informal industry – seizing here the opportunity provided by a process of decentralization of informal industries (largely tolerated by the public authorities) driven by increasing land prices in the central area of the metropolis.

Roll-Out Neoliberalism: Reinserting the Pirate Suburb into the Entrepreneurial Metropolis

The main effect of the succession of embargoes and administrative divisions and of the consequent structuration of a "pirate suburb" has been the segmentation of Lahraouiyine into socially polarized enclaves, each governed by a form of sociability based on regional origins, political affinities, or traffic networks. These extremely localized forms of sociability progressively extended, became structured, and ultimately led to the creation of neighbourhood associations. While the state softened the regulation of illegal suburbs, these associations became the main interlocutor of the inhabitants, finally allowing a controlled reintegration of the pirate suburbs within the metropolis. The 2002 legislative elections pointed out a rupture in the punitive administrative treatment of the inhabitants of Lahraouiyine. This first election took on an exemplary dimension for the new regime, which aimed to break with the authoritarian practices of previous decades. This was a question of displaying a high participation rate and transparent elections. Registration on the electoral roll was made mandatory. These dispositions allowed the inhabitants of Lahraouiyine to gain access to a national identity card, a residence certificate, and a family record book. The involvement of local elected representatives in the following elections further expanded the voters' lists. Viewed as electoral reservoirs, the illegal suburbs are considered strategic places by the elected local representatives.

However, Lahraouiyine has benefited above all from an evolution of the regulation of the illegal suburbs since the turn of the century, with a substitution, for the former repressive treatment, of an integrative logic aimed at solving the worst social and political consequences of the "roll-back" neoliberal period – a new phase in the neoliberalization of Morocco's regulation of space that can be labelled as "roll-out neoliberalism." The key instrument in this evolution, which explains the progressive (although still incomplete) reintegration of the pirate suburbs into Morocco's main metropolis, is the National Initiative for Human Development (INDH). This government program has its roots in the deficiencies of public-private partnerships and the consequent promotion of the principle of participation of local residents and associations in access to basic services in cities – and especially in urban peripheries – of developing countries (Navez-Bouchanine 2001). Established by the state in 2005, the INDH initially aims at empowering local actors, and especially associations, by constraining the strategies of private operators in the poorest areas. It officially aims to encourage income-generating activities, support social, cultural and sporting activities and, in the end, support local initiatives and strengthens the aspects of "good governance". Through its modes of intervention in areas that have been long ignored and marginalized by the state, the INDH has used its "social entry" to initiate changes in peripheral neighbourhoods, aiming at reinvesting some previously disinvested lands, and using financial and institutional support to grant access to basic services, education, and micro-finance. With regards to the so-called illegal neighbourhoods – that is to say, the slums and substandard housing – the INDH plans to grant them basic amenities, infrastructure, and services. This new mode of regulation also aims at fostering the development of the local associations and increasing the participation of the inhabitants in the management of their territories. Nevertheless, it raises controversies regarding its effectiveness and its capacity to initiate changes in the poorest rural and suburban areas.

In Lahraouiyine's case, two types of associations exist. The first are national associations, often institutionalized, which operate as non-profits and aim for social regularization, micro-finance, and knowledge, and plead for a better recognition of a local context of marginality that is often identified with a closed and dangerous microcosm. These national associations (such as the Association marocaine des droits humains, which aims at protecting human rights and which is present in Lahraouiyine) operate in a sectorial, not a territorial manner, and as such have (almost) no relationship with the second type of associations, the local ones.

By contrast, the local associations emerged in the illegal peripheries in the 1990s under the leadership of local institutional actors of the douars. They focus mainly on areas where needs are still unmet, such as water access and environmental management (Baron and Belarbi 2010).

Most of the local associations in Lahraouiyine declare questions of literacy, environmental protection, sport, and cultural events as their priorities. Nevertheless, in reality, they rarely develop activities related to the objectives of the INDH. In Lahraouiyine, the most efficient and visible associations are those which focus on the basic issues of water, sanitation, and garbage collection. The associations operating in Douar Mediouni are the most dynamic in this field, such as Al Wafae ("fidelity"), which was created by the municipality and a group of primary-educated inhabitants in the early 2000s in order to manage the two water wells of the douar. Characterized by its demographic weight, and dominated by the local elites of the landowners but also by the arrivals of 1996, Douar

Mediouni constitutes the nucleus of Lahraouiyine, in terms of associative mobilizations. However, these associations, which seek above all to benefit from the aid of the local authorities and also from their agreement to carry out activities, follow, in their official objectives, the trend of calling for a social upgrade to the so-called vulnerable neighbourhoods.

The current associative dynamics in Lahraouiyine are therefore fuelled above all by concrete needs. As a member of an association explains, "The one who has power over the water, has it also over the inhabitants" (interview with a resident). But they are also fueled by a strong desire for official recognition, in a territory which has long been qualified as a pirate suburb. In the mind of the members of the associations, it appears a privilege to get closer to the most influential local institutional actors. The president of another association explains: "By doing the associative work, we became recognized. We sign agreements with the governor, we attend the official meetings, we are asked [by the public authorities] to be present in the demonstrations. Before that, we were ignored" (Interview with the president of an association).

The phase of roll-out neoliberalization therefore appears to have softened the political regulation of Lahraouiyine, allowing the ongoing reintegration of the previous pirate suburb into metropolitan society. But such a reintegration is still carefully controlled through new political instruments such as the INDH. For example, the majority of the associations that operate in the field of services and amenities are controlled by the local elected representatives under the supervision of the CQS (the Centre of Social Qualification, which depends on the Ministry of the Interior). As the director of the CQS explained: "We are trying to extend the associative fabric to Lahraouiyine. For this, we create a kind of competition between associations. The trick is simple, one provokes things and one says that they are initiated by the associations. The governor is preparing the relay. They are politically framed associations. They are working towards the state control of society. Without saying that it is the state that is behind" (Interview with the director of the CQS).

Conclusion

In Morocco, the neoliberal agenda, although it was initially brought by external forces (the structural adjustment plan of 1982), did not lead to a collapse of the traditional style of political and economic domination organized around the Makhzen. On the contrary, the authoritarian state proved remarkably able to adapt itself to this new international agenda

(Bayart 2007; Zemni and Bogaert 2011). Nevertheless, the first phase of neoliberalism, characterized by the rolling back of the developmentalist state in a time of massive rural exodus, paved the way for the emergence of "grey spaces" over which the control of the Makhzen was uneasy. These grey spaces were mainly located in the new suburbs inhabited by former peasants. To the state, facing strong social opposition to the implementation of the neoliberal agenda and to the growing inequalities it implied, the illegal peripheries soon threatened to turn into a reminiscence of "*bled siba*" ("ungoverned spaces," a historical term in pre-colonial Morocco in opposition to "*bled el makhzen*," the spaces over which the state exerts its full power).

In neoliberalizing Morocco, the consequent strategy of containment and punitive regulation of the suburban "grey areas" ultimately led to the creation of the pirate suburbs, in which the inhabitants, deprived of urban services and subjected to several embargoes, developed new forms of organization in order to enable the survival and the further development of illegal housing. Subsequently, the structuring of these networks of sociability provided the basis for the return of the state under the new regime. This new phase of "roll-back neoliberalism," which aims to bind the excluded from the "modernization" of Casablanca to the metropolitan society, is accompanied by a filtering through which certain types of sociability appearing in the pirate suburbs are co-opted by the state, while others are sanctioned.

Today in Lahraouiyine, a new official development plan predicts the construction of new buildings on several hectares. These housing projects are intended for the resettlement of inhabitants of the slums in the city centre, where the bidonvilles are being razed in order to clear the ground for new developments that fit the uses of a solvent population. Lahraouiyine is henceforth recognized by the state, but only as a reserve for the population excluded from the globalized metropolis, the showcase of a "modern" Morocco. Such a controlled and limited reintegration of Casablanca's "Chechnya" into the metropolitan area does not eradicate the many problems affecting the former "pirate suburb." Lahraouiyine is still considered by the national press as the hub of drug trafficking in Casablanca (Chennaoui, 2015), and the 2016 legislative campaign, during which residents claimed their right to basic urban services (sewage, drinking water, transport, and security), highlighted once again the local political dysfunctions when some inhabitants reported "elected officials exploiting misery by selling water from the well" (Ait Akdim, 2016).

NOTES

1 The Maghreb traditionally refers to the western part of the Arab world. It encompasses most of the Arab-Berber region of western North Africa (Morocco, Algeria, Tunisia, Libya, and Mauritania).
2 What we describe by "illegal" in this chapter refers to the official French expression used by the Moroccan administration to design the areas in which forms of "non-regulatory housing" ("habitat non-réglementaire"), including slums ("bidonvilles"), have proliferated by bypassing the official planning documents.
3 This research is based on various interviews realized in Lahrarouiyine and with civil servants in Rabat. The quotations in this chapter are taken from these interviews.
4 The Casablanca bombings were a series of bombings in May 2003 and March and April 2007 in which 53 people were killed. The suicide bombers came from the illegal suburbs of Casablanca.

REFERENCES

Aalbers, M. (2014). "Do Maps Make Geography? Part 1: Redlining, Planned Shrinkage, and the Places of Decline." *ACME: An International Journal for Critical Geographies* 13, no. 4, 525–56. https://www.acme-journal.org/index.php/acme/article/view/1036

Abouhani, A. (2012). *Gouverner les Peripheries Urbaine. De la Gestion Notabilaire a la Gouvernance Urbaine au Maroc* [Governing the Peripheries: From Notables' Management to Urban Governance in Morocco]. Paris, France: L'Harmattan.

Ait Akdim, Y. (2016). "Au Maroc, quand les élus prospères du PAM font campagne dans les bidonvilles" [In Morocco, When the PAM's Prosperous Elected Representatives Are on the Stump in the Slums]. *Le Monde*, 5 October 2016. https://www.lemonde.fr/afrique/article/2016/10/05/au-maroc-quand-les-elus-prosperes-du-pam-font-campagne-dans-les-bidonvilles_5008805_3212.html

Ameur M., and Filali-Belhaj, A. (1997). *Développement urbain et dynamiques associatives: Rôle des amicales dans la gestion des quartiers urbains* [Urban Development and Associations: The Role of Local Clubs in the Management of Urban Neighbourhoods]. Rabat, Morocco: Agence Nationale de Lutte contre L'Habitat Insalubre.

Barthel, P.A., and Planel, S. (2010). "Tanger-Med and Casa-Marina, Prestige Projects in Morocco: New Capitalist Frameworks and Local Context." *Built Environment* 36, no. 2, 176–91. https://doi.org/10.2148/benv.36.2.176

Baron, C., and Belarbi, W. (2010). "Gouvernance participative et rôle des associations pour l'accès à l'eau dans la périphérie de Casablanca (Maroc)" [Participatory Governance and the Role of Associations for Access to Water in Casablanca's Peripheries]. In Schneier-Madanes, G. (ed.), *L'eau mondialisée,* 381–401. Paris, France: La Découverte.

Bayart, J.-F. (2007). *Global Subjects: A Political Critique of Globalization.* Cambridge, UK: Polity Press.

Béal, V., Fol, S., and Rousseau, M. (2016). "De quoi le 'smart shrinkage' est-il le nom? Les ambiguïtés des politiques de décroissance planifiée dans les villes américaines" [What Lies behind Smart Shrinkage? The Ambiguities of the Politics of Planned Degrowth in American Cities]. *Géographie, Économie, Société* 18, no. 2, 211–34. https://doi.org/10.3166/ges.18.211-234

Belarbi, W. (2011). "Mobilisations des habitants et régulations territoriales dans la périphérie sud de Casablanca – Cas de Lahraouiyine" [Residents' Mobilizations and Territorial Regulations in the Southern Peripheries of Casablanca]. PhD diss. Faculté des lettres et des sciences humaines de Rabat.

– (2015). "Les mobilisations sociales dans les territoires périphériques de Casablanca pendant les années 1990" [Social Mobilizations in the Peripheral Territories of Casablanca in the 1990s]. *L'Année du Maghreb* 12, 137–53. https://doi.org/10.4000/anneemaghreb.2423

Ben Othman Bacha, H., and Legros, O. (2015). "Introduction: Politiques urbaines et inégalités en Méditerranée" [Introduction: Urban Politics and Inequalities in the Mediterranean Area]. *Les Cahiers d'EMAM,* 27. https://doi.org/10.4000/emam.1168

Berry-Chikhaoui, I. (2007). "Les citadins face aux enjeux d'internationalisation de la ville: Casablanca et Marseille, où est le Nord, où est le Sud?" [Urban Residents Confronting the Stakes of Internationalization: Casablanca and Marseille, Where is the Global North, Where is the Global South?]. *Autrepart* 41, no. 1, 149–63. https://doi.org/10.3917/autr.041.0149

Bogaert, K. (2011). "The Problem of Slums: Shifting Methods of Neoliberal Urban Government in Morocco." *Development and Change* 42, no. 3, 709–31. https://doi.org/10.1111/j.1467-7660.2011.01706.x

Chennaoui, S. (2015). "Lahraouiyine, fief du trafic de drogue" [Lahraouiyine, Stronghold of Drug Trafficking]. *Le360.ma*, 7 February 2015. http://fr.le360.ma/societe/lahraouiyine-fief-du-trafic-de-drogue-31735

Escallier, R. (2001). "De la tribu au quartier, les solidarités dans la tourmente" [From the Tribe to the Neighbourhood, the Threatening of Solidarities]. *Cahiers de la Méditerranée* 63, 13–40. http://journals.openedition.org/cdlm/9

Florin, B., and Semmoud, N. (2014). "Introduction: Marges en débat" [Introduction: Discussing the Margins]. In Semmoud, N., Florin, B., Legros,

O., and Troin, F., *Marges urbaines et néolibéralisme en Méditerranée*, 15–41. Tours, France: Presses Universitaires François Rabelais.
Iraki, A. (2003). *Des notables du Makhzen à l'épreuve de la « gouvernance »: Élites locales, territoires, gestion urbaine, et développement* [Notables of the Makhzen and the Test of Governance: Local Elites, Territories, Urban Management, and Development]. Paris, France: Inau/L'Harmattan.
Le Tellier, J. (2009). "Accompagnement social, microcrédit logement, et résorption des bidonvilles au Maroc" [Social Accompaniment, Housing Microcredit, and Slum Reduction in Morocco]. *Les Cahiers d'EMAM* 17, 55–70. https://doi.org/10.4000/emam.248
Leveau, R. (1976), *Le fellah marocain, défenseur du trône*. [The Moroccon Fellah, Defender of the Throne]. Paris, France: Presses de Sciences-Po.
Navez-Bouchanine, F. (2001). "Villes, associations, aménagement au Maroc" [Cities, Associations, and Planning in Morocco]. *Les Annales de la Recherche Urbaine* 89, no. 1, 112–19. https://doi.org/10.3406/aru.2001.2387
– (2003). "Les chemins tortueux de l'expérience démocratique marocaine à travers les bidonvilles" [The Difficult Paths to Morocco's Democratic Experience through Slums]. *Espaces et Sociétés* 112, no. 1, 59–81. https://doi.org/10.3917/esp.g2003.112n1.0059
Peck, J., and Tickell, A. (2002). "Neoliberalizing Space." *Antipode* 34, no. 3, 380–404. https://doi.org/10.1111/1467-8330.00247
Rachik A. (2002). *Casablanca: L'urbanisme de l'urgence* [Casablanca : Emergency Urbanism]. Casablanca, Morocco: Imprimerie Najah El Jadida.
Rousseau, M., Amarouche, K., and Salik, K. (2017). "L'urbanisation des périphéries rurales de Rabat entre autoritarisme et néolibéralisme" [The Urbanization of Rabat's Rural Peripheries between Authoritarianism and Neoliberalism]. In Berger, M., and Chaléard, J.-L. (eds), *Villes et campagnes en relations: Regards croisés Nord-Sud*. [Cities and Countrysides in Relationship: Northern and Southern Crossed Views] Paris: Karthala, 141–54.
Simone, A. (2006). "Pirate Towns: Reworking Social and Symbolic Infrastructures in Johannesburg and Douala." *Urban Studies* 43, no. 2, 357–70. https://doi.org/10.1080/00420980500146974
Zaki, L. (2005). "Pratiques politiques au bidonville (Casablanca, 2000–2005)" [Political Practices in the Slums (Casablanca, 2000–2005)]. PhD diss., Institut d'Études Politiques de Paris.
Zemni, S. and Bogaert, K. (2011). "Urban Renewal and Social Development in Morocco in an Age of Neoliberal Government." *Review of African Political Economy* 38, no. 129, 403–17. https://doi.org/10.1080/03056244.2011.603180

12 State-Led Housing Provision Twenty-Five Years On: Change, Evolution, and Agency on Johannesburg's Edge

MARGOT RUBIN AND SARAH CHARLTON

Introduction

In South Africa, new-built government-provided housing settlements, often on the edges of urban areas, are consolidating and maturing differently from what was envisaged by the state. Echoing adjustments made by occupants to government-built housing elsewhere (see, for example, Ghannam 2002; Tipple 2000), adaptations in South Africa include the addition of outside rooms or the introduction of businesses on the property, many of these unsanctioned by authorities (Rubin and Gardner 2013; Turok and Borel-Saladin 2016). By and large these changes discomfort and unsettle the state, provoking a variety of responses (Charlton, Gardner, and Rubin 2014).

Viewed collectively, these neighbourhood transformations, with their mix of autoconstruction and state infrastructure, resonate with Caldeira's (2017) notion of peripheral urbanization, but instead of resident-led incrementalism[1] our focus is on state-conceived and produced developments in a context without a history of autoconstruction on a large scale. In this different sequencing of actors in the production, consumption, and transformation of space – state first, residents later, as unforeseen "agents of urbanisation" (Caldeira 2017, 5) – our focus is on further phases and forms of action and reaction by the authorities. In this chapter we thus consider these responses by the state and others, focusing specifically on the issue of added external rooms, labelled "backyarding." Using settlements in the vicinity of Johannesburg (figure 12.1) we discuss different types of responses, including enforcement of regulations; conditional facilitation of backyard dwellings using local level

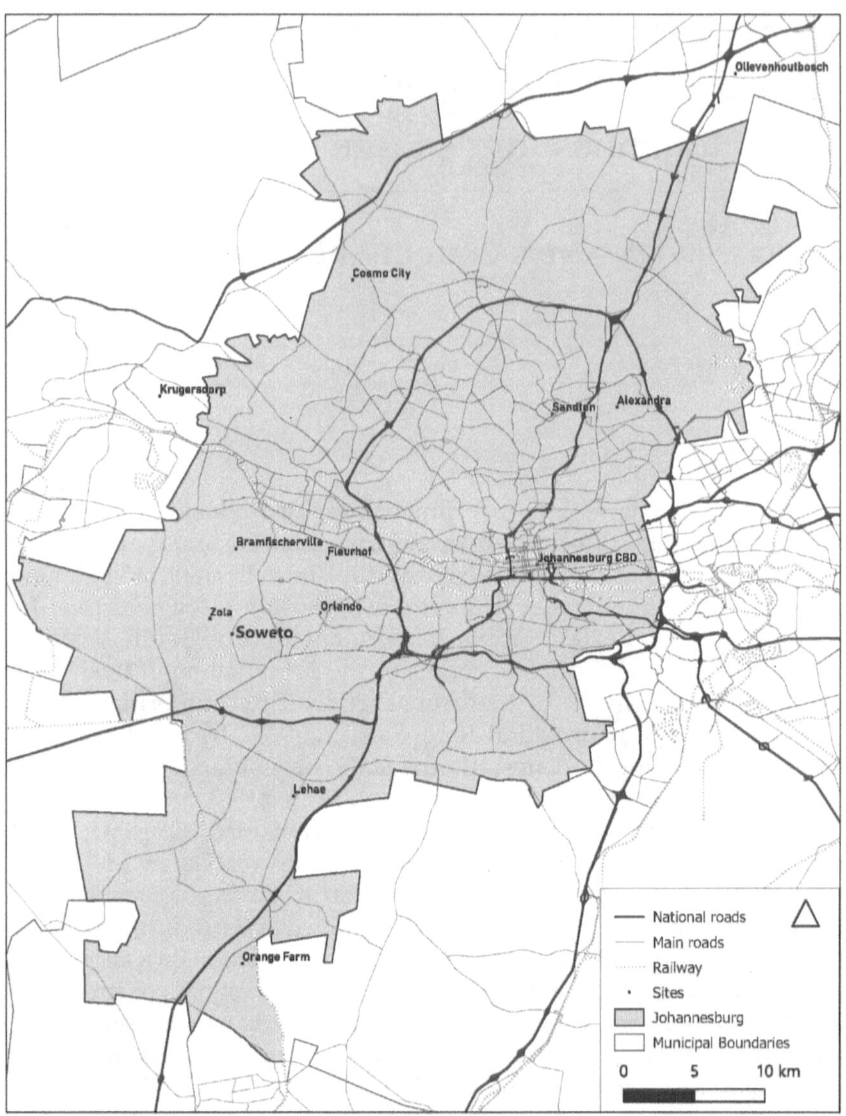

12.1 Location map of selected housing sites in or around Johannesburg (shaded), the centrally located metropolitan area in the province of Gauteng. Map created by Miriam Maina, 2018.

agents; the direct supply of external rooms; supportive policy formulation as well as a laissez-faire, hands-off approach. We argue that the cases, taken together and put alongside government discourse and practices, show an awkward process of adaptation by sections of the state to an unforeseen and unwelcome reality, though at times they also reveal paralysis and avoidance. The process of adaptation by the state is not a linear progression towards a resolved position, but rather reveals aspects of a complex process of bricolage, understood as "a matter of borrowing and copying bits and pieces of ideas from elsewhere, drawing upon and amending locally tried and tested approaches, cannibalising theories, research, trends and fashions and not infrequently flailing around for anything at all that looks as though it might work" (Ball 1998, 126).

Although bricolage has historically been used to describe the complex iterations between interest groups that go into policy construction and the translation and translocation of policy from elsewhere, here we extend the idea of bricolage and policy-making to demonstrate that where policy "ends up" is due to the agglomeration of political interests, capacity constraints, conflicted political positions, and a rough pragmatism (Wilder and Howlett, 2014; Stone, 2017). The form of bricolage as policy-making through practice we discuss in this chapter also resonates with the call for "more pragmatic approaches and responses to the forms of urbanity" (Mosselson 2015, 17) that reflect the daily practices of residents. Here we demonstrate that residents' and homeowners' adaptations are making their way into policy, albeit haltingly and unevenly, echoing the processes discussed in Rubin (2018). The chapter is thus also a response to Ekers, Hamel, and Keil's (2012, 405) call to arms, in their comment that "suburban governance thus is about accounting for both the converging and diverging patterns of peripheral development. Doing so requires paying attention to the varied agents, methods, relations and institutions through which development is managed. Together, these can be viewed as the mechanisms of suburban governance." Thus, we begin to unpack the making and remaking of these sites, the various actors and the manner of policy practice that has evolved to govern these spaces.

The case we discuss, however, does not sit comfortably within many conceptualizations of suburbia. Harris (2010) identifies three defining features of suburbia: peripherality, physical and social newness, and low-density construction. However, many of the sites in our discussion are not peripheral, and while they are certainly not dense by the standards of Southeast Asia or Latin America, key to understanding the practices that we are describing is that they are densifying..Once again this is in a way that differs from the cases that Mabin, Butcher, and Bloch (2013) have examined in sub-Saharan cities, which are older middle class or "colonial" suburbs densifying through retrofitting with residential and commercial activities.

A further layering of the idea of suburbanism is the important conceptual contribution of Ekers et al. (2012, 418) who note that "suburban governance can be viewed through the distinct but complementary modalities of the state, capital accumulation and private authoritarianism" with close ties between the financialization of suburbs by the private sector and the state. The case we are offering surfaces a hybrid moment, a combination of older state investments in housing and new autoconstruction by households and individuals who are densifying these environments. But even state investment differs from other contexts, as in our case the state invests in the construction of the full range of housing and services, not just infrastructure as in Indian cities (Kennedy 2007), or finance as in the North American case (Ekers et al. 2012). These settlements are not entirely autoconstructed, or the result of private entrepreneurship, nor are they solely public or state-subsidized units or the work of large private developers (Mabin et al. 2013); they have combined a kind of quiet, private property development, led and undertaken by households who are beneficiaries of state subsidized housing with the intention of some small capital accumulation (Bayat 2010; Charlton 2013). As such, what we will demonstrate below is slightly different from the current work on suburban growth, and offers a form of what Bank (2011) identifies as a fractured urbanism: a clinging on to and use of state-led infrastructure in ways that are important and beneficial to poorer people, and then in turn the state's response to this repurposing.

Our research on bricolage and backyarding took place in Gauteng, a province of about 13 million people,[2] with South Africa's largest city Johannesburg as a central component. The national statistics bureau, StatsSA, reports that although Gauteng is the smallest province by size, it is considered the country's economic powerhouse, is responsible for a third of South Africa's GDP, and is home to three of the country's largest metropolitan areas (BusinessTech, 2018). The bricolage evident in approaches to backyarding also makes visible

12.2 Original starter/core house. Photo by Sarah Charlton, 2011.

the inconsistencies and incoherence that exist within and between the various departments and spheres of government. Such thinking rebuts any conceptualization of the state as a unified entity, and complexifies the notion of networked governance by introducing elements of informality, confusion, inconsistency, and personal agendas into the analysis (Klijn 2008). It also helps to explain the variable morphology of housing settlements on the periphery, where, due to mixed response, neighbourhoods have evolved in various and highly heterogeneous ways.

Context and Background

South Africa's post-apartheid low-income housing program is known for the scale and extent of its intervention, with the South African government estimating that over 20 million people have benefited from some type of house, housing subsidy, or, as the government terms them, "housing opportunities" (2016). Many of the 4.3 million houses in the program consist of a house on a plot of land for ownership

12.3 Evolution of RDP housing post-2001. Photo by Sarah Charlton, 2010.

in one of many new-built neighbourhoods of predominantly low-rise and detached houses that have substantially extended the urban footprint of cities and towns. During the 1990s the first generation of state-sponsored housing was basic "starter" or "core" housing, requiring the end user (mostly very low-income or no-income households) to add to and complete it over time using their own labour or savings (see figure 12.2). The specific form this process might take was not well articulated in policy documents, but at least in the early stages the look and feel of these settlements was likely to be one of houses under construction in an evolving and incrementally changing area.

In 2001 the national government introduced a significant change to both the housing delivery mechanisms and the end product. This placed far greater emphasis on state-driven and managed housing development, effectively bypassing the various forms of private-sector and community developers, and, significantly, requiring the delivery

12.4 Current version of state-supplied housing. Photo by Sarah Charlton, 2011.

of a complete house of specified dwelling size and materials (Charlton and Kihato 2006; see figure 12.3 and figure 12.4). This policy change served to downplay individual and community self-build initiatives, and signalled an increased emphasis on formal, complete, and orderly neighbourhoods.

Yet, despite these efforts by government to construct modern, fully serviced, and comprehensive suburbs, many changes have emerged through residents' individual adaptions and appropriations, by and large unanticipated, unauthorized, often non-compliant, and largely unwelcome in the government's view. Discussed by scholars over the years, these changes include the proliferation of backyard dwellings of all kinds of construction quality (Robins 2002; Lemanski 2009; Rubin and Gardner 2013; Shapurjee and Charlton 2013; Zweig 2015; Gardner 2015), many forms of home-based economic activity (Robins 2002; Wariawa 2014; Gardner 2015), and other activities in public space such as taxi stands, car washes, trader's stalls, and animal grazing.

Scholars have discussed these changes in different ways: as a reversion from formal living to informality (Robins 2002; Lemanski 2009); as necessary adaptations for households to generate income to support the costs of living formally (Robins 2002; Lemanski 2009; Charlton 2013); as contributing to important forms of cheap rental accommodation (Watson 2009); as significant supplementation to landlords' income (Watson 2009; Gardner and Rubin 2016); as a means of densification of neighbourhoods realized initially as problematically low density areas (Shapurjee and Charlton 2013; Gardner 2015); as important contributors to the transformation of areas from mere housing developments to human settlements (Charlton et al. 2014); and as "a rude reminder of the failure of planners, policy makers and developers to acknowledge the complexity and heterogeneity of everyday social life and lived experience" (Robins 2002, 511). Less documented in published work have been the practical and attitudinal responses by authorities (though see Charlton 2018).

Historically, official pronouncements tended to be strongly disapproving: statements by provincial or national housing politicians in the 2000s were critical of shack-like backyard dwellings and of certain businesses apparent in residential areas, in particular those owned by foreigners and those where the original beneficiary appeared to have surrendered the entire house to a retail or commercial undertaking. In parts of South Africa, the construction of backyard dwellings by residents was forbidden, and units were demolished and landlords reprimanded (Charlton et al. 2014). In 2010, Ruby Mathang, the portfolio head of housing for the City of Johannesburg, announced that "the City wants to eradicate corrugated iron shacks by 2014, and no backyard dwellings will be tolerated because they contravene by-laws and exert pressure on its resources," reflecting national responses and opinions (City of Johannesburg 2010). According to Turok and Borel-Saladin (2016, 387–8, citing the Presidency in 2010), the Presidential Delivery Agreement for Sustainable Human Settlements dismissed the importance of backyarding, simply noting that "whilst informal rental is an accommodation provider these units are illegal, do not conform to minimum standards and thus cannot be accounted for in this document."

In 2006, there was an attempt by provincial government to thwart the construction of external rooms for rent, through a directive to the provincial Gauteng Department of Human Settlements to ensure that toilets and amenities were built inside the state-subsidized units: "The intention is to frustrate new homeowners who build backyard shacks.

When the toilet is inside the house it will make it harder for them to have tenants using it" (Victor Moreriane, the housing department's spokesman, quoted in Mabuza 2006, n.p.). Although there was departmental acceptance that housing beneficiaries rented out backyard shacks to supplement their income, apparently the practice of backyarding derailed the government's plan to eradicate shacks in the province and undermined the state-provided housing: "There are other ways to make an income that do not devalue the new houses that government has built to empower communities. These houses must be treated as an asset, not only as shelters" (Victor Morerian in Mabuza 2006, n.p.). At the same time, diverse practical responses by the state started to emerge, such as the pilot scheme to improve the quality and living conditions in backyard dwellings in two areas of the older township of Soweto, which we discuss below.

More recently there has been official recognition of the needs of backyard tenants: "Government has always focused more on providing houses to beneficiaries residing in informal settlements while overlooking those dwelling in backyards. Backyard dwellers must also be prioritised," pronounced Nocawe Mafu, member of the Parliamentary Human Settlements Portfolio Committee (Ramothwala 2015, n.p.). As a consequence, of late, there has been a bifurcated response to backyarding. On the one hand, led by the National Department of Human Settlements, there has been a change in the official discourse concerning backyarders: Minister Sisulu announced in her 2016 Budget Speech that backyarders – who less than a decade ago were illegal law-breakers – are now the state's prime concern: "In allocating subsidised housing we will be prioritising backyard dwellers whose concerns about queue jumping by urban new comers are legitimate. These are people who are law abiding and have stuck by the rules" (Sisulu, 2016, n.p.). However, while foregrounding the people living in backyard dwellings, this approach is still premised on the eradication of informal structures and the relocation of backyarders into formal settlements. The other strand of the bifurcated approach is the idea of an in situ incremental upgrading of neighbourhoods congested with backyard rooms through infrastructure provision – implicitly accepting the existence of backyard units into the future. This infrastructure focus seems to have emerged from the experience of Gauteng Department of Human Settlements, whose interventions into backyarding over the last few years were termed "suicidal and dangerous" by one senior official because of the policy distortions and social tensions they created (provincial housing official, personal communication with author, 1 February 2017). Prompted by experiences in Alexandra K206 in northern Johannesburg, and Zola and Orlando in Soweto (details below), there is now a policy directive that subsidies for second dwellings will not be provided, but rather infrastructure interventions will be the key focus. In addition, the South African Local Government Association (SALGA) has been pushing for an agenda of differentiated responses and limited intervention in what is seen as a functional market (Rubin and Gardner 2013).

Responding to Backyards: Public-Private Partnerships, Community Management, and State Intervention

Up to this point, we have outlined shifting perspectives at the national and to some extent the provincial sphere of government; the cases below discuss responses to backyarding that mainly involve local government in new settlements. We discuss three types of responses in the greater Johannesburg area in post-apartheid state-provided housing settlements, also summarizing the provincial state interventions made in older existing townships as a counterpoint to help explain some of the attitudes and changes in practice that have been seen.

(1) Maintaining and Losing Control: Local Authorities and the Private Sector in Cosmo City, Olievenhoutbosch, and Fleurhof

Within the array of responses to backyarding, one strand takes a fairly authoritarian approach aimed at stopping some forms of this practice and shaping others. The cases of Cosmo City, Olievenhoutbosch, and Fleurhof demonstrate this approach. Cosmo City, begun on site in 2004, is considered by government as a flagship post-apartheid mixed-income and mixed-housing-typology development, including 8,000 fully or partially subsidized free-standing houses and 3,300 mortgage-linked houses (City of Johannesburg 2008a). The project is a public private partnership between the provincial authorities, the City of Johannesburg (CoJ), and a joint-venture private company, Codevco (Lategan 2012). From the beginning there was a clear policy on backyard dwellings that initially shaped the growth of extensive but formally constructed secondary units (Rubin and Gardner 2013; see figure 12.5). The policy had two aspects: first, new residents went through an induction process informing them that backyarding was only possible if permanent materials of brick and mortar were used and if they

12.5 Formal double-storey backyard rooms in Cosmo City. Photo by Sarah Charlton, 2015.

received planning approval (Kotzen and Suttner, 2018). Second was surveillance and demolition: Codevco-employed scouts cycled through the settlement and reported by-law infringements, including non-conforming backyard structures which were then demolished.

However, the urban management of the settlement was then handed over to the City of Johannesburg. As a consequence, the network of scouts is no longer operating, and the city has not been able to control by-law infringements and non-conforming backyard units (see figure 12.6). Reports suggest backyarding is now "mushrooming" in Cosmo City, with the backyard population estimated to be as high as 30,000 people (Kriegler and Shaw 2016), out of official provincial Human Settlements Department estimates of an overall population of 70–80,000. Many units now consist of temporary materials and are crowded into the yards of the main houses, far exceeding city guidelines. The high demand for accommodation proliferates in this form because of the city's lack of capacity to police and enforce by-laws, as well as fierce

12.6 Unsanctioned backyard dwelling in Bram Fischerville, similar to those in Cosmo City. Photo by Sarah Charlton, 2011.

resistance from residents to the demolition of non-compliant backyard dwellings. The city succumbed to this community pressure and left the units in place. With the change in political leadership in the city council in 2016 (from the African National Congress [ANC], the liberation party and dominant party, to a Democratic Alliance and Economic Freedom Fighters coalition), the new politician in charge of planning has promised to change the response to backyarding, contending that the ANC's "demolitions of backroom structures were a lazy solution to a complex issue that rise [*sic*] from socio economic challenges experienced by affected residents" (Cosmo City Chronicle 2015, n.p.).

A joint approach to backyarding by a local authority and private developers can also be found in Olievenhoutbosch, a 5,000-unit mixed-income development begun in 2009, close to the periphery of Johannesburg in the neighbouring municipality of Tshwane (Lategan 2012). Private developers have argued that forms of backyard rooms, additions, or unscheduled home businesses that contribute to "a sense of informality" in the area lower property values and affect the sale of mortgage housing (Charlton, 2013). In 2010 the local authority was responding actively when developers reported backyard shacks starting to emerge in the yards of bonded housing. However, by 2016 protests by backyard dwellers over their lack of formal, owned housing revealed extensive backyarding across the settlement, including, it seems, informal construction: "Almost every third house in Olievenhoutsbosch Extension 2, for instance, has an average of seven shacks in its yard" (Ndlazi 2016, n.p.). It is not clear what has happened to the private developer and local authority's efforts at control.

Fleurhof, to the west of Johannesburg's Central Business District (CBD) and on land that has historically lacked development, is now seen as a further emblematic mixed-income, mixed-use housing project. Begun in 2008, the 10,411-unit project comprises a mix of fully and partially subsidized houses, and rental and ownership. Fleurhof is a partnership between Gauteng Province, the City of Johannesburg, the listed company Calgro M3, and a variety of funding partners, including International Housing Solutions and the four big private banks. Given the experience of other mixed-income housing developments, including Cosmo City and Olievenhoutbosch, the developers and the city have chosen a zero-tolerance response to backyarding. The developers are still active on site and constantly monitor the settlement, protecting their investment and making sure that by-laws are observed and enforced.

(2) Community Self-Management: Bram Fischerville, Orange Farm, and Lehae

Moving from direct control of backyarding by the private sector and the state, we discuss a second broad response, in the form of community self-management in some state-funded housing projects. The iconically named Bram Fischerville[3] is a 22,000-unit development on the western edge of Soweto, with subsidized houses constructed between 1997 and 2002 and more recent phases of mortgage housing (Gardner 2015). Initiated by the Gauteng Provincial Department of Housing, Bram Fischerville's eroded gravel roads and insufficient stormwater management demonstrates significant differences between the province and the city on the appropriate quality of infrastructure (Charlton, 2013). More than twenty years after its establishment, some social facilities remain undeveloped, yet the

12.7 Main house and backyard units in Bram Fischerville. Photo by Margot Rubin, 2013.

location is in demand as a place of residence, with rents generally higher than elsewhere in Soweto (Gardner 2015). Rent is a significant form of income for individual households: on average one in every four houses has a backyard unit accommodating either new residents or family members splitting off from original resident households, and some houses have as many as ten backyard units. By 2011 about 16 per cent of the population were living in informal-material shacks (Gardner 2015). A range of home-based businesses can also be found throughout the neighbourhood.

Gardner (2015) claims there was strict control by the City of Johannesburg to prevent the construction of backyard dwellings in the early years, echoing national government policy. In addition, the local councillor Gardner cites refers to the "ANC" not allowing backyard shacks, although it is not clear if this refers to the party or the ANC government. However, the councillor also noted that both of these forms of oversight, "political and town planning control," were relaxed after 2009, and there is seemingly "no control" happening now.

Reference to local implementation of the wishes of the ANC is echoed in the neighbourhoods of Lehae and Orange Farm, some distance to the south of Bram Fischerville (Charlton 2013). Originating as an informal settlement, Orange Farm grew rapidly in the 2000s (Murray 2008) and has since received considerable infrastructure, social facility, and housing investment from government in response to its poverty-stricken character. Despite significant investment, city officials have also seen the area as having a "below the radar" dimension (Charlton 2013, 300), as inadequate policing services apparently enable migrants without all necessary documentation to gain a foothold, including through accessing rental accommodation. While many backyard dwellings were evident in 2010, people interviewed in Orange Farm's Extension 6 reported that they and their neighbours were in the process of taking down shacks made of precarious materials to comply with an agreement they had made amongst themselves as residents (Charlton 2013). An interviewee cited in Charlton (2013, 319) noted that they understood this to be the wishes of leadership: "[Former ANC and South African president] Thabo Mbeki wants the shacks [taken] down." This community found more formally constructed backyard dwellings acceptable, however.

Similar anti-shack sentiments were apparent amongst residents of Lehae, a much newer neighbourhood on farmland south of Soweto some twenty-five kilometres from central Johannesburg planned to accommodate approximately 10,000 sites (City of Johannesburg 2008b). An interviewee cited in Charlton (2013, 318) noted that shacks are not wanted in the new housing development by either government or community members: "We don't want the backyard rooms." Another interviewee observed that "other ... communities are allowed to have backyard shacks" while they in Lehae were not, though her prediction was that in due course people would just start building rooms anyway (cited in Charlton 2013, 319).

(3) Government-Approved Backyarding: Alexandra, Orlando, and Zola

Our third form of response to backyarding is the pilot schemes by the Gauteng Provincial Department of Housing (now the Department of Human Settlements) to improve and regulate, or indeed create, backyard dwellings. These initiatives view backyarding as part of the housing environment, and perhaps even desirable, but needing to conform to standards both in terms of material quality and the relationship between landlords and tenants.

The K206 or Extension 9 project in the East Bank area of Alexandra (Alex) was a pioneering initiative to build an external room for rent along with each house, in recognition that rental activity is key to the

Alex economy. Alex is an old township now very centrally located a few kilometres from the upmarket business hub of Sandton. Established in 1912 prior to the withdrawal of Black people's land rights through the 1913 Land Act (Johannesburg Development Agency 2014), Alex is unusual in that Black residents had freehold tenure until the 1960s when these rights were revoked (City of Johannesburg 2012b). Over the decades the area became increasingly congested as demand for housing grew. Of the approximately 90,000 residential structures in the wider Alex area in 2001, 52,000 or 57 per cent were backyard dwellings (Johannesburg Development Agency 2014). In 2001 major infrastructure investment and built environment improvements were announced, including a significant amount of new government-subsidized housing on adjacent land, among other things to de-densify the incredibly crowded township. Self-built backyard dwellings that soon started to emerge in the newly developed areas were frowned upon by Alex Renewal Project (ARP) staff as undermining improvement initiatives (Shapurjee and Charlton 2013).

The K206 new-build housing project acknowledged that subletting and outside rooming was inevitable, and with appropriate design could achieve a higher density and more efficient use of very well-located land, as well as also offer "economic opportunity to those in need" (Mokonyane 2007). As the deputy director of the ARP wrote in a 2016 letter cited by Ratau (2017, 32), "K206 ... replicated what already took place within a majority of yards in Old Alexandra ... [with] one landlord owning multiple rentals within their yard." Two rental rooms with a shared bathroom were built along with each house, intended also "to answer the complex question of how to house people who do not qualify for a housing subsidy" (Mokonyane 2007), a problem faced when attempts to de-congest "old Alex" ran into existing residents who were not eligible for the government housing grant.

However, this experimental project encountered problems when authorities selected and moved tenants into the rental rooms in order to hurriedly free up land for development in advance of the 2010 World Cup, apparently "without permission or knowledge" of the beneficiary of the primary dwelling (Cox 2012). Tenants reportedly had a range of grievances and were unwilling to pay rent to the primary owner; as Wilson Makoba, a tenant cited in Cox's article (2012, n.p.) said, "These so-called landlords got their houses for free, so why should we pay rent when we were forcibly removed?" Landlords in turn proposed evictions of tenants or illegally evicted tenants, resulting in a tense and dysfunctional private rental situation. At the same time, more precarious shack-like rooms emerged in the development, despite beneficiary education and training to the contrary. An ARP Community Liaison Officer, cited by Ratau (2017, 49), said: "Clear instruction was given about the prohibition of the construction of shacks. Extending your house is allowed, however no shack is allowed." A K206 landlord with a shack structure explains how in her case the two issues in K206 are connected: "I had to extend my house because we were promised a main house and rentals but ... our rentals were given out to other residents." In other cases, shacks have been built to accommodate family members (cited in Ratau 2017, 41).

A similar experiment has been tried by the Gauteng Provincial Government in partnership with Emfuleni Local Municipality in the Golden Gardens project, adjacent to Johannesburg on its southern border. Golden Gardens was planned for 666 houses, each with "equally beautiful backyards units which can be used for rental purposes" (Gaba, n.d.). This reflects an awareness of circular and temporary migration, as one goal is that "some people who are in this region for work and educational purposes only [can] ... have some decent accommodation." However, Gauteng Department of Human Settlement officials report difficulties: some tension between landlords and tenants, confusion in the state itself over the source of funding for the backyard units and whether this constituted double subsidization of the owner-beneficiary, and that the expectation that tenants would come from a particular hostel has been thwarted by landlords who have put in family members instead (Provincial Housing Official 2, personal communication with Margot Rubin, 1 February 2017).

The Backyard Pilot Project in the old suburbs of Orlando and Zola in Soweto provides a different example of the Gauteng Provincial Government's efforts to directly intervene in backyarding. This project aimed to replace informally constructed backyard structures with rooms compliant with building regulations and by-laws (Rubin and Gardner 2013). Two or three rooms and a shared ablution facility per yard were funded and constructed by the provincial Department of Housing (figure 12.8). By 2013 between 2,000 and 3,000 back rooms had been constructed. However, this was far fewer than the original number of backyard dwellings, resulting in the program "dislodging more households than it accommodated" (Rubin and Gardner 2013, 22). There were further reasons why a number of the original tenants did not benefit from the new improved accommodation: landlords

12.8 State-upgraded housing in Orlando. Photo by Margot Rubin, 2013.

wanting equivalent income from the better-quality but less numerous new improved units raised rents; some landlords put family members in the new units rather than tenants; and, against policy, some units were used as businesses or shops instead of accommodation. The consequence was the displacement of an estimated 70,000 people, sentiments of dissatisfaction from backyarders who felt that landlords had benefitted twice from the state, and reduced housing provision.

Explanations and Conclusions

Three types of responses have been described in the examples above: control of permitted forms of backyarding by local authorities in partnership with private developers, community self-management echoing political and government wishes, and direct provision of backyard structures and regulation of landlord-tenant relationships.

In the "control" cases, all three settlements discussed show that private developers have been key in the response to backyard developments, educating, training, and advising homeowners and providing a local monitoring and facilitation service. Yet they are dependent on municipal authorities such as City building control units to issue fines and demolish units. In all cases, the private sector's involvement is limited, and after a period the urban management gets handed over to the municipality. In the case of Cosmo City this has meant that there is now a very reduced and uneven response to units that contravene the City's by-laws.

In the second type of response, community self-regulation in Bram Fischerville, Orange Farm, and Lehae indicates residents' desire for dignified and decent suburban living (Charlton 2018) and improved shelter circumstances for all. Less clear is the extent to which their opposition to informal-looking construction is a "bottom-up" initiative – articulations of community desires – or, alternatively, a case of central government or ANC party ideas which have filtered down to ground level (Charlton 2013). Residents' efforts seem fragile, however: in Bram Fischerville informal shacks have become widespread despite these earlier attempts, while at the same time it appears that the state is largely absent in this aspect of urban management.

The third form of response shows a more directly interventionist state experimenting with delivering compliant backyard units, yet struggling to find the mechanisms to do this effectively. Despite difficulties, this suggests a receptive and adaptive state exploring ways to respond to a de facto situation. Having run into difficulties both in direct provision and in attempting to formalize tenancy arrangements, the Gauteng provincial government's backyarding improvements have now shifted to improving underlying infrastructure. Yet efforts are limited and lack coherence, directed at backyarding in old townships with little evidence of proactive up-sizing of infrastructure capacity in new housing developments (Provincial Housing Official, personal communication with Margot Rubin, 1 March 2017).

Currently, there seems to be an overwhelmingly laissez-faire approach to backyarding across the city: few plans for intervention, limited urban management, and little if any building control. Evident in the examples above is a change over time as well as space in response to backyarding, and a differential response between local government and the provincial authorities. The Gauteng Provincial Department of Housing's response has evolved over the years, from no recognition of backyarding and its potential role in settlements in the 1990s to some direct supply in the first decades of twenty-first century, to a retreat from the complexities of this into a focus on infrastructure capacity expansion in older townships. Despite their involvement in establishing new housing settlements, there a sense that there is "nothing that binds them to the project[s]" post-delivery, viewing ongoing management as the responsibility of the local municipality (Provincial Housing Official, personal communication with Margot Rubin, 1 March 2017). Proactive preparation for backyarding in forthcoming developments through expanded infrastructure provision is also not evident. Further, provincial officials report that the once-energetic interaction with the National Department of Human Settlements' proposed backyarding guidelines has gone silent.

Tracking the official stance of the City of Johannesburg reveals a change of attitude and thinking over time, but very little of the official policy has been reflected in practice. Beginning about 2011–12 with the Sustainable Human Settlements Urbanisation Plan (SHSUP), there was "a plan advocat[ing] a conceptual shift towards supporting and facilitating backyard rental stock" (City of Johannesburg 2012a, n.p.). SHSUP planned for the inclusion and improvement of existing backyard stock in an incremental manner, as well as careful planning in new settlements through "designs / capacity of bulk infrastructure to anticipate the additional rental component." SHSUP never became official policy, however, and was largely ignored. Once again, backyards were overlooked and not seen as a central concern until a few years later: in 2015 Ros Greef, member of the mayoral committee responsible for development planning, said that "our policy ... advocates a conceptual shift towards supporting and facilitating backyard rental stock" (The Star 2015, n.p.). This was followed by requests for proposals from innovators to formalize and legalize structures to meet specifications set out by the town planning division, so that they would comply with building regulations, generate rental income, and add to the value of the property (Dlamini 2016, n.p.). However, this project has not made much progress, and it appears almost nothing has been done, although the issue is still a matter of concern for the land use management department, and also attracts attention from private financiers of housing. A CoJ housing official has noted that backyarding no longer features in city internal discussions from either of the two potential lead departments, planning or human settlements, and no budget has been allocated to backyarding (CoJ Housing Official, personal communication with the authors, 3 March 2017).

Overall, the different responses reflect particular moments in state thinking, starting with a lack of acceptance of backyarding and thus pursuing eradication and formalization, and ending up in a kind of pragmatic acceptance and official support for compliant forms, but in practice showing little recent on-the-ground engagement with the phenomenon. This endpoint seems to be related to changes in the national policy milieu, an acceptance of the persistence of backyarding and thus its adoption and incorporation, and also the limited success of the interventions to date. Added to this has been both passive and active resistance by communities protesting against demolition, or continuing to build despite fines and demolitions. These resistances, within the right to housing framework and delivery programme embraced by many in South Africa, have not been the kind of rights-claiming, organized mobilizations Caldeira (2017) sees in many resident-developed settlements elsewhere, but rather a pragmatic push-back at impractical constraints, stimulating in turn more pragmatic accommodations by the authorities. At the local level, a lack of urban management capacity, increased numbers of houses and housing settlements, and reduced budget has meant that council has few tools to manage infringement. Thus, the current de facto "policy" and on-the-ground response of acceptance and adoption seems to be a case of policy bricolage: a status made up of changed attitude, a sense that intervention is difficult and costly, and a lack of resources. Such bricolage holds a great deal of hope, especially for peripheral settlements, in that it allows for the adoption of current practice and community needs, while recognizing the limitations of the state. Yet it does call for government infrastructure support and other guidance to ameliorate potentially poor living conditions. Such an approach begins to demonstrate an evolving state and suitable interventions distilled from the best parts of local everyday practice and official expertise and enforcement. Significantly, rather than the progression from irregularity and illegality that Caldeira (2017) emphasizes through the engagements of many actors consolidating self-made settlements over time, this discussion flags that irregularity and illegality may be periodic or episodic conditions in city formation, as the limits of formally built and idealized spaces in conditions of poverty are tested, transgressed, and, eventually, adjusted.

NOTES

1 For Caldeira (2017, 5), the concept of peripheral urbanization "does not involve spaces already made that can be consumed as finished products

before they are even inhabited," as in the South African situation. Her focus is on people-driven settlements, although she sees state-provided housing as significantly impacting city formation in areas previously characterized by autoconstruction, such as Chile.

2 According to the Gauteng City Region Observatory; see http://www.gcro.ac.za/about/the-gauteng-city-region/.

3 Named after the anti-apartheid defence lawyer in the Rivonia treason trial of the early 1960s.

REFERENCES

Ball, S.J. (1998). "Big Policies/Small World: An Introduction to International Perspectives in Education Policy." *Comparative Education* 34, no. 2, 119–30. https://doi.org/10.1080/03050069828225

Bank, L. (2011). *Home Spaces Street Styles*. Johannesburg, South Africa: Wits University Press.

Bayat, A. (2010). *Life as Politics: How Ordinary People Change the Middle East*. Palo Alto, CA: Stanford University Press.

BusinessTech. (2018). "If Gauteng Was a Country Its GDP Would Be the Seventh Largest in Africa." *BusinessTech*, 20 April 2018. https://businesstech.co.za/news/business/239347/if-gauteng-was-a-country-its-gdp-would-be-the-seventh-largest-in-africa/

Caldeira, T. (2017). "Peripheral Urbanization: Autoconstruction, Transversal Logics, and Politics in Cities of the Global South." *Environment and Planning D: Society and Space* 35, no. 1, 3–20. https://doi.org/10.1177/0263775816658479

Charlton, S. (2018). "Confounded but Complacent: Accounting for How the State Sees Responses to Its Housing Intervention in Johannesburg." *Journal of Development Studies* 54, no. 12, 2168–85. https://doi.org/10.1080/00220388.2018.1460465

– (2013). "State Ambitions and Peoples' Practices: An Exploration of RDP Housing in Johannesburg." PhD diss., University of Sheffield. http://etheses.whiterose.ac.uk/7412/listing/c/sarahcharltonwitsacza/#sthash.oSlHY48V.dpuf

Charlton, S., and Kihato, C. (2006). "Reaching the Poor? An Analysis of the Influences on the Evolution of South Africa's Housing Programme." In Pillay, U., Tomlinson, R., and du Toit, J. (eds), *Democracy and Delivery: Urban Policy in South Africa*, 252–82. Cape Town, South Africa: HSRC Press.

Charlton, S., Gardner, D., and Rubin, M. (2014). "Post-intervention Analysis: The Evolution of Housing Projects into Sustainable Human Settlements." In *From Housing to Human Settlements: Evolving Perspectives*, 75–94. Braamfontein, South Africa: South African Cities Network.

City of Johannesburg. (2008). "Cosmo City is a Thriving Suburb." http://www.cosmocityfunrun.co.za/cosmo-city-joburg/

– (2008b). "Lehae Takes Mixed Income Route." Accessed 16 March 2013. http://www.joburg.org.za/index.php?option=com_content&task=view&id=2863&Itemid=198

– (2010). "Fresh Look at Housing Plans." 1 December 2010. Accessed 16 March 2013. http://www.joburg.org.za/index.php?option=com_content&view=article&id=5979:fresh-look-at-housing-plans&catid=123:housing&Itemid=204

– (2012a). "Sustainable Urban Human Settlements Urbanisation Plan (SHSUP)." Presentation to Human and Social Mayoral Sub-Committee, 3 August 2012, City of Johannesburg Development Planning.

– (2012b). "Alex Renewal Makes Progress." 23 February. http://www.urbanlandmark.org.za/downloads/clipping_jhb_feb2012.pdf

Cosmo City Chronicle. (2015). "Back Rooms Creeping into Bonded Houses." *Cosmo City Chronicle*, 12 August. https://www.facebook.com/Cosmocitychronicle/posts/984645624932416:0

Cox, A. (2012). "Whose Home Is It Anyway?" *The Star*, 2 August. http://www.iol.co.za/the-star/whose-home-is-it-anyway-1354623

Dlamini, P. (2016). "Turning Shacks into Assets." *The Times* (Johannesburg), 29 February 2016.

Ekers, M., Hamel, P., and Keil, R. (2012). "Governing Suburbia: Modalities and Mechanisms of Suburban Governance." *Regional Studies* 46, no. 3, 405–22. https://doi.org/10.1080/00343404.2012.658036

Gaba, S. (n.d.). Massive housing development goes ahead in Golden Gardens. Emfuleni Local Municipality. http://www.emfuleni.gov.za/index.php/emfuleni-news/515-massive-housing-development-goes-ahead-in-golden-gardens.html

Gardner, D. (2015). "Bram Fischerville." In Todes, A., Harrison, P., and Weakley, D. (eds), *Resilient Densification: Four Studies from Johannesburg*, 38–70. Johannesburg, South Africa: University of the Witwatersrand, Gauteng City-Region Observatory.

Gardner, D., and Rubin, M. (2016). "The 'Other Half' of the Backlog: (Re) Considering the Role of Backyarding in South Africa." In Cirolia, L., Görgens, T., van Donk, M., Smit, W., and Drimie, S. (eds), *Upgrading Informal Settlements in South Africa: A Partnership Based Approach* (77–94). Cape Town, South Africa.

Ghannam, F. (2002). *Remaking the Modern: Space, Relocation, and the Politics of Identity in a Global Cairo*. Berkeley: University of California Press.

Harris, R. (2010). "Meaningful Types in a World of Suburbs." In Clapson, M., and Hutchison, R. (eds), *Suburbanisation in Global Society*, 15–47. Bingley, UK: Emerald.

Johannesburg Development Agency. (2014). "Alexandra Renewal Project Improving Residents' Lives. 27 June. http://www.jda.org.za/index.php/latest-news/news-2014/151-june/1544-alexandra-renewal-project-improving-residents-lives

Klijn, E.-H. (2008). "Governance and Governance Networks in Europe." *Public Management Review* 10, no. 4, 505–25. https://doi.org/10.1080/14719030802263954

Kennedy, L. (2007). "Regional Industrial Policies Driving Peri-Urban Dynamics in Hyderabad, India." *Cities* 24, no. 2, 95–109. https://doi.org/10.1016/j.cities.2006.06.001

Kotzen, B., and Suttner, S. (2018). "Infrastructures of Transformation: A Contextual Study of Housing in Cosmo City." In Harrison, P., and Rubin, M. (eds), *Urban Innovations: Researching and Documenting Innovative Responses to Urban Pressures*. Pretoria, South Africa: Department of Planning, Monitoring and Evaluation.

Kriegler, A, and Shaw, M. (2016). "Comfortably Cosmopolitan? How Patterns of 'Social Cohesion' Vary with Crime and Fear." *SA Crime Quarterly* 55, 61–71. https://dx.doi.org/10.17159/2413-3108/2016/v0n55a46

Lategan, L.G. (2012). "A Study of the Current South African Housing Environment with Specific Reference to Possible Alternative Approaches to Improve Living Conditions." MA thesis, North-West University, South Africa.

Lemanski, C. (2009). "Augmented Informality: South Africa's Backyard Dwellings as a By-Product of Formal Housing Policies." *Habitat International* 33, no. 4, 472–84. https://doi.org/10.1016/j.habitatint.2009.03.002

Mabin, A., Butcher, S., and Bloch, R. (2013). "Peripheries, Suburbanisms, and Change in Sub-Saharan African Cities." *Social Dynamics* 39, no. 2, 167–90. https://doi.org/10.1080/02533952.2013.796124

Mabuza, K. (2006). "Pulling the Chain on More Shacks." *Sowetan Live*, 13 November.

Mokonyane, N. (2007). Budget speech by Gauteng MEC for Housing Nomvula Mokonyane for the 2007/08 Financial Year. Delivered on 15 June. http://polity.org.za/article/mokonyane-gauteng-housing-prov-budget-vote-200708-15062007-2007-06-15

Mosselson, A. (2015). "'Johannesburg Has Its Own Momentum': Towards a Vernacular Theorisation of Urban Change" Paper presented at the RC21

International Conference, Urbino (Italy) 27–9 August. http://www.rc21.org/en/wp-content/uploads/2014/12/I2-mosselson.pdf
Murray, M. (2008). *Taming the Disorderly City: The Spatial Landscape of Johannesburg after Apartheid*. Cape Town, South Africa: UCT Press.
Ndlazi, S. (2016). "Backyard Dwellers See Red over Rent." *Independent Online*, 19 May. http://www.iol.co.za/news/south-africa/gauteng/backyard-dwellers-see-red-over-rent-2023521
Ramothwala, P. (2015). "Fix Backyard Units – Salga." *The New Age (KwaZulu-Natal)*, 5 June.
Ratau, F. (2017). "The Spatial Impacts of the Alexandra Renewable Project (Arp) on the Management of Housing Density: The K206 Housing Development." Undergraduate research report, University of the Witwatersrand, Johannesburg. Available from the authors.
Robins, S. (2002). "Planning 'Suburban Bliss' in Joe Slovo Park, Cape Town." *Africa: Journal of the International African Institute* 72, no. 4, 511–48. https://doi.org/10.3366/afr.2002.72.4.511
Rubin, M. (2018). "At the Borderlands of Informal Practice of the State: Negotiability, Porosity, and Exceptionality." *Journal of Development Studies* 54, no. 12, 2227–42. https://doi.org/10.1080/00220388.2018.1460466
Rubin, M., and Gardner, D. (2013). *Developing a Response to Backyarding for SALGA*. Final Report, The South African Local Government Association (SALGA). http://pmg-assets.s3-website-eu-west-1.amazonaws.com/150602BackyardingFinal.pdf
Shapurjee, Y., and Charlton, S. (2013). "Transforming South Africa's Low-Income Housing Projects through Backyard Dwellings: Intersections with Households and the State in Alexandra, Johannesburg." *Journal of Housing and the Built Environment* 28, no. 4, 653–66. https://doi.org/10.1007/s10901-013-9350-9
Sisulu, L. (2016). Minister of Human Settlements Budget Speech & Responses by DA. Briefing, 3 May. https://pmg.org.za/briefing/22463/
South African Government. (2016). "Human Settlements Pocket Guide, 2015/2016." https://www.gcis.gov.za/content/resourcecentre/sa-info/pocket-guide/pocket-guide-south-africa-20152016
Stone, D., 2017. "Understanding the Transfer of Policy Failure: Bricolage, Experimentalism and Translation." *Policy & Politics* 45, no. 1, 55–70. https://doi.org/10.1332/030557316X14748914098041
The Star. (2015). "Add Value to Your Property by Putting Up a Backyard Shack." *The Star*, 20 February 2015. http://www.iolproperty.co.za/roller/news/entry/add_value_to_your_property

Tipple, G. (2000). *Extending Themselves: User-Initiated Transformations of Government-Built Housing in Developing Countries*. Liverpool, UK: Liverpool University Press.

Turok, I., and Borel-Saladin, J. (2016). "Backyard Shacks, Informality, and the Urban Housing Crisis in South Africa: Stopgap or Prototype Solution?" *Housing Studies* 31, no. 4, 384–409. https://doi.org/10.1080/02673037.2015.1091921

Wariawa, A. (2014). "Investigating the 'Regulation' of Economic Activities in Mohlakeng Extension 7." Planning Honours Research Report 2014, Wits University, Johannesburg, South Africa.

Watson, V. (2009). "Strategic Literature Assessment for Informal Rental Research Project." Report to the Social Housing Foundation, Johannesburg, South Africa, March.

Wilder, M., and Howlett, M. (2014). "The Politics of Policy Anomalies: Bricolage and the Hermeneutics of Paradigms." *Critical Policy Studies* 8, no. 2, 183–202. https://doi.org/abs/10.1080/19460171.2014.901175

Zweig, P.J. (2015). "Everyday Hazards and Vulnerabilities amongst Backyard Dwellers: A Case Study of Vredendal North, Matzikama Municipality, South Africa." *Jàmbá: Journal of Disaster Risk Studies* 7, no. 1. https://doi.org/10.4102/jamba.v7i1.210

13 From Informal Settlements to Harmonious Communities: Professional Squatters and the Many Actors of Urbanization in Metro Manila

ABIDEMI COKER

Much of the literature on cities in the Global South of today leans towards eclectic and deep descriptive analyses with perspectives on the specific conditions that bring personality to these cities. What were once descriptions of lack and chaos are now, more and more, considerations of vibrancy and connections. This chapter looks at the megacity of Manila in the Philippines through an open lens, in order to understand the processes and actors engaged in the urban housing sector, particularly in the provision of low-cost housing for the "poorest of the poor." The poorest here are squatters, identified as persons or families living in informal settlements, where they have built illegally on public or privately-owned land or rented units from landlords who cannot provide security of tenure. In the Philippines, such families are officially called "informal settler families." The examples of housing provision outlined here are a synthesis of the multitude of social programs for relocating poor families from danger zones and other precarious alcoves of metropolitan Manila.

As in many cities in the Global South, the politics surrounding land and its distribution can have severe impacts on the implementation of wide-ranging urban policies. The repercussions on the poor pose challenges so significant that a system of constant negotiation of housing rights has been established, whereby communities must defend or pursue their rights while government and non-governmental actors attempt to reconfigure housing delivery mechanisms. The result in Manila has been the development of several housing programs for the poorest city inhabitants, yet challenges remain for those who live at the peripheries of the city, in terms not only of physical locality but also of access to infrastructure and information.

This chapter discusses the factors that promote or impede access to housing by examining the relationships between the various actors and the expectations of the beneficiaries of the programs. The overall purpose of the research was to identify the ways in which families in Manila organize themselves or are organized by other actors to gain access to housing. One of the discoveries during the fieldwork, which was conducted in informal communities located in danger zones as well as newly relocated communities in Manila, is that poor urban residents could qualify for various housing programs from government and NGOs. The condition was that families in informal areas come together and form associations, through which they could apply to be beneficiaries of the housing programs. Furthermore, the findings highlight professional squatters (families with arguably more income than others in the area) as important – though controversial – members of the communities, in terms of formalized housing access. Discussion of professional squatters appears mainly in national housing policy[1] and indicates the need for socio-political analysis, starting with identifying who professional squatters may be and how they relate to the communities and the housing programs. These complex dynamics in communities that are not often geographically peripheral, but are certainly socially marginal, create both a challenge and an opportunity in the city.

The Poverty of Informal Settler Families

Squatters or informal settler families are persons living in informal settlements and other unregulated areas, or on illegally occupied land. Informal settlers are often seen by NGOs and government agencies in the Philippines as underprivileged persons who are often informally employed and represent the poorest 30 per cent of the urban population. However, as Neuwirth (2005) recognized, some squatters are formally employed, and thus informality cannot be seen as a defining characteristic. In Manila, I found that some squatters I interviewed were active in trained roles such as teaching and cooking and had spouses or family members who worked in registered companies or government agencies. Therefore, in my work I look at informal settlers as "very low-income families who do not have access or resources, or are unwilling to access formal housing for which they would have tenure security, and who live illegally in unregulated areas or structures without [adequate]

13.1 Seaside homes in Ulingan, Manila, considered a danger zone. Photo by Abidemi Coker.

amenities such as potable water, sanitation, drainage, paved roads and waste management" (Coker 2016, 89).

Informal settler families living in danger zones, for example on the edge of waterways (including the sea), along rail tracks or under bridges, face exceptionally precarious conditions, including extreme weather conditions, lack of security, fire, pollution, and other daily hazards. Although some families have resided in specific communities for decades, the conditions are just as bad in these long-established informal areas, such as Ulingan, where some families engage in the essential economic activity of charcoal production, which results in an increase in health issues from air pollution.

Other squatter settlements are distributed around the fringes of the city, sometimes in small pockets such as in abandoned buildings or along the heavily polluted Pasay River and its tributaries. The city of Manila, part of the National Capital Region that forms Metro Manila, has the largest percentage of poverty in the country, at 11.9 per cent in 2015. There, many families live along the Manila bay where they are vulnerable to regular weather events such as typhoons. Many of the

13.2 Games and charcoal production in Ulingan, Manila. Photo by Abidemi Coker.

areas in Manila where the poorest live have been designated as danger zones, due to their proximity to the sea. According to the Borgen Project (Quigley 2018), there are 526 such impoverished communities scattered around the cities and municipalities that comprise Metro Manila – with 2.5 million people living in these conditions.

Despite economic growth in the Philippines of 5.9 per cent in 2015 and 6.8 per cent in 2016 (Asian Development Bank 2018), poverty and informality have remained constant. The Philippine population was estimated by the United Nations at over 103 million in June 2017. Metro Manila, which serves as the capital region and is comprised of sixteen cities, had 12.8 per cent of the national population according to the 2015 national census, and is the most densely populated administrative region in the country. With exorbitant land prices and minimal space for housing development in one of the world's most densely populated cities, many families simply cannot afford to rent formally and are forced to occupy open areas, even those in danger zones. The stark reality is that many poor families reside in close proximity to the gated communities of upper-middle-income and high-income families. Remaining in the city due to work and social ties appears to be important for poor families, and perhaps more so for chronically poor families, who have faced poverty for extended periods of time or throughout their lifetimes. Such families have in the past been least affected by development interventions (Hulme, Moore, and Shepherd 2001) – but things have changed over the past two decades. Despite, or perhaps in tandem with, the challenges of governance and the lackluster interest in low-income housing supply from the formal market, several NGOs have sprung up alongside government housing initiatives to improve living conditions in the city.

The objective of low-income housing provision for the "poorest of the poor" in Manila is to relocate families living in danger zones into safe and secure homes within or outside the metropolitan area. However, the means of doing so are varied, depending on the type of organization coordinating and funding a particular program. Government housing programs that target poor families in particular, such as the Community Mortgage Program (CMP) and, especially, projects by the National Housing Authority (NHA), tend to move families outside the city, and their processes can take several years to be completed. The programs of these government agencies also require small – but at times arguably unaffordable – payments for the housing. Programs run by NGOs and urban poor groups in Manila were more likely to provide housing within the city and at minimal or no cost, as the number of families in each project was relatively small. Also, through advocacy, sponsorships, and campaigns, modest homes were built in shorter periods of time than the government programs. Such programs have been carried out by the well-known NGO Gawad Kalinga, which operates nationwide, as well as smaller groups such as St Hannibal Empowerment Center (SHEC) and Urban Poor Associates (UPA). These organizations have been able to rehouse families within the districts in which they originally resided.

Community Housing Associations and the Challenges of Local Governance

The actors providing low-income housing programs in Manila do so at various scales, from housing a few families at a time to settling hundreds of families in new housing areas. The geography of new housing is also different in the various schemes, as large government housing programs tend to move families outside the metropolitan area, while NGOs often acquire small pockets of land to relocate families within Manila. Inasmuch as there are numerous housing programs, the efforts of well-meaning stakeholders in the housing scene were found to be duplicated and to follow very complex processes. What ties together these two types of housing provision processes is the intent to house the poorest and the focus on participation as an essential element. It is important to note that participation here refers mainly to practical arrangements that enable the project to take shape. The need for participation is to ensure that beneficiaries are genuine squatters – not professional squatters – and to improve ownership of the project during and following relocation, which is perceived as a measure to help prevent families returning to squatting. The housing delivery mechanisms require several layers of governance institutions and partnerships.

Housing associations are important features of the housing scene due to the requirement for informal settlers interested in the programs mentioned above to join a community association. It is through these associations that residents apply for housing programs; some programs run by NGOs, on the other hand, request beneficiaries to create an association following their application, since the squatters may come from various parts of the district or sub-cities of the metropolitan area. Whereas previously the linkages between communities and local government or larger government institutions were weak, due to an emphasis on poverty reduction since the 1990s, some direct engagements have been seen, for example in the programs run by the CMP and NHA that allow interaction between the two actors around identifying the community's needs and the government's plans for the community. By positioning housing institutions closer to local government and to low-income areas in the city, state organizations imitate initiatives of civil society (Mitlin and Satterthwaite 2004).

This linkage between community and state does not mean that their relationship is fluid. Rather, due to challenges, NGOs and urban poor groups have retained their roles as mediators and coordinators for state housing projects, as they are known to have better relations with and understanding of the communities. Linkages between the communities and the NGO potentially enhance the communities' access to local government, which is most relevant where relocation is desired within the local government area. Curiously, despite the presence of NHA or CMP offices in certain local government areas, the government's relocation plans were mostly to areas outside the city, and getting consensus from community associations has proven all but straightforward. This may also be due to the designation of available land, if any is available for housing purposes. In addition, as the local government serves mainly as a landholder tasked with maintaining master lists of residents, it can be sidestepped during other stages of the housing process as it is not involved in developing the participatory processes and is minimally, if at all, involved the planning or actual construction of housing.

Although NGOs (for example, Habitat for Humanity Philippines) have had minimal roles in building new houses, they have had supportive functions in rebuilding following disaster, or in providing technical expertise and harnessing manpower for the construction of homes. The core connection between the international NGOs, local NGOs, and community is in their call for participatory processes, which hinges on the former Millennium Development Goals. In addition, governments have been tied to decentralization policies that call for programs to alleviate poverty by reducing vulnerability and improving assets, while engaging people through various participatory approaches that bring the government closer to the people (Post and Baud 2002; Mitlin 2003).

In principle, decentralization means local government is given greater responsibility to manage and implement the Philippines' Urban Development and Housing Act (UDHA). However, meeting this responsibility is a challenging feat for local government, as they tend to be allocated insufficient resources, and in poorer areas of the city the tax base is too small for local government to supply housing and services without the concerted inputs of other governance agencies. Many local government units (LGUs) lack the financial and technical resources to engage in housing provision. Hence, the role of local government in housing delivery is stunted, limited to creating and maintaining master lists of residents, supplying amenities such as water, and, when possible, providing small plots of land for housing in their areas. According to Ballesteros (2009), LGUs often lack the political will to undertake their roles, when in fact shelter programming and planning has become more localized.

On the other hand, with NGOs stepping in as mediators with government, as well as providing housing themselves, they have become notable actors in housing provision for the poorest. The successes of the NGOs are based on their relations with the community members, local government, and private supporters, and in the case of SHEC and Gawad Kalinga, land was sourced through donations. Thus, the role of the NGOs discussed here cannot be undermined; they have a wealth of knowledge about the processes and politics of housing programs, including those of the state, and influence on public authorities to allow for families to be rehoused within Manila. What is perhaps most impressive about the NGOs' housing efforts lies in the process, as all the NGOs included in the research followed participatory methods known as "values formation" and "sweat equity." These participatory methods were developed to improve the morale as well as the skills of the families prior to, during, and following relocation, and to enable the families to be a "community" in their new homes. Values formation and sweat equity are discussed in the following section.

Participation in Housing Provision

The process of housing provision is depicted in table 13.1. The nine stages of the housing process depicted here are presented from the perspective of the funding or implementing organization, and highlight the various roles and activities of the key stakeholders.

Although the table begins with an identification of families to participate, it is important to note that some form of participation occurs in various forms in most stages of the process, even prior to and during identification – especially in NGO-led projects. The NGOs Gawad Kalinga and St Hannibal Empowerment Center (SHEC) involved families throughout the process, by providing training, workshops, and couples counselling to educate and "empower" the families to become weavers of their own destiny. Sweat equity is a component of values formation in which families contribute hours of labour to the physical construction of the building.

The idea, supported by evidence from Gawad Kalinga, is that through values formation the families get to know each other well and develop good camaraderie, as well as pride in their contributions, so that once they move into their new homes they will feel bonded and have a well-functioning community. Also, the objective is to help families transition "from trainee to trainer," to develop the leadership qualities of community members. The NGOs were particularly concerned with the ways the community would function to maintain their housing and environment and to transform the families' mindsets and improve their circumstances. Social preparation is much the same, but is a lighter model of participation as it mostly occurs before and during construction, mainly to engage families in the process so that values are improved and ownership attained, and to identify professional squatters and ensure they are not included in the program. In the NHA and

13.1 The nine stages of the housing provision process

Stage	Description
1. Identification of families/receiving of applications	Families apply through the local government or are identified through interaction with the NGO's staff, who have been immersed in the community for some time.
2. Checking of master lists	Local government checks applicant names against a master list of residents of the area to ensure they have lived in the area for a long enough period of time and have not availed of another social housing programme. This stage can take time, as practiced by the NHA for example, as the names are checked against lists of beneficiaries in various programmes, and it is a process used by both state and non-state housing programmes.
3. Formation of community association	For the NHA, an association has to be registered with the Housing and Land Use Regulatory Board. Associations availing of NGO housing identify themselves and their members and board to the NGO in question.
4. Identification of needs	The community and partners discuss their needs, negotiate about options (for example, cost), and the association is informed about the process and the area identified for relocation, are encouraged to pay dues, etc. NGOs are known to invite the local government into meetings at this stage to have them better understand their residents and to legitimize the process and improve relations for current and future projects.
5. Identification and purchasing of land	Land has often already been identified prior to the commencement of the process, but it is at this stage that the challenge of having the association accept the land becomes evident, and issues to do with professional squatters affect the process, sometimes tremendously.
6. Agreement on housing plans	For many programs, housing plans need to be approved by the housing association to legitimize it and give ownership of the process to the families involved. This is in order to minimize problems following construction and relocation.
7. Social preparation and values formation	Social preparation in the form of meetings with NHA district officers occurs throughout the process, but appears to intensify as the process advances. Values formation involves training, meetings, and couples counselling throughout the process.
8. Commencement of building	This stage involves construction on site with sweat equity as the physical and social contribution of the families to be relocated.
9. Completion, relocation, and values formation	The building construction is finalized, families are moved, and values formation occurs throughout this process, including after families have moved into the new houses. Post-relocation values formation occurs in NGO programs only. Once families are relocated, the local government is tasked with demolishing and clearing former informal settlements.

CMP programs, NGOs tend to be present either due to the communities' needs, in the case of the former, or the requirement of a "mobilizer," in the case of the latter. NGOs and their participatory techniques thus feature to some degree when low income housing is being developed, even by the state.

There are some differences in the ways the housing process works for state agencies and non-state organizations. Participatory approaches in housing provision have been used not only by NGOs but also by in government housing schemes. The backbone of participatory housing mechanisms is for families who qualify to organize themselves or to join an existing community association or an association applying for a project run by a particular NGO (the latter is more common with NGOs than with government programs). These associations are formalized, with elected officers including presidents and treasurers tasked with collecting dues to contribute to the program, or negotiating and liaising on behalf of the community. For some NGOs, such as SHEC, the dues paid over a period of time were seen as a show of commitment from the families, and those dues were used in purchasing land and building materials, or hiring skilled labor. Gawad Kalinga, however, did not expect participants to pay for any part of the housing, as their programmes are based on donations from the local government, politicians, and business leaders. This ability of an NGO to provide housing for free indicates their access to considerable funds and ability to negotiate land prices when necessary. The increased links between Gawad Kalinga and local government and other powerful actors relates to the assessment by Porio et al. (2004) that NGOs have a significant role in the decentralization of housing, in such a way that the NGO becomes a mediator between community associations and government authorities.

Whereas the intended impact of values formation is to create communities that are secure, clean, and harmonious, social preparation does not lean towards "empowerment." Instead, social preparation was developed as a means of attaining a consensus of families and to facilitate a smoother housing process. However, I argue in my dissertation (Coker 2016) that social preparation is to some extent an imitation of values formation, based on the results of some government housing programs following which relocated families returned to the city to squat. The notion of values formation is, however, very specific to housing organizations in Manila, although the goal of empowerment is rooted in NGO and poverty discourse in general. The concept of empowerment may be contested, but generally refers to the development of people's abilities to make strategic life choices (Kabeer 2001) or to have the freedom to take action to shape their lives (Narrayan 2002).

Those families in areas proclaimed as danger zones and in the process of relocation who did not want to participate in housing associations faced eviction. This exclusionary practice was tied to the principle that all squatters want a decent living space, and those who did not want to relocate were suspected of being professional squatters. An analysis of professional squatters and their influence on the housing provision process is discussed in the following section.

The outcomes of participation can extend outside the housing project, such as in the case of Ulingan, where the community association was able to collect enough money, in addition to the assistance provided by the city councillor, to bring water to their community. Over time, the association was able to pay back to the councillor the amount of the down payment made on their behalf. Although the process of savings and repayment took five years, it was an important improvement for the community and showed the relevance of savings. The community association in Ulingan continued to save, as the money collected also helped members in times of personal crises, such as a death in the family.

The Challenge of Professional Squatters

Professional squatters have appeared to be a major concern in Philippine housing policy, although less so in academic texts, since the 1990s. Yet this group can be difficult to pinpoint in a given community. Generally speaking, professional squatters are individuals or families who live in very poor informal settlements, but who may not be regarded as the "poorest of the poor." In my interviews with NGOs and state housing actors, professional squatters were discussed as families who live in danger zones, for example, but have several poorly built houses in the area that they rent out, while their own houses are made from concrete and are air-conditioned. Erhard Berner (1997, 210) was one of the first scholars to mention professional squatters in reference to the Philippines' 1992 Urban Development and Housing Act (UDHA), which defines professional squatters as "slum dwellers with income above the poverty line." The Act itself states:

> "Professional squatters" are individuals or groups who occupy lands without the express consent of the landowner and who have sufficient income for legitimate housing. The term shall also apply to persons who

> have previously been awarded homelots or housing units by the Government but who sold, leased or transferred the same to settle illegally in the same place or in another urban area, and non-bona fide occupants and intruders of lands reserved for socialized housing. The term shall not apply to individuals or groups who simply rent land and housing from professional squatters or squatting syndicates. (Republic Act No. 7279, 1992)

The presence of professional squatters in legislation and their regular mention by all housing providers interviewed for my study shows the propensity of their impact on informal housing and housing provision in Manila. In the strictest sense, professional squatters are landlords who have no formal ownership of the land on which they have built rental housing, therefore offering no tenure security and little or no amenities. One can suppose that these illegal landlords pay no taxes on their "properties" or from income from the rentals, thus retaining the informality of such housing services.

Pertinent to the housing programs discussed here, professional squatters cause confusion and hinder values formation and social preparation by causing conflict in the community housing associations, absconding from meetings, refusing proposed housing plans, or changing their positions at important stages of the process. The NHA views such behaviour by so-called incalcitrants as deplorable, and implements measures requesting the community themselves to be vigilant and out suspected professional squatters. The assumption is that the incentive for whistle-blowing is the reduction of delays and other complications (Lee Kuan Yew School of Public Policy 2012). However, where the suspected professional squatter is a neighbour or landlord, or a leader in the community, the socio-political dynamics in the community may make it challenging for community members to out their peers. Shared history and feelings of solidarity between community members who have lived together for years, even decades, cannot be underestimated; and as Gilchrist (2009) discusses, strong feelings of community are often most apparent when people are threatened. In addition, issues of security and safety may hinder willingness to identify another person in the community as a professional squatter if that person is known to be a gang member or affiliated with a gang. The enforcement of anti-professional squatting measures has therefore been weak, as coordination is insufficient and

13.3 Mixed housing in Ulingan, Manila. Photo by Abidemi Coker.

it is not clear who whistle-blowers should report findings to or what happens when they do.

The UDHA definition, as has been used by other scholars, including Shatkin (2007) and Ballesteros (2009), does not take into account the nature of informality itself, whereby incomes may fluctuate and people may be able to access small plots informally, which they then rent out. As an example of this complexity, one informal settler in Ulingan had informally bought the house of her neighbour, who had moved out, and opened a small "sari-sari" store behind which she rents out a room. Such entrepreneurial families could also be labelled as professional squatters under the official definition provided above. However, it has been found that income generation is a foremost preoccupation of the urban poor (Grant 2004), and investment in housing allows for spaces for home-based work. This is especially challenging during housing projects, because the presence of supposed professional squatters can elongate the process. From this perspective, it is necessary to consider in what ways some families that fit the definition of professional squatters actually contribute to urbanization in general by providing housing as a commodity for low-income families who would otherwise have no place to live.

Adding to that, squatting cannot be seen as a short-term housing solution, and hence the networks and shared histories mentioned earlier may mean that some squatters have invested not only financially but also socially in relationships and community associations that have value for the wider community. As Berner (1997) describes, the value of kinship within localities allows for the formation of organized groups and collective action, which can, essentially, contribute positively to the housing process presented here. Professional squatters, if seen as potential leaders who have not previously availed of low-income housing, can be instrumental members of the community. The challenge of identifying "real" professional squatters and allowing genuine community members to access housing lies in the ability of the local government, NGOs, and regulatory authorities to coordinate themselves and prevent syndicates from taking housing meant for poor families. It would thus be important to redefine professional squatters and develop more robust means of fishing them out, especially in long-standing communities where these families or syndicates may have ties to the local government and other actors. The politics of major players in the context of decentralization and rapid urbanization are profound in the megacity, where the struggle over land remains complex.

Peripheral Urbanization in Manila

The empirical case of Manila presented here highlights two areas of tension common in urbanization in the South. One revolves around the responsibilities and relationships between powerful actors, which hinge on the struggle over land. It is the lack of land – or the unaffordable expense of it – that pushes poor families into informal settlements at the peripheries of the city. It is also that unaffordability that engages NGOs to take on the urban challenge of providing adequate housing for the poor. The example of NHA housing owners who move back to the city shows that low income housing outside the city is undesirable to informal settlers – people do not want to live far away, for various reasons, particularly livelihood. So even though families may be marginalized living in informal settlements, moving farther from the city increases the social and economic marginalization of the poor.

While NGOs are looking to provide decent urban spaces for the poor within the city, they are also helping poor families claim a stake in inclusive urban processes, which is justified when considering that families have lived in the city for many years, even decades, and poverty

13.4 New housing community in Pasay, Manila. Photo by Abidemi Coker.

alone should not be the deciding factor in where they should live. As the long-winded government housing schemes that relocate families to the outskirts of the city have proven, many poor families are unable to settle for tenured housing when their livelihoods become distant or require longer travel distances and increased transportation costs. The NGO model thus shows more promise as a reasonable way of reducing marginalization, although the true impacts of values formation have not been ascertained. However, considering the scale of urbanization, increasing land prices, and congestion, meeting housing needs requires continued innovative approaches that take into account the other needs of the urban poor.

The second area of tension concerning urbanization in the South is the sustained notion that the poor need to be organized and need to participate. Whereas urbanization and suburbanization of middle- and higher-income groups demand individual financial ability to participate and freedom to choose based on one's income, low-income housing developments are organized "for" the poor rather than "by" the poor in Manila. The requirement of participation has potential benefits during and following the housing process in terms of ensuring a degree of ownership of the process. However, housing successes in Manila have not been easy – the politics and powers at play have hindered civil society participation (Shatkin 2007). Thus, the participatory methods developed by local NGOs that have been included – though to a lesser degree – in government housing programs points to a shift in the way informal settlers and urban governance are being approached.

The chapter has highlighted initiatives to house informal settlers through participatory processes in an effort to make the programs sustainable. What it demonstrates, in relation to the rest of this book, is that peripheral urbanization already exists in Manila, as neighbourhoods or communities that are located in danger zones are socially marginal. It is this marginality that makes them peripheral, as their formal linkages with the city are limited even when they provide services and goods of economic value to the wider society. Ulingan is a case in point as the charcoal-producing community provides an important commodity to the city, but one that is specialized and not recognized as such. Ulingan residents do not reap the benefits of a more structured urban economy and have no significant returns over the long term. This is an unfortunate aspect of informality and the informal economy, as despite their contribution to overall city life, their socio-economic lives remain isolated or minimal in relation to the rest of the city.

Thus, it is not useful merely to look at communities at the periphery as suburbs or new subdivisions, as this would require engaging mostly with formalized built environments, when in fact cities in developing countries may exhibit peripheral qualities within the city proper. It is more useful, in the case of Manila, to look at the socio-economic flows, attachments to the city, and histories of informal settlements and new residences created through the housing processes discussed here as part of the city and its overall evolution. Maybe then the contributions of the urban poor will be better appreciated and their needs incorporated more seriously in urban development.

NOTE

1 Relevant acts include Urban Development and Housing Act of 1992 (Republic Act No. 7279); National Urban Development and Housing Framework (NUDHF), 1993-1998; Socialized and Low-Cost Housing Loan Restructuring and Condonation Act of 2008 (Republic Act No. 9507); Comprehensive and Integrated Shelter Financing Act of 1994 (Republic Act No. 7835). See http://www.hudcc.gov.ph/content/laws-and-executive-issuances

REFERENCES

Asian Development Bank. (2018). "Philippines: Economy." https://www.adb.org/countries/philippines/economy

Ballesteros, M.M. (2009). *Housing Policy for the Poor: Revisiting UDHA and CISFA*. Philippine Institute for Development Studies Policy Notes, 2009-04. https://dirp4.pids.gov.ph/ris/pn/pidspn0904.pdf

Berner, E. (1997). *Defending a Place in the City: Localities and the Struggle for Urban Land in Metro Manila*. Manila, The Philippines: Ateneo de Manila University Press.

Coker, A. (2016). "Negotiating Informal Housing in Metro Manila: Forging Communities through Participation." PhD diss., Jyväskylä University.

Gilchrist, A. (2009). *The Well-Connected Community: A Networking Approach to Community Development*. 2nd ed. Bristol, UK: Policy Press.

Grant, U. (2004). "Surviving the City: Livelihoods and Linkages of the Urban Poor." In Devas, N. (ed.), *Urban Governance, Voice, and Poverty in the Developing World*, 53–67. London, UK: Earthscan.

Hulme, D., Moore, K., and Shepherd, A. (2001). *Chronic Poverty: Meanings and Analytical Frameworks*. Chronic Poverty Research Centre, Working Paper No. 2. http://dx.doi.org/10.2139/ssrn.1754546

Kabeer, N. (2001). "Resources, Agency, Achievements." In *Discussing Women's Empowerment: Theory and Practice, Sida Studies* No. 3, 17–57. https://www.sida.se/contentassets/5e45d330e16743179cefc93de34e71ac/discussing-womens-empowerment---theory-and-practice_1626.pdf

Lee Kuan Yew School of Public Policy. (2012). *Manila's Poor: Bridging Service Gaps and Strengthening Mental Resilience*. Asian Trends Monitoring Bulletin 17. https://issuu.com/nuslkyschool/docs/atm-17

Mitlin, D. (2003). "Addressing Urban Poverty through Strengthening Assets." *Habitat International* 27, no. 3, 393–406. https://doi.org/10.1016/s0197-3975(02)00066-8

Mitlin, D., and Satterthwaite, D. (2004). "Addressing Deprivations in Urban Areas." In Mitlin, D., and Satterthwaite, D. (eds), *Empowering Squatter Citizen: Local Government, Civil Society, and Urban Poverty Reduction*, 244–77. London, UK: Earthscan.

Narrayan, D. (ed.) (2002). *Measuring Empowerment: Cross-Disciplinary Perspectives*. Washington, DC: The World Bank.

Neuwirth, R. (2005). *Shadow Cities: A Billion Squatters, a New Urban World*. New York, NY: Routledge.

Porio, E., with Crisol, C.S., Magno, N.F., Cid, D., and Paul, E.N. (2004). "The Community Mortgage Programme: An Innovative Social Housing Programme in the Philippines and Its Outcomes." In Mitlin, D., and Satterthwaite, D. (eds), *Empowering Squatter Citizen: Local Government, Civil Society, and Urban Poverty Reduction*, 54–81. London, UK: Earthscan.

Post, J., and Baud, I. (2002). "Evolving Views in Urban and Regional Development Debates in Africa, Asia, and Latin America: Introducing the Key Themes." In Baud, I., and Post, J. (eds), *Realigning Actors in an Urbanizing World: Governance and Institutions from a Development Perspective*, 1–23. Farnham, UK: Ashgate.

Quigley, A. (2018). "10 Important and Little-Known Facts about Poverty in Manila." *The Borgen Project*, 6 May 2018. https://borgenproject.org/10-facts-about-poverty-in-manila/

Shatkin, G. (2007). *Collective Action and Urban Poverty Alleviation: Community Organizations and the Struggle for Shelter in Manila*. Farnham, UK: Ashgate.

14 Suburbanisms of Ethnocracy: Building New Peripheries in Israel/Palestine

ODED HAAS

State-led urban development aimed at specific ethnic or religious groups is an enduring strategy in the production of Israel's peripheries and in the Zionist colonization of Israel/Palestine as a whole. Grounded in the combined "conquest of labour" and "conquest of land" ideology of early twentieth-century pragmatic Zionism, modern suburbs continue to establish Jewish control over national territory. This chapter discusses the planning of and resistance to a future "Arab city" in the western Galilee region of northern Israel and considers some of the risks and possibilities the new suburban development carries for Palestinian communities. Unlike other chapters in this collection, the space examined here has yet to be produced outside of planning documents and the contending imaginaries of the state and local inhabitants. Therefore, the chapter relies, among other sources, on conversations with local inhabitants as well as planners involved in the project. The focus is on the battle of Palestinian citizens of Israel (PCI) against the plan of the new suburb, Tantour, which unites opposition to urban development through anti-colonial struggle. It thus reveals how a neoliberal "solution" to a "housing crisis," selling state land for developing private housing, is situated within continuous marginalization and oppression of Palestinians in Israel. This chapter suggests an "ethnic logic of space" as explanation for how both the housing crisis and the solution for it are conceived by the state and challenged by local inhabitants.

First, the chapter reviews some key moments in the making of the plan for Tantour and in its relationship to the existing local Palestinian community, wherefrom the main resistance to the project originates. The case study is then situated within the historical and ideological context of an ethnic logic of space, which refers to the explanatory power of Lefebvre's ([1974] 1991) *Production of Space* to elucidate the marginalization of an "Arab sector" in Israel to a socio-spatial periphery under an "ethnocratic" regime. Finally, by focusing on a local activist group, resistance to the new suburb and its position as a new "Arab city" is understood as a struggle for the right to the city: for appropriation of space; for the production of space by inhabitants as an alternative to state-led urban development; and therefore – in this particular context – as an anti-colonial struggle.

Still, since the capitalist production of space is a dialectical process in which space is produced and reproduced simultaneously by the state and by inhabitants, the process always also includes the emergence of new possibilities for resistance (Lefebvre [1974] 1991). Therefore, by looking at a local struggle against a new housing development in the periphery, this chapter aims to unravel the dynamics of current suburbanization as an ideological process. Exploring the ways in which the state may manipulate a "housing crisis" for (attempted) de-politicization of the periphery helps to interpret a wider process of suburbanization, and at the same time consider new manifestations of a particular ongoing spatial practice in the current worldwide urban condition.[1]

"We Want the Revolution": Plagues, Plans, and Planners

Since the 1970s, the State of Israel has sanctioned several plans to be made for Tantour hill in the Galilee region, in the outskirts of ancient Acre in northern Israel. Only in 2014 did the government approve the establishment of "a new urban zone for the non-Jewish population" in the area as a means to increase housing supply for Palestinians in the Galilee.[2] The goal of the new "Arab city" is to offer a modern, planned alternative to the marginalized, underserviced Palestinian localities that have been systematically neglected by the state in general, and by its planning authorities in particular. The National Master Plan 44 is the main planning document outlining the project, and it lays out approximately 10,000 housing units to be built along with public spaces, main commercial streets, and an industrial area. Its planners perceived it originally as an extension to the adjacent Palestinian town of Judeida-Makr, adding to it much needed open public spaces, commercial activity, and employment opportunities that would improve overall living conditions in the poor, neglected community, and connect it better to other communities in northern Israel. However, as the plan moved forwards and upwards in the convoluted Israeli central planning mechanism, it was altered, reduced, and reshaped to an almost exclusively residential suburb, deliberately and bluntly disconnected from the existing town.

National Master Plan 44 is yet another stage in a local history of top-down decisions imposed on the Palestinian town of Judeida-Makr, described by local activists as "plagues" inflicted upon their community (J and F, local activists, personal communications with author, 9 September 2016 and 15 September 2016, respectively). Now amounting to about 20,000 residents, Judeida and Makr were originally two separate villages whose story illustrates the history of Palestinian localities in Israel. In the 1948 Nakba – the catastrophe of over 700,000 Palestinians who fled or were expelled from their homes during the Israeli War of Independence – both villages surrendered in order to survive. In the aftermath of the Nakba, internal refugees (Palestinians who were displaced within the new State of Israel) arrived in the area and settled among the landowners, creating a new landless class in these 700-year-old communities of Muslim and Christian Arabs. In the early 1970s, the state built public housing estates in the village of Makr, intended for Palestinian families from the nearby Arab-Jewish "mixed" city of Acre. This initiative that supplied housing for families in need was in fact part of a broader strategy to secure a Jewish majority in "mixed" cities; in this case, by housing some of the poorest Palestinian residents in an exclusively Palestinian community, thus not only ethnically segregating the population but at the same time marginalizing Palestinian communities in comparison to neighbouring Jewish ones.

These public housing estates, the only modern, planned neighbourhoods in Makr, quickly became the poorest, most crime-ridden parts of the village. In the coming decades, the villages of Judeida and Makr became a mosaic of spatially differentiated populations, where different groups typically remain in their own neighbourhoods. Later, in 1976, the state confiscated lands from the two villages as part of a general scheme to block the development of Palestinian communities. And in 1984, Judeida and Makr were officially consolidated into one municipality without consulting local residents, thus exacerbating rifts within the community. Nation-wide systemic issues that reflect the state's approach towards Palestinian communities, such as extremely insufficient infrastructure, reduced public services, and oppressive planning policies, along with the particular social complexity of Judeida-Makr render the town one of the poorest in the country. Similarly, to many other Palestinian communities in Israel, the town suffers from high rates of unemployment and acute shortage in social services and usable public spaces. Consequently, residents have become completely disillusioned with the municipal government and its inability to cope with the severe lack of resources. In fact, municipal elections in Judeida-Makr have turned into an extremely violent arena for contention between different groups of residents, perpetuating fear among inhabitants from challenging local government and its cooperation with the state's planning authorities (J, local activist, personal communication with author, 9 September 2016).

In the midst of this everyday reality in the Galilee, the declared purpose of the government's decision to build a new "Arab city" was to provide a housing solution specifically for the Palestinian "middle-class" in the region. But in order to attract a highly educated, well-off population, the planning authorities decided the project must be planned as a separate urban area, disconnected from Judeida-Makr and its bad reputation; this, despite the planners' professional recommendation to integrate the new plan into the existing urban fabric. Moreover, National Master Plan 44 is devised to facilitate semi-private compounds, in terms of urban design and lot configuration. The planners' intention was to mitigate private market reality in the traditional Palestinian society that traditionally builds family compounds on private land that expand as the family grows. If Plan 44 were to be executed, groups of families from the same village could purchase adjoined lots and create semi-private neighbourhoods in the new "Arab city." The vision was to allow an incremental transition to a modern, urban setting. However, as the planning process progressed, the state pushed for higher density and for the selling of land to multiple private developers, which would hinder such design sensitivity. In addition, while Plan 44 is pending approval by various levels of government, the project has been granted legal status under the National Committee for Planning and Construction of Special Housing Areas (Ministry of the Interior 2018): a government body that was legislated in 2014 as a tool for coping with the national housing crisis. Once the government declares certain state land as a priority area for housing, the committee promotes plans for that area on a fast track that is essentially cleared from public scrutiny. This alternative route intended for promoting massive construction of housing units, which are mainly sold in the private market, is simplified by eliminating the hierarchy of planning procedures – local, regional, and national – each of which involves feedback from the general public as well as from various social and environmental organizations. Room for opposition is thus significantly diminished.

Including Tantour in a Priority Plan for Housing (Ministry of the Interior 2018) enabled the government to sanction a new plan for the construction of 25,000 housing units (two and a half times the number in Plan 44). According to architect Amir Kolker, a senior partner in the firm that holds the contract to design this plan, the new "Arab city" would house approximately 120,000 residents (approximately six times the current population of Judeida-Makr) and its area would span not only the lands originally meant for establishing Tantour but also the existing town of Judeida-Makr itself and additional lands surrounding it (A. Kolker, personal communication with author, 27 September 2016). It will have extensive parks, a large industrial zone to service the whole region, several commercial streets, and multiple public and civic buildings. The street layout is based on a modern orthogonal grid pattern that would encompass the old villages; and the planners envision it will facilitate the continuity of urban space between old and new neighbourhoods and between religious or family groups that will organize to purchase parcels of lots together. The grid was chosen for its "enabling" character that ostensibly does not impose certain spatiality on the community, and as a spatial form that would also counter the negative stigma of the local community. The planned density of sixteen housing units per dunam in multi-storey buildings, as in many large Israeli cities, and an expected average housing price as high as in some predominantly Jewish urban areas in the region, will make Tantour undeniably different from existing Palestinian communities in Israel.

The architects and planners working on the project of Tantour approach it with a sense of mission: devising a plan that will fulfil a growing demand among the "highly educated Palestinian middle-class" for a vibrant urban environment. While such a demand may be evident in the sacrifice that some Palestinian families are willing to make – albeit out of lack of choice and housing options – in moving into Jewish or "mixed" cities where they suffer from racial discrimination in exchange for urban life, no input from the local population has been sought in shaping this planning vision. On the one hand, this plan determines the expansion of the existing town rather than building a separate suburb, which could potentially better the living conditions of local residents. On the other hand, current residents who experience housing precariousness will not be able to afford moving to the new project and could be displaced as their town transforms into a desirable location. Professor Rassem Khamaisi, a Palestinian geographer and planner who has researched urbanization processes among PCI and who has been deeply involved in the planning process for the new "Arab city," says that "some people may want to cling to a small village mentality, but the urbanization process of the 'Arab sector' cannot be stopped" (R. Khamaisi, personal communication with author, 11 September 2016). Architect Kolker adds, "The local residents want to keep building private homes, but we [the planners] want a revolution!" (Personal communication with author, 27 September 2016). This statement, while offering an opportunity for modern, planned space for the "Arab sector," suggests that the potential to stop the ongoing oppression of local residents is in their own hands and depends on their willingness to change their housing habits. Along with the urban character of the plan itself, it implies that the project of Tantour is not meant to alleviate local housing distress, but rather to provide housing options for a particular type of Palestinian population – one that can afford and is willing to transition from traditional village life to a model of homeownership that is detached from land ownership, and thus away from disrupting the "Judaization of the Galilee" and into the housing market, as explained in the following parts of the chapter.

Urbanization in/of the Galilee: An Ethnic Logic of Space

The planners' claim that the new "Arab city" will be "revolutionary" must be situated within the particular ideology that engendered both the project of Tantour and resistance to it. Jabareen (2014a) refers specifically to the "urban" in Lefebvre's ([1974] 1991) theory of the production of space as a useful explanatory framework for the question of Palestinian communities in Israel. The "urban" is a mediating level between state power and everyday life, and urbanization is accordingly the non-neutral, capitalist production of commodified space for controlled consumption that is segregating and marginalizing differences for creating a clear spatial order (Lefebvre [1970] 2003). Hence, spatial segregation is imperative in urbanization (Kipfer 2011), and in that sense, the Palestinian struggle for national liberation is part of global uneven development (Hanieh 2013). In addition, the state is at the core of Lefebvre's spatialization of capitalism, since it has the power to produce its own space of national territory – state-space – out of material space, through uneven development (Brenner and Elden 2009). Colonization is then understood not as a metaphor for colonial relations, but rather as a particular statist form of organizing hierarchical territorial relations that facilitate political domination (Kipfer and Goonewardena 2013).

In Israel's settler-colonial society, state-space is produced by an ethnocratic regime according to ethnonational ideology – in this case, Zionist aspiration to a nation-state – and through an "ethnic logic of capital": an "ethnoclass" system that determines the distribution of political and economic power among different ethnic groups with different levels of inclusion in the nation. The settler group of Ashkenazi Jews is the dominant ethnoclass, while indigenous Palestinians are the most oppressed (Yiftachel 2006).[3] Accordingly, the Zionist colonization of Israel/Palestine as a whole can be referred to as the "Judaization of space," which embodies the production of Israeli national territory as the space of the Jewish state.[4] Judaization of space consists of establishing Jewish presence throughout the national territory, alienating Palestinians from the landscape, and depriving the landscape its Palestinian identity (Tzfadia and Yacobi 2011; Hanafi 2009). This is achieved through various spatial strategies such as an ethnicized legal land system, ethnically biased urban development agendas, and appropriation of vernacular architecture, all manifesting an ethnic logic of space.

Judaization of space began with Zionist settlement in Palestine through the ideology of "conquest of labour" and "conquest of land" that advocated purchasing lands, working the land, and forming an all-Jewish working class as a means for building a modern Jewish society in the Land of Israel. It evolved to an ethnically bifurcated economy in the State of Israel and a spatialized "Arab sector" (Hever 2012; Peled and Shafir 2002). Palestinian communities are systematically marginalized; their territorial jurisdiction and urban development are limited in order to prevent non-Jewish control over land; and since the majority of PCI are employed by Jewish employers but effectively excluded from the Jewish housing market they must commute daily into Jewish urban spaces (Jabareen 2014b; Hever 2012; Yiftachel 2006; Peled and Shafir 2002; Falah 1989). The social, economic, and geographical peripheralization of PCI along with blatant ethnic discrimination has essentially, although not formally, created a separate housing market with extremely limited housing options. Consequently, the state is able to define a "housing crisis" within the "Arab sector" and to imagine an "Arab city" as a solution for it.

Significantly, the periphery has a crucial role in Judaization of space. In the production of space, the socio-spatial centre and the periphery produce each other through a constant struggle over the power to organize social space and meaning, and over the ideology of urbanization (Lefebvre [1970] 2003). Following the ethnic logic of space, the state disperses ethnoclass enclaves across the national territory, thus fixing separations in space; at the same time, Jewish communities in the geographical peripheries are rendered guards of the national territory, and hence marginalized groups among them can gain centrality by inhabiting the periphery. Peripheral communities are regarded front-line defenders of the landscape from Arab takeover, defeaters of wilderness, and albeit cynically, the reason for waging war when their communities are attacked (Tzfadia and Yacobi 2011).[5] Therefore, the periphery is central for reproducing the Jewish state as such. And, indeed, since the state was founded it has been building new cities for certain groups in the periphery; from attempted forced urbanization of Bedouin communities in the 1960s to establishing Jewish-Orthodox suburbs for increasing Jewish presence next to the Green Line in the 1990s. Specifically, in the Galilee region, suburbanization has been highly ethnically contested. In the 1970s, the state confiscated Palestinian lands in areas where Jewish majority was at risk, to facilitate the construction of new exclusively Jewish communities. These suburban developments were designed as an intentional sprawl, blocking Palestinian-controlled territorial continuity (Falah, 1989). However, since the periphery is produced and reproduced through an ongoing struggle over the power to produce space, it is also the space where resistance is generated (Lefebvre [1970] 2003). The response to the massive land confiscation in the Galilee region was a protest against dispossession and for recognition in Palestinian nationality, which peaked on 30 March 1976 as the Day of Land general strike and mass demonstrations in the "Arab sector." It has since been marked by annual demonstrations in Palestinian cities against Judaization of space and for Palestinian liberation. Nevertheless, in 1979–80 the state introduced an official plan, popularly referred to as "Judaization of the Galilee" (Sofer 1992: 24), to establish exclusively Jewish gated communities, Mitzpim (in Hebrew: outlooks), which were located strategically to prevent Palestinian towns from spreading out on the Galilee landscape and creating territorial continuity. This Day of Land protests marked the beginning of annual demonstrations nation-wide but especially in the Galilee against Israel's discriminatory land system and the spatial injustice of confiscating Palestinian lands and turning them to state land. It also marked a shift in the state's response towards PCI protest to a militarized police control. Along with the Mitzpim in the Galilee and with calls from the far right for a population transfer, these practices signify a "creeping apartheid": practices that were developed for dispossession and alienation of Palestinians from their lands in the

Occupied Territories since 1967 have been penetrating Israel (Hanafi 2009; Yiftachel 2006; see also Agnoti 2013).

Revolution would therefore require undermining the Judaization of the Galilee as part of the Zionist colonization of Israel/Palestine as a whole. On the contrary, Tantour is planned on lands confiscated from Palestinians mainly from Judeida-Makr, where most residents will not be able to afford living in the new city. Moreover, architect Eran Mebel, who is the leading planner of National Master Plan 44, describes how the plan takes its inspiration explicitly from the Jewish suburbs in the Galilee, following his admittedly surprising realization that "everyone wants to live in the same kind of space, regardless of their ethnicity" (E. Mebel, personal communication with author, 7 September 2016). Meaning, although the plan for Tantour may express sensitivity to some local issues such as multi-purpose public spaces, attention to pedestrian circulation, and so on, the core concept of the new residential suburb references the very spaces that were produced as intentional Jewish sprawl to block the spread of Palestinian communities in the region. Furthermore, in order to localize the new city in the landscape, the planners incorporated vernacular architectural elements such as arches and inner courtyards in their renderings of the detailed design. This has become common in Jewish spaces in the Galilee, especially the Mitzpim, since the 1980s. Architecture that references local tectonic identity represents cultural attachment to the landscape, to the place that is the Land of Israel (Eretz Israel), whereof Zionist ideology derives historical legitimacy for Jewish self-determination in Palestine. Such elements become essential in authenticating Judaization of space and thereby concurrently alienated from their capacity to symbolize Palestinian nationalism and incorporated in Israeli state-space (see Yacobi 2008). In this respect, as the first "Arab city" to be built since 1948, the project of Tantour seemingly marks a shift towards legitimizing Palestinian presence in Israeli state-space. For the first time, a state-led large-scale urban development is targeting specifically the Palestinian population, promising better, modern infrastructure, housing and urban life. Nonetheless, such inclusion in state-space is conditioned on continued ethnic segregation and controlled presence through state-sanctioned land grabbing and spatialization of the "Arab sector" as separated. The project of Tantour then becomes instrumental in reproducing settler-colonial relations in Israel/Palestine.

However, once we recognize space as produced, it becomes a site for political struggle. And so, it happened that despite – and precisely because of – the project's goal to increase (landless) private homeownership among PCI, the project instigated resistance among the local community. Local activists relate their struggle against the new suburb to the broader Palestinian struggle against ongoing Zionist colonization through "conquest of land," thus indicating that the new "Arab city" may be revolutionary not by imposed modernization of Palestinian society nor by increased participation in the housing market but rather by opening a new site for a struggle against capitalist production of politically dominated spaces. Their struggle is a political (ethno-class) struggle: a struggle for decolonization.

From Solution to Resistance

"Pioneers of urbanization" are expected to emerge out of the "Arab sector," in the form of highly educated middle-class families that will become homeowners in Tantour (A. Kolker, personal communication with author, 27 September 2016). The "Arab city" is then supposed to resume a Palestinian urbanization process that was halted in 1948 by the Nakba. This solution to the housing crisis assumes the main problem of spatialized Palestinian society in Israel – marginalized through the ethnic logic of space – is lack of a viable housing market. The objective of the new suburb is therefore to increase housing supply, facilitate more housing choice for Palestinians, who are effectively excluded from Jewish spaces, and enable greater participation of the "Arab sector" in the market (A. Kolker, personal communication). A neoliberal approach to housing as a commodity combined with an ethnic logic of space justifies the investment of state resources, most notably the allocation of land, in an "Arab city." In other words, rendering spatial segregation a market solution to a "housing crisis" facilitates enduring ethnic discrimination under the democratic façade of Israeli ethnocracy. However, it is precisely the production of Tantour as an acceptable Palestinian space in Israeli national territory that situates it in conflict with existing Palestinian communities in Israel in general and in the Galilee region in particular. The role of the project in Zionist colonization is revealed by the attempt to urbanize a Palestinian middle-class into state-conceived suburbanization at the expense of local housing needs, thus sustaining power in the hands of the ruling ethnoclass. By increasing private homeownership in the "Arab sector" the state promotes consumerism that hinders the Palestinian national struggle – echoing Israel's strategy to encourage urban development in certain parts of the West Bank (Hanieh 2013).

Resistance to the project therefore came quickly after the government's decision to establish a new city for the "non-Jewish population." In particular, a self-organized group of activists from the Palestinian town Judeida-Makr – the local Khirak (in Arabic: movement or unrest) – has been fighting the plan in multiple arenas, from appearing before planning institutions to demonstrating on the ground. The Khirak started as a grassroots political movement in 2009, when plans for Tantour were only in the early phases of devising what would become the new "Arab city." It was then driven by residents of Judeida-Makr who had claims for the lands that were included in National Master Plan 44, but from its inception the movement took upon itself to fight against corrupt local government. After failing in the 2013 municipal elections, the Khirak was re-established in 2016 but this time refraining from any affiliation to specific contenders in local politics. Replacing short-term goals of public office with long-term opposition to state-led development projects such as highways that would circumference and smother the town, the group of activists gained the faith and trust of local residents (Khirak Group, personal communication with author, 15 September 2016). In its current, more principled form, the Khirak activist group crosses multiple lines of division in the local Palestinian community: it includes religious and secular members; internal refugees alongside land owners from locally deep-rooted families; various political views, from Communist Party supporters to the recently disallowed northern faction of the Islamic movement. Lack of women participation is explained by women's fear of voicing opinions, enforced by the community's own traditional views that preclude women engagement in politics. However, group members claim that many women in the community privately express their support in the group's actions.

Despite its small number of about ten devoted activists, the group is in fact vastly condoned by the local community, as expressed in personal communication and over social media. Evident of the Khirak's popularity is that since they began fighting against National Master Plan 44 they have also become a de-facto shadow municipality in Judeida-Makr, confronting the deficient local government and looking after everyday matters in the name of local citizens who put their trust in the small group. While the Judeida-Makr Khirak occasionally collaborate with similar grassroots organizations in other Palestinian communities in the Galilee, they are the only group who managed to form an alternative to the official local government. Moreover, although some NGOs advocating spatial justice in Israel have been helping the activists to the extent of expressing support in their struggle, the main resources for action remain the time and labour of the activists themselves, who now contemplate lending their experience to other PCI activist groups that focus on the issues of land and development (Khirak Group, personal communication with author, 15 September 2016).

In the eyes of the Khirak, the group is mainly fighting against fear: fear from the corrupted local government and the violence of elected officials; from traditional conservatism that keeps people from engaging in political activity; from being displaced by the new development in Tantour; and, mainly, from expressing opposition to state actions. The activists describe that many people in the community continue reliving the trauma of the Nakba and refrain from resisting state-sanctioned planning so to not be marked as a threat by the state (Khirak Group, personal communication with author, 15 September 2016).[6] Therefore, although the struggle against the project of Tantour is ongoing, the activists claim success in helping the local community overcome some of its fears in supporting the Khirak. From mundane local issues to an anti-colonial struggle, from garbage disposal to reversing land confiscations, the Khirak is fighting for appropriation of space. Indeed, the Khirak makes actual use of the space when its members gather on Tantour hill for social activities and sometimes for discussing resistance.

The Khirak's main argument for rejecting the plan for Tantour is that the state is trying to solve a "sectorial" issue at the expense of taking care of needs of Palestinians already living in that space. Their objection to the new development stems from the plan's goal to attract a Palestinian "middle-class," thus neglecting those whose housing distress is most urgent. The activists point out that no real public participation was incorporated into the planning process, and that the planners mainly consulted with groups they perceived as potential homeowners in Tantour. They argue the resources invested in Tantour should have been used for improving basic living conditions in their extremely poor town. For the Khirak, the political question of the "Arab city" is inseparable from urban design: while they concur with the planners' impression that everyone wants a modern, well-planned space, the high density proposed in Tantour is perceived as forced urbanization and an attack on their right to plan for themselves on their land. In addition, the local residents are sceptical about the project's ability to generate employment and warn that future residents of Tantour will have to continue commuting into large cities or nearby exclusively Jewish communities for work. In that sense, as they articulate it, "the future of the project is already inscribed in the history of Palestinian towns in the Galilee" and it reproduces a spatially segregated "Arab sector" (Khirak Group, personal communication with author, 15 September 2016).

Furthermore, while the activists' long-term aspiration is a decolonized housing market free from ethnic discrimination, they agree that in the immediate future a new "Arab city" would increase housing choice for some. However, if the project is actually successful in attracting young professional Palestinians from all over the Galilee, then, first, it will solve none of the pressing issues that Palestinian communities are coping with; second, it will only deepen the marginalization of other communities that will lose some of their economically stronger population; and third, the process of Palestinian families penetrating Jewish communities – albeit out of lack of housing choices – will be reversed, and hence Judaization of the Galilee prevails. This last point is especially telling of the role of the "Arab city" in the current neoliberal variant of Judaization of space. National aster Plan 44 emulates the Jewish spaces that PCI reportedly "like" (E. Mebel, personal communication with author, 7 September 2016); but with this all-new development, the state is effectively attempting to reverse the "mixing" of such spaces by creating an alternative where middle-class PCI can concentrate in a separate city of their own through self-segregation that is facilitated by the "Arab sector" housing market. This strategy is crucial in light of urban ethnocracy – the articulation of ethnoclass discrimination in urban regimes – that is relatively open, since cities require certain diversity and fluidity in order to function and grow (Yiftachel 2016). Establishing a new "Arab city" potentially assists in minimizing the inevitable tendency of ethnocratic urban space to open up possibilities for conflict and resistance on grounds of liberal citizenship. As an alternative, the Khirak and local Palestinian planners suggest that the state invests in existing Palestinian communities and encourage their development. This would enable a more locally produced urbanization process that could potentially include housing development that is more attuned to inhabitants' needs.

In response, the Jewish and Palestinian planners involved in the project argue that being "stuck" in the trauma of the Nakba prevents PCI from recognizing the benefits of the new "Arab city," and that resisting the project for the sake of resisting Zionist colonization is useless (A. Kolker, personal communications with author, 11 September 2016, 27 September 2016, respectively). Clearly, articulating the problem of marginalized Palestinian communities in Israel as a "housing crisis" warrants an asset-based solution that is focused on increasing Palestinian homeownership rates. However, the state is building the periphery by increasing housing supply within the ethnoclass-controlled market. Tantour is planned on lands that were confiscated from Palestinians and that will be sold to private developers, to be then sold back to Palestinians who can afford to move into the project. The local "housing crisis" is thus manipulated by the state to shape a Palestinian middle-class detached from the land, and "pioneers of urbanization" become agents of colonization.

Conclusion

In the project of Tantour, Israel is investing unprecedented resources in its Palestinian population. However, local resistance unveils the role of the state's asset-based approach to housing in the reproduction of Israeli state-space as well as Zionist colonization of Israel/Palestine as a whole. The new "Arab city" will undoubtedly improve housing conditions for those Palestinian households who can afford it, but at the cost of displacement from the marginalized, stigmatized Palestinian communities that are the socio-spatial peripheries of the Jewish state, the spaces of Palestinian resistance. The state is thus attempting to displace the Palestinian struggle itself from an anti-colonial demand for desegregation to a market-based demand for affordable homeownership in ethnically exclusive suburbs. In that way, suburbanization in the periphery becomes a strategy for depoliticization of the periphery. Certainly, the Israeli planning institutions do not expect to solve all the problems stemming from the deep oppression of the "Arab sector" by building one suburb in the Galilee (S, Palestinian planner from the Galilee, personal communication with author, 23 September 2016). Instead, the case of Tantour suggests that by treating urban development as separate from the Palestinian national struggle, the state is sustaining its ethnic logic of space. Specifically, a neoliberal ideology that seeks to solve a "housing crisis" by increasing the rate of homeownership is utilized as counter-resistance strategy.

Still, the dialectics of the capitalist production of space open "interstices" in the "urban" opportunities for resistance (Lefebvre [1974] 1991). In this case, the rebuilding of the periphery opens a path for opposition to urban development, which could serve as a legitimate route for resistance to colonial subjugation. Indeed, the government's announcement on the Tantour project engendered local protest that now connects everyday matters with a determinedly anti-colonial struggle to participate in the production of space.[7] This sheds light on the state's interest in declaring Tantour a "Special Housing Area," ostensibly for the benefit of the local community but essentially a less democratic planning process.

The struggle of the Khirak activist group implies that current suburbanization processes cannot be disconnected from local and historical particularities. In that sense, the perspective of the production of space reveals, on the one hand, how current colonial relations may be advanced through neoliberal urban development, and on the other hand, how specific ideologies intertwined with capitalism shape urbanization processes. In a time when many capitalist nations devise market-solutions to "housing crises," the inherent contradictions of local political economy elucidate the impact of such strategies on the reproduction of socio-spatial peripheries. In the context of recent claims for Planetary Urbanization, a perspective of constant struggle is crucial for a critical understanding of suburbanization beyond capital accumulation, as a process tangled in local politics. And the more global suburbanisms rebuild peripheries around the world, the more possibilities for resistance emerge.

NOTES

1 Recent writings about Planetary Urbanization argue that the "urban revolution" Lefebvre predicted is now complete, as the world is fully urbanized into a capitalist society through the production of space. Urban/non-urban and other binaries thus become obsolete as analytical categories for explaining the ongoing process of urbanization. See Brenner and Schmid (2015).

2 The wording comes from the instructions of the National Committee for Planning and Construction, which was first published on 3 February 2009. On 4 November 2014 the National Committee gave the first approval to National Master Plan 44, which then moved forward to the next step in the planning process – objections from local planning authorities – where it remained indefinitely. Information on the Plan can be found on the Israeli Ministry of the Interior website at http://mavat.moin.gov.il/MavatPS/Forms/SV9.aspx?tid=91&esid=10

3 According to Yiftachel (2006), the three "ethnoclasses" of Israeli society are the settler group of Ashkenazi Jews; the later immigrants, predominantly Mizrahi Jews (from Arab nations) but also Jews from the former Soviet

Union and Ethiopian Jews; and indigenous Palestinians. Since 1948 there has been an ongoing spatial segregation of the different ethnoclasses. The middle ethnoclass is included in the nation but simultaneously excluded from centres of power and decision-making (e.g., deserving public housing in the 1950s but mainly in peripheral communities), as evident in the dialectical production of socio-spatial periphery referred to in this chapter.

4 The concept of "Judaization of space" refers to the production of the national territory of the Jewish nation (Falah 1989; Tzfadia and Yacobi 2011) rather than to Jewish religion (although religion has an important role in Zionist colonization and the ethnocratic regime). For discussion on "Judaization of space," see also Yiftachel (2009) and Monterescu (2011).

5 For example, Israeli communities located in the country's southern periphery in close proximity to the Gaza Strip are generally neglected and marginalized; however, when they are suffering attacks from Gaza, the government puts them in the spot light of public debate and uses their suffering as excuse to wage war on Gaza (Tzfadia and Yacobi 2011).

6 Khirak activists describe how residents of Judeida-Makr do not distinguish the actions of the Ministry of Construction and Housing – or any other government agency – from the actions of the Shin-Bet, the Israeli security agency. They distrust any state-sanctioned action and are afraid to resist them (Khirak Group, personal communication with author, 15 September 2016).

7 Demonstrations against Tantour included calls such as "This is our land!" alluding to Day of Land protests and emphasizing the land-grabbing that facilitated the establishment of the new "Arab city."

REFERENCES

Angotti, T. (2013). "Orientalist Roots: Palestine and the Israeli Metropolis." In *The New Century of the Metropolis: Urban Enclaves and Orientalism*, 75–94. New York, NY: Routledge.

Brenner, N., and Elden, S. (2009). "Henri Lefebvre on State, Space, Territory." *International Political Sociology* 3, no. 4, 353–77. https://doi.org/10.1111/j.1749-5687.2009.00081.x

Brenner, N., and Schmid, C. (2015). "New Epistemology of the Urban?" *City* 19, no. 2–3, 151–82.

Falah, G. (1989). "Israeli 'Judaization' Policy in Galilee and Its Impact on Local Arab Urbanization." *Political Geography Quarterly* 8, no. 3, 229–53. https://doi.org/10.1016/0260-9827(89)90040-2

Hanafi, S. (2009). "Spacio-Cide: Colonial Politics, Invisibility, and Rezoning in Palestinian Territory." *Contemporary Arab Affairs* 2, no. 1, 106–12. https://doi.org/10.1080/17550910802622645

Hanieh, A. (2013). *Lineages of Revolt: Issues of Contemporary Capitalism in the Middle East*. Chicago, IL: Haymarket Books.

Hever, S. (2012). "Exploitation of Palestinian Labour in Contemporary Zionist Colonialism." *Settler Colonial Studies* 2, no. 1, 124–32. https://doi.org/10.1080/2201473X.2012.10648829

Jabareen, Y. (2014a). "Jaffa-Haifa-Beirut." The Arab City in Israel. Nalaga'at Centre, Jaffa. Lecture. 12 June.

– (2014b). "'The Right to the City' Revisited: Assessing Urban Rights – The Case of Arab Cities in Israel. *Habitat International* 41, 135–41. https://doi.org/10.1016/j.habitatint.2013.06.006

Kipfer, S. (2011). "The Times and Spaces of (De-)colonization: Fanon's Counter-Colonialism, Then and Now." In Gibson, N. (ed.), *Living Fanon: Global Perspectives*, 93–104. Basingstoke, UK: Palgrave Macmillan.

Kipfer, S., and Goonewardena, K. (2013). "Urban Marxism and the Post-colonial Question: Henri Lefebvre and 'Colonization.'" *Historical Materialism* 21, no. 2, 76–116. https://doi.org/10.1163/1569206x-12341297

Lefebvre, H. (1974) 1991. *The Production of Space* (Donald Nicholson-Smith, trans.). Oxford, UK: Cambridge, MA: Blackwell.

– (1970) 2003. *The Urban Revolution*. Minneapolis, MN: University of Minnesota Press.

Ministry of the Interior (2018). *Priority Plan for Housing No. 1058-1059*. http://mavat.moin.gov.il/MavatPS/Forms/SV4.aspx?tid=4

Monterescu, D. (2011). "Estranged Natives and Indigenized Immigrants: A Relational Anthropology of Ethnically Mixed Towns in Israel." *World Development* 39, no. 2, 270–81. https://doi.org/10.1016/j.worlddev.2009.11.027

Peled, Y., and Shafir, G. (2002). "The Frontier within: Palestinians as Third-Class Citizens." In Shafir, G., and Peled, Y., *Being Israeli – The Dynamics of Multiple Citizenship*, 110–37. Cambridge, UK: Cambridge University Press.

Sofer, A. (1992). "The Outlooks in the Galilee – A Decade since Their Establishment." *Karka* 34, 24–9.

Tzfadia, E., and Yacobi, H. (2011). *Rethinking Israeli Space: Periphery and Identity*. New York, NY: Routledge.

Yacobi, H. (2008). "Architecture, Orientalism, and Identity: The Politics of the Israeli-Built Environment." *Israel Studies* 13, no. 1, 94–118. https://doi.org/10.2979/isr.2008.13.1.94

Yiftachel, O. (2006). *Ethnocracy: Land and Identity Politics in Israel/Palestine*. Philadelphia, PA: University of Pennsylvania Press.
– (2009). "Theoretical Notes on 'Gray Cities': The Coming of Urban Apartheid?" *Planning Theory* 8, no. 1, 88–100. https://doi.org/10.1177/1473095208099300
– (2016). "Extending Ethnocracy: Reflections and Suggestions." *Cosmopolitan Civil Societies: An Interdisciplinary Journal* 8, no. 3, 30–7. https://doi.org/10.5130/ccs.v8i3.5272

15 The Making of Cairo's Vast Planned Periphery: Particularities and Parallels Revealed through an Examination of Four Suburban Cultural Assemblages

KARL SCHMID

Metropolitan Cairo has leapt in geographical size since the early 1990s, due to the expansion of informal settlements built on the agricultural periphery of the city and the transformation of the adjacent desert into privatized communities marketed to middle- and upper-class Egyptians. These communities are mostly gated, with aspirational names like Beverley Hills, Uptown, Al Rehab ("Spacious"), Dreamland, and Evergreen. This chapter explores this planned, privatized, and massive suburbanization of Cairo since the 1990s. I follow Aihwa Ong in attempting to see urban transformations through "the complexity of particular engagements rather than subject them to economistic or political reductionism" (Ong 2011, 3). Suburban Cairo was created through a range of forces and actors, and analysis should "open up to the multiplicity of events, interrelationships, and factors that, in ways both chaotic and strategic, expected and unforeseen, are in play in the formation of particular urban environments" (10).

An integrative anthropological perspective is used here to focus in on four cultural assemblages contributing to the production and transformation of Cairo's peripheries. Each of these assemblages encompass interconnected political, economic, and cultural processes. The first assemblage involves the entanglements between the planning and patronage practices of the state and military. The second assemblage centres on the neoliberal paradigm shift, with relevant implications found in the transformation of urban consumption and governance practices. The third assemblage focuses on "inter-referencing" (Ong 2011, 17) carried out by private national and transnational developers as they create speculative and spectacular spaces. A final assemblage is located in the lived experience of these new areas. What is known about the people consuming these peripheral urban lifestyles, and what is their relationship with the rest of the city? Once these urban cultural assemblages have been considered together, the particularities of Cairo's development as well as parallels with regional and global suburbanization can be identified more clearly.

Cairo's recent urban transformation intersects with many dramatic political developments that will be briefly summarized in order to identify key actors and contextualize what follows. For almost all of the nearly six decades between the 1952 and 2011 revolutions, three presidents with military backgrounds ruled over Egypt through the National Democratic Party (NDP): Gamal Abdel Nasser (1956–70), Anwar Sadat (1970–81), and Hosni Mubarak (1981–2011). The most important national urban development projects for all three presidents were the "New Towns" projects. These were large government-planned communities that consisted of freestanding or satellite cities adjacent to Cairo and Alexandria. Their purpose was to redirect population growth away from valuable agricultural land. By 1996, the public projects had failed to create viable economic cities: the combined population of five of Cairo's New Towns had reached only 165,741 residents out of a target population of 1,850,000 residents (Fahmi 2008, 272). From the mid-1990s onward, these desert cities were mostly privatized and repurposed as affluent suburban communities. These, too, have supplied only a small fraction of Cairo's housing needs, leaving informal housing the only option for an estimated 50 to 75 per cent of residents (Mitchell 2002, 287; Singerman 2007, 86). Therefore, most Cairenes live in ashwaiyats or "spontaneous" communities, and Cairo's ashwaiyats include four of the twenty largest slums in the world (Davis 2006, 28).

The ashwaiyats are a crucial part of the political base of the Muslim Brotherhood, a moderate Islamic movement that has been for decades the popular alternative to the government regime. With the near-abandonment by the government of the informally housed population in terms of infrastructure and services, the Muslim Brotherhood and its networks of food cooperatives and other services took on the nature of a parallel state. Following the 2011 revolution, Mohamed Morsi, the Muslim Brotherhood candidate, became Egypt's first democratically elected president. But his weak control over the state contributed to the ease with which he was deposed by the military a year later. The coup solidified the ascendency of the military that outlawed and violently suppressed the Muslim Brotherhood and further marginalized the revolutionaries and the fulool, or remnants of Mubarak's regime (Al-Amin 2015); General Abdel Fattah el-Sisi has ruled as president of Egypt since then. In doing so, the military, long kept at a distance from political office, added direct power over the country to its diverse economic activities and control of Cairo's desert periphery. The military's economic contribution to gross domestic product is estimated at 25 per cent, but no one is quite sure, as their activities and holdings are deliberately opaque (Dorman 2013, 1588). The exploration of the following four assemblages reveals the many urban entanglements of these actors, as well as intersecting economic and cultural forces that have shaped the desert peripheries of Cairo.

First Assemblage: The Entanglements of Land and State and the Repurposing of the New Cities

Cairo has long been seen as ungovernable and even as an urban failure in terms of planning. Nevertheless, since the 1952 revolution, considerable resources have been expended on urban development, planning expertise has been offered by non-governmental organizations, and outside capital has been attracted to the city. Cairo's high population densities have guided most planners to look at the eastern and western desert peripheries so as to avoid expanding into Egypt's scarce agricultural land base, located mainly to the north and south. However, most of the housing expansion of Cairo has occurred informally on this agricultural land as a direct outcome of other interests and objectives of the government and the military, both key owners of desert land.

Desert towns or cities initiatives began after the 1952 revolution but coalesced by 1979 into a broad project to build up to eighteen freestanding and satellite cities in the deserts around Cairo and other cities in order to redirect population growth (Singerman 2007, 84; Stewart 1999, 14). These projects follow a long history of imposing order on the city by creating new highly defined areas and subdivisions, like the garden suburb of Heliopolis, built in 1905, the 1950s middle-class residential area of Muhandisin ("The Engineers"), or the administrative Nasr City suburb of the 1960s, located near Cairo's army bases and the airport. The new cities approach also reflected the limited ability of the state to penetrate and engage the existing Egyptian urban fabric effectively (Dorman 2013, 1586–9). These desert city projects were to supply housing for ordinary low-income citizens, but while one of the first new cities, called 6th of October, was eventually successful in providing affordable housing, industry, and jobs, the remainder of the new cities failed to achieve their population targets despite draining as much as a fifth of state capital spending by 1991 (1591). For prospective residents, the new cities frequently lacked sufficient economic planning, posed higher mobility and living costs, and strained social connections with families and friends (Feiler 1992, 308).

How and why did these cities transform into the affluent subdivisions and trendy gated communities for the upper middle class and elite, starting in the early 1990s? The answer is found in the way that both the regime and army put up barriers to, or outright rejected, the use of the desert for assisted informal development. Several major Western aid agencies, including the World Bank, the US Agency for International Development (USAID), and the Institut d'aménagement et d'urbanisme de la région d'île-de-France (IAURIF), proposed semi-formal development in the desert to divert lower-income groups from agricultural land. Elements of the government were appalled at these proposals, which were then criticized as an acceptance of informality and a potent threat to the authority of the state. An Egyptian general articulated the military's alarm that these plans might enact a potentially dangerous informal slum encirclement of the city (Dorman 2013, 1597). By the early 1990s there was a tremendous infusion of foreign aid and investment, as well as remittances from Egyptians living and working abroad. This made these areas enormously attractive for speculative public and private development for a widening base of affluent Egyptians and foreign investors. As the failures of the new cities became apparent, both parties encouraged the transformation of the desert peripheries into a private investment "wild west" from the 1990s onward.

The military's shared dominance of the desert was part of the outcome of the 1973 war with Israel, when it occupied defensive positions in the desert around the cities and held on to these areas, estimated in the 1980s to approach half of metropolitan Cairo (Dorman 2013, 1603). It maintained control of these spaces even as other state agencies fought for access and ownership. Mubarak's strategy of mollifying and depoliticizing the army by "allowing it to develop a parallel economy and set of social-welfare institutions dedicated to its own corporate aggrandizement" (1588) contributed to its ability to maintain its land holdings. By the 1980s the military had already created vast army communities to provide housing for all professional serving and retired soldiers

(Mitchell 2002, 274). The military has its own construction companies, factories, and other enterprises that allowed it to rapidly and effectively construct housing and infrastructure. After rejecting or blocking semi-informal development in areas proposed by international donors, it subsequently sold or developed them into affluent subdivisions (Dorman 2013, 1595).

Second Assemblage: Neoliberalism and the New Cities

To what extent have shifts in global capitalism and the rise of neoliberal imperatives shaped the suburbanization of Cairo? International neoliberal institutions in the 1990s promoted the idea that Egypt was following a "model" neoliberal path of reform, one spearheaded by a younger generation of regime insiders. Principal among them was Gamal Mubarak, the president's son, who wielded influence as Deputy Secretary General of the ruling party and head of its policy committee. Yet, in Egypt, while neoliberalization processes contribute to the restructuring the city, in practice "actual existing neoliberalism" often operates contrary to the ideology of "open, competitive, and unregulated markets, liberated from all form of state interference" (Brenner and Theodore 2002, 350). The government in the 1990s was lauded as leading in terms of the privatization of industry and finance; however, this masked the way that most privatized enterprise remained tethered to state institutions through financial and infrastructural dependencies, as well as through substantial ownership shares controlled by public institutions and government office holders (Mitchell 2002, 275–81).

The military remained in principle one of the few powerful social actors in Egypt that often directly opposed the privatization of public-sector enterprises, following its role as a trustee of statist policies coming out of the 1952 revolution (Marshall and Stacher 2012). Wikileaks revealed cables of two United States ambassadors to Egypt noting the strong opposition the military had to Gamal Mubarak's reforms, which were perceived as threatenening their interests and social stability in the country (Marshall and Stacher 2012, n.p.). After the 2011 revolution, the army is presently in near competitor-free economic expansion through collaboration and joint investments in numerous sectors, including large oil, gas, renewable energy, computer, communications, housing, and infrastructural projects. Its collaborators include private and sovereign wealth funds, Western multinationals, and Gulf conglomerate partners. Today, "Military Inc." offers "the same system of benefits that

Egyptian oligarchs have enjoyed, including preferential treatment in bidding for state contracts, privileged access to infrastructure and services, and advance notice of pending projects," with the addition of soldiers to deliver security in an unstable environment, "a type of insurance no other state actor can provide" (Marshall and Stacher 2012, n.p.). Actually existing neoliberalism in Egypt is a capitalism of public–private partnerships operating in a framework of autocracy that has changed little with the shift from the Mubarak regime to the military.

Alongside these private-public partnerships, two key socio-cultural aspects of neoliberalism – segregation and speculation – interact to create the gated suburbs. Historically, Cairo's very vibrant street life was generated by high-density living and the mixing of classes, notwithstanding the subservient roles of the poor. The new gated communities on the peripheries, on the other hand, are extreme in their distancing and almost complete exclusion of non-affluent classes outside of direct employment. Neoliberal economic practices in Egypt help accelerate the development of segregated spaces because they have promoted the retrenchment of public spending that helped to level differences, contributing to the exacerbation of income inequality (Brenner and Theodore 2002). In turn, various forms of security and safety panics can arise as greater disparities take hold and conflict is heightened. One example of this in Cairo are the Islamic insurgencies emanating from the ashwaiyats in the 1990s. The city became dangerous for the ostensibly respectable classes, and degraded, too, as neglected buildings collapsed, and sewage, electrical, and water infrastructure failed due to corruption and insufficient investment.

It is important to identify how the proliferation of enclosed malls in Cairo in the late 1980s and 1990s signalled the crisis of neoliberal retrenchment and provided a cultural gateway for the creation of gated communities (Abaza 2006). The new malls turned their backs on open streets and traditional consumption spaces like Cairo's Khan-el-Khalili souk. Undesirable classes were now easily denied entry or expelled from these new consumption spaces; as little as a fifth of the population could be seen as legitimate consumers of the expensive goods and services within (199). For the affluent classes, the attraction of the mall was found in other parallel cultural paradigms of separation; as watched spaces they were considered to be places where people had better manners and women could feel less intimidated. Moreover, businesses saw the value of locating in malls to appeal to both expatriate workers who had become used to Saudi and Gulf mall culture, and luxury consumers who had the option of travelling abroad to places like Dubai (Abaza 2006; Mitchell 2002).

Speculation is another neoliberal process made possible by privatization and the financialization of assets. The value of desert suburbanization in Egypt was made attractive to speculation because capital was able to "transcend social-spatial infrastructures and [a] system of class relations that no longer provide a secure basis for sustained accumulation" (Brenner and Theodore 2002, 355). The desert land allowed for a deregulated, essentially carte blanche space at very low cost. This land could therefore increase substantially, even exponentially in value. Another factor in speculation was that, by their exclusivity, the developments assembled in one place Cairo's most affluent base of consumers, and the only ones likely to produce enough marginal returns to interest foreign capital in a poor country. Finally, the gated communities provided defensible space (Blakely and Snyder 1997); the privatized nature of the space and the new governance structures meant that it was less affected by perceived loss of property value due to problematic or incompatible uses, people, and other buildings. Other forms of speculation drove more housing speculation, as leveraged positions on the booming Egyptian stock market in the 2000s flowed into the gated communities (Mitchell 2002). There was also a strong "cash" element to these investments, both because of the overall wealth of some of the consumers, and because of investments from the accumulated savings of Egyptians working in Saudi Arabia and Gulf States following the First Gulf War. As Denis (2008, 1107) argues, cash fuelled new neoliberal sensibilities of speculation and segregation: "At the other end of the social spectrum, the requirement to pay in cash encourages ostentatious buildings that magnify, in their adornment, the financial power of the owner. The ostentatious may be all the better expressed given the building of new prestige residential centralities in the desert; the sense of propriety needed in cohabitation in the metropolis, in proximity to others, is no longer felt necessary. The ways of living in, of experiencing the metropolis have become exclusive."

Third Assemblage: Transnational Developers and Inter-Referencing

The third assemblage focuses on inter-referencing carried out by national and transnational developers as they design and build peripheral "communities," responding to the ripe conditions of speculation. Alongside the benefits of exclusion and security, media discourses and developers' promotional campaigns have also elevated concern about Cairo's pollution as a health hazard, and the deleterious stress effects of noise pollution (Denis 2008). In the promotion of gated communities, developers advise the abandonment of the crowded city for green space, healthy living, and peace and quiet. The communities are designed from an international palette of models that include Western suburban lifestyles and regional, Arab, and Islamic influences. Through this practice of "inter-referencing," developers cite "what constitutes urban success and achievements in a world of circulating city symbols" (Ong 2011, 18). Examining some of these models in theme and design, as well as their developers, will help us understand their intersections with the spatial reconfiguration of Cairo, as well as ground broader neoliberal paradigms in specific projects.

Dreamland, begun in 1995, is among the first of the aspirational communities, and has been promoted as one of the largest private urban developments in the Middle East, with a marquee location only a few kilometres west of the Great Pyramids of Giza (Mitchell 2002, 274–82). Its design reflects several trends of inter-referencing. Two international design firms were employed by the Bhagat Group developers to integrate the shopping mall culture of Egypt and Dubai with American golf and leisure lifestyles. The US–based Development Design Group created the overall plan that now includes a golf course, conference centre, Dream Mall, and the Hilton Pyramids Golf Resort. This firm's portfolio includes similar multi-purpose private developments, such as the Dubai River Walk, Zonk'izizwe in South Africa, Commerce City in Istanbul, a resort community in Spain, and the Easton Town Center in Columbus, Ohio (Development Design Group 2018). Toronto firm FORREC, designers of Universal Studios in California, translated the American Disneyland model into Dreampark, billed as the "leading amusement park of the Middle East" (Wikipedia Contributors n.d.). After the relative success of Dreamland, the hybrid American-Dubai lifestyle was echoed in other gated developments like Beverley Hills, Palm Hills, and Golf City.

In many ways these communities are not so much hybrids of Western and Middle Eastern cultures, but can rather be understood as a new development form: neoliberal simulacra projects. Simulacra are signs that no long really refer to anything, except through a kind of superficial recognition (Douglass and Raento 2003). Like the themed and juxtaposed casino-hotels of Las Vegas, where décor is playful but secondary to the function of gambling and shopping, the themes of the gated community are supportive of the functions of security, speculation, and consumption. The development of Hyde Park references London in name, but in it are Italian-style villas and modernist townhouses. Many gated communities suggest an interpretation of California via Dubai, but this somewhat misses the point that both Dubai and California are successful producers of simulacra themselves. And what frees up the communities to be designed in this way is the neoliberal abandonment of any meaningful integration with their surroundings. Neoliberal projects tend to eschew systemic or collective development approaches in favour of defined and defensible projects linked to private beneficiaries. Just as in Las Vegas, where the Egyptian-themed Luxor Hotel shares the strip with Treasure Island and Caesar's Palace, Egyptians cruising the Ring Road of Cairo can follow their desires to Dreamland, El Lotus, and Future City. Each space is designed to encapsulate a new collection of aspirations.

The developers of these spaces grapple both with aspirational designs and with the functions of these spaces, which need to serve mobile and cosmopolitan lifestyles. Buyers often include foreigners, Egyptians who work for transnational corporations, and the middle classes who migrated for work abroad and returned with savings to invest. At Katameya Heights Golf and Tennis Resort, a marketing strategy was to entice Egyptian expatriates with a landscape and design that reflected cosmopolitanism and the good life through leisure (Fahmi 2008, 276). Inter-referencing is therefore a competitive as well as creative practice. Competition includes providing residents opportunities for the accumulation of cultural capital in lifestyle and education (Bourdieu 1986). International language schools and, if possible, universities are a desired feature of most communities; along with other international schools and universities, the American University in Cairo, branded as "Egypt's Global University," recently moved from Midan Tahrir in central Cairo to a 260-acre campus in New Cairo City (American University in Cairo 2016).

While these projects aim to be cosmopolitan and international, the developers capitalizing on this inter-referencing are more provincial. Many of the large property developers have diverse portfolios of developments, often all over the globe, but they are largely based in Egypt or the Gulf States. It appears likely that national and regional elite networks are at work, forging partnerships that are based on trust and mutual benefit. Dreamland was the ambitious project of Dr Ahmed Bhagat of Bhagat Group, who was "reputed to be a front man for unpublicized moneymaking by the presidential family" (Mitchell 2002, 185). The Bhagats are one of the oligarch families of Egypt, involved in a diversity of economic activities including real estate, manufacturing, and internet services. Another developer, Talaat Moustafa Group (TMG), the country's largest publicly traded real estate company, was revealed to have obtained some 33 million square metres without auction and far below market value for its 120,000-unit residential subdivision of Madinaty ("My City"). Gamal Mubarak, the president's son and neoliberal czar, held stakes in holding funds invested in TMG (Fattah and Kassem 2011). Following the revolution, SAE, the second largest public developer in Egypt, had its purchase transaction annulled, and former Housing Minister Ahmed El Maghraby was charged with related profiteering. SODIC, Egypt's third largest developer, also faced scrutiny over its land bank, and charges were laid against its chair, Magdi Rasekh. Mubarak's oldest son, Alaa, is married to Rasekh's daughter (Fattah and Kassem 2011).

Notably, corporations with ties to the military did not suffer these types of investigations after the 2011 revolution. An example of ongoing Gulf State collaborations is the military's partnership with Dubai-based Emaar Properties, ranked as the top Middle East property developer by market capitalization (Arabian Business 2014). It constructed Dubai's icon – and the world's tallest building – the Burj Khalifa. Emaar claims to be "focused on creating one of the largest real estate companies in Egypt," and one of its newest projects is a highly exclusive mall/golf/housing complex called Uptown (Emaar Misr 2016). The project reveals close collaboration with the military on a "sensitive deal" on land owned by the military and used for military training, and will include a private road from Cairo's highways to the complex that could involve the clearance of a nearly slum. The "Dubai-sation of Cairo" replicates some of the same relations in Dubai "between political power and capital: namely, that they are usually one and the same" (Elshahed 2014, n.p.).

The Fourth Assemblage: The Lived Experience of These Spaces

There are more than sixty gated communities in and around Cairo, and some non-gated public desert communities. Media reports and an ethnographic lens can begin to fill in an impression of how these spaces are inhabited and how residents newly relate to the old city. The most obvious general observation about the Cairo gated communities is that most are currently underpopulated, and some appear almost deserted.

They contribute very little to the nation's housing shortages. Despite lofty government occupancy projections, "the total 2006 population of all new towns in Egypt had not reached 800,000 inhabitants, of which 600,000 were to be found in the new towns around Cairo. This represented just over 1 percent of the nation's total population" (Sims 2015, 141). New Cairo, which contains the Al Rehab City development, for example, had 118,678 total residents in the 2006 census, less than 10 per cent of its target for 2022 (Sims 2015). Primary among the reasons for non-occupancy is the affluent consumer base, the speculative nature of the investments, the way that this gated housing figures into cultural norms about planning for the future, and employment and commuting challenges.

Non-occupancy among the affluent consumer base is a critical factor. While housing varies in price, with apartments in Golf City averaging about $200,000 USD, and villas $350,000 USD (Fahmi 2008, 275), the average cost of housing puts it out of reach for all but upper-class Egyptians. They often own more than one property, so a new city acquisition may be in addition to a family home in central Cairo, a vacation property on the Red Sea or Mediterranean coasts, and possibly a home abroad. A new city home may be used for weekend retreats, which is why the many leisure amenities such as golf, tennis, and swimming are so important to the communities.

A second important reason for non-occupancy is that many properties are bought for speculation by Egyptians or foreigners. These speculators hope to sell them at a profit in the near future, or longer-term, when the developments become more occupied and desirable (Fahmi 2008, 277–84). There was a property boom in the new cities from 2007 to 2011, beginning with a 116 per cent price increase in 2007. While Egypt lacks a housing index to track prices, the new cities' price percentage increases are estimated to have outpaced central Cairo properties by three to one, even though prices are still higher per square metre in the centre. The 2007 price jump came as restrictions on foreign ownership were relaxed, making Egypt far more open to speculation than even Dubai, which maintains ownership restrictions outside a few designated areas. Encouraging the trend even more, President el-Sisi's new investment law in 2015 dispensed with any remaining restrictions on foreign ownership of land and property (Shawkat 2015).

A third factor in non-occupancy is the Egyptian cultural norm of parents practising housing planning for their children. The affluent have the choice of buying homes in the new cities for when their children get married, while ordinary Egyptians in informal neighbourhoods continue to build "up" for their future generations, adding floors and apartments to existing buildings as money becomes available. Finally, the lack of jobs in the mainly residential enclaves means that commuting is often necessary for work or school. Egypt is a country where only 11 per cent of families own a vehicle (Sims 2010), and even families that have one vehicle likely have several family members that will still need to use limited public transit options or expensive private mass transportation options to reach the core of the city or other gated communities. For those commuting by car, there are onerous traffic jams on Cairo's Ring Road to face.

As with the new cities of the twentieth century, the affluent new cities of the twenty-first century demonstrate varying levels of success, although it would be premature to suggest that they will suffer the same fate. While some might fail, unlike the older desert projects, the new affluent gated communities generally have government backing and considerable funds allocated for infrastructure development (Kirk 2016), and private developers who have a reputational stake in their success. They have also experienced almost constantly rising prices, and ride on a wave of media and advertising campaigns about their desirability. As well, several of the gated communities have been increasingly successful at attracting businesses and organizations, such as technology companies and some government ministries and international organizations. Many of these have relocated from central Cairo. Al Rehab City is one of the older and more diversified of the gated communities, with five foreign language schools, a private hospital, and two shopping malls. As more of the workplaces of potential owners relocate, the gated communities will have more success in attracting permanent residents (Marafi 2011, 43–5).

Al Rehab's residents articulate their desire to live in their gated suburbs in relation to the older city. The attraction of the gated community is found in what it offers over the more traditional middle-class and upper-class neighbourhoods of Cairo. Mid-city neighbourhoods like Heliopolis and Muhandisin are expensive, despite providing the benefits of excellent access to the rest of the city. They are also inferior as defensible space (Newman 1972). Muhandisin, for example, is very close to the ashwaiyat of Imbaba. Public intermixing with the lower classes who stroll or window shop is a source of anxiety. Many Al Rehab City residents express a class-based moral panic about such intermixing and equate the environmental pollution of the streets with the moral pollution of the lower classes, who are characterized as being rude, lawless, uncivilized, and likely drug users (Marafi 2011). In other words, a general environmental and moral degradation is perceived as having irredeemably afflicted even the middle- and upper-class neighbourhoods of Cairo (Denis 2008, 1102–3).

An "othering" process occurs through persistent media representations linking the ashwaiyats, crime, and the moral degradation of society (Marafi 2011, 33). It is being reinforced, in the case of Al Rehab City, by the possibilities of defensible space and a siege-like mentality. Despite business concerns, residents have fought for and won increased restrictions on the ability of outsiders to enter for shopping or other purposes. The moral panic has been internalized in many ways as the walls, gates, and separation reinforce the idea that the workers who come in and sometimes reside inside the walls are themselves a security risk. Residents concern themselves with whether the low pay of the private security guards will motivate them to steal, and whether the deliverymen know too much about their comings and goings. Servants are often described in derogatory ways as living in "nests" (ishash in Arabic) where drugs and crime proliferate. Despite having walled home compounds, many residents are installing bars on their windows and doors and extra security lighting, and purchasing guard dogs, a previously unheard-of practice in Cairo. At the same time, with shopping options expanding inside, residents feel less need to leave, reinforcing this perception of living in a better world (Marafi 2011, 51–62). While urban class differentiation has always existed in Cairo, the agency of the residents is being used to take segregation to a new, unprecedented level.

Conclusion

The approach of examining assemblages of political, economic, and cultural contributions to the making of Cairo's periphery has served to ground the role of broader processes of globalization and neoliberalism, and to demonstrate interconnections, trends, and some contradictions between several categories of actors. The question can now be asked about what is particular in Cairo's suburbanization, and what has parallels to other cities. Unlike many gated communities in Europe and the United States, it is not the influx of immigrants or racial minorities which contributes to the drive for separation in Cairo, but very strong class divisions. Dubai's social divisions are different from Cairo in this respect – it is the four-fifths of the population who are guest workers and have no secure status who are the excluded "Other." However, in common with many other cities, the Cairo and Dubai enclaves have the tendency to reinforce or even exacerbate divisions. This is also true in the United States (Blakely and Snyder 1997), and in regional mega-cities like Istanbul, where the poor in informal developments "now invoke feelings of 'otherness' and social disorder" (Erkip 2000, 375).

Unlike Dubai and Istanbul, and many other regional cities, the pace at which the most affluent citizens are rushing to abandon the core and older suburbs of Cairo is remarkable. The movement of universities, language schools, and international organizations to the new cities is aiding this momentum. Moreover, the push to abandon old Cairo to the poor may have taken a new turn. In May 2015, Housing Minister Mostafa Madbouly announced the building of an entirely new capital at a cost of $45 billion USD, with a timeline of seven years (Guerin 2015; RT News 2015). The Egyptian government, as well as departments, ministries, and embassies, would be relocated to the new capital, projected to hold 5 million citizens. Superseding Dreamland is a planned theme park that would also be seven times larger than Disneyland in California, while the new capital airport would be larger than Heathrow. This is a mega-project managed by United Arab Emirates billionaire Mohammed Alabbar, already noted as a collaborator with the Egyptian military through Emaar (Smierciak 2015). Given Alabbar's track record of completed projects, the new capital may avoid being added to a long list of never-realized Egyptian mega-projects. Even if it is never built, its design is highly symbolic of the many converging trends of neoliberalism, authoritarianism, and private-sector partnerships. Cairo's massive desert suburbanization is fixing urban space in ways that will make it difficult to reintegrate a highly divided society in the future.

REFERENCES

Abaza, M. (2006). "Egyptianizing the American Dream." In Singerman, D., and Amar, P. (eds), *Cairo Cosmopolitan: Politics, Culture, and Urban Space in the New Globalized Middle East*, 193–220. Cairo, Egypt: American University in Cairo Press.

Al-Amin, E. (2015). "Egypt's Destiny between Four Political Forces and Four Dates." *Counterpunch*, 19 June. http://www.counterpunch.org/2015/06/19/egypts-destiny-between-four-political-forces-and-four-dates/

American University in Cairo. (2016). "About." *The American University in Cairo*, 1 October. http://www.aucegypt.edu/about

Arabian Business. (2014). "Revealed: GCC's Top 10 Developers." *Arabian Business*, 9 October. http://www.arabianbusiness.com/revealed-gcc-s-top-10-developers-567120.html

Blakely, E., and Snyder, M.G. (1997). "Divided We Fall: Gated and Walled Communities in the United States." In Ellin, N. (ed.), *Architecture of Fear*, 85–101. New York, NY: Princeton Architectural Press.

Bourdieu, P. (1986). "The Forms of Capital." In Richardson, J. (ed.), *Handbook of Theory and Research for the Sociology of Education*, 241–68. New York, NY: Greenwood.

Brenner, N., and Theodore, N. (2002). "Cities and the Geographies of 'Actually Existing Neoliberalism.'" *Antipode* 34, no. 3, 349–79. https://doi.org/10.1111/1467-8330.00246

Davis, M. (2006). *Planet of Slums*. London, UK: Verso.

Denis, E. (2008). "Cairo between Traces and Liberal Re-Foundation: Demographic Stabilization, Extension of the Metropolitan Area, and the Renewal of Forms." In Jayyusi, S.K., Holod, R., Petruccioli, A., and Raymond, A. (eds), *The City in the Islamic World*, 1091–120. Boston, MA: Brill.

Development Design Group. (2018). "Projects." *Development Design Group*. Accessed 21 November 2018. http://www.ddg-usa.com/projects/#international

Dorman, W.J. (2013). "Exclusion and Informality: The Praetorian Politics of Land Management in Cairo, Egypt." *International Journal of Urban and Regional Research* 3, no. 5, 1584–610. https://doi.org/10.1111/j.1468-2427.2012.01202.x

Douglass, W.A., and Raento P. (2003). "The Tradition of Invention: Conceiving Las Vegas." *Annals of Tourism Research* 31, no. 1, 7–23. https://doi.org/10.1016/j.annals.2003.05.001

Elshahed, M. (2014). "From Tahrir Square to Emaar Square: Cairo's Private Road to a Private City." *The Guardian*, 7 April. https://www.theguardian.com/cities/2014/apr/07/tahrir-square-emaar-square-cairo-private-road-city

Emaar Misr. (2016). "Real Estate Projects in Egypt." *Emaar Misr*, 5 August. https://www.emaar.com/en/what-we-do/communities/egypt

Erkip, F. (2000). "Global Transformations versus Local Dynamics in Istanbul." *Cities* 17, no. 5, 371–7. https://doi.org/10.1016/s0264-2751(00)00033-0.

Fahmi, W.S. (2008). "The Right to the City, Stakeholder Perspectives of Greater Cairo Metropolitan Communities." In Jenks, M., Kozak, D., and Takkanon,

P. (eds), *World Cities and Urban Form: Fragmented, Polycentric, Sustainable*, 269–92. New York, NY: Routledge.

Fattah, Z. and Kassem, M. (2011). "Egypt's Developers Pay the Price for Their Ties to Mubarak." *Bloomberg*, 7 June. http://www.bloomberg.com/news/articles/2011-06-06/mubarak-s-legacy-clouds-future-of-egypt-s-property-boom-as-deals-unravel

Feiler, G. (1992). "Housing Policy in Egypt." *Middle Eastern Studies* 28, no. 2, 295–312. www.jstor.org/stable/4283494

Guerin, O. (2015). "Egypt Unveils Plans to Build New Capital East of Cairo." *BBC News*, 13 March. http://www.bbc.com/news/business-31874886

Kirk, M. (2016). "The Extraordinary Privilege of the Wealthy Suburbs in Egypt." *Citylab*, 1 July. http://www.citylab.com/tech/2016/07/egypt-new-cities-infrastructure-spending/489608

Marafi, S. (2011). "The Neoliberal Dream of Segregation: Rethinking Gated Communities in Greater Cairo, a Case Study, Al-Rehab City Gated Community." MA thesis, American University in Cairo.

Marshall, S., and Stacher, J. (2012). "Egypt's Generals and Transnational Capital." *Middle East Report* 262. http://www.merip.org/mer/mer262/egypts-generals-transnational-capital

Mitchell, T. (2002). *Rule of Experts: Egypt, Techno-Politics, Modernity*. Berkeley, CA: University of California Press.

Newman, O. (1972). *Defensible Space: Crime Prevention through Urban Design*. New York, NY: Macmillan.

Ong, A. (2011). "Introduction: Worlding Cities, or the Art of Being Global." In Roy, A., and Ong, A. (eds), *Worlding Cities: Asian Experiments and the Art of Being Global*, 1–26. Oxford, UK: Blackwell Publishing.

RT News. (2015). "Egypt Unveils Blueprints for New $45bn Capital City." *RT News*, 14 March. https://www.rt.com/news/240793-egypt-cairo-new-capital

Shawkat, Y. (2015). "Egypt's Deregulated Property Market: A Crisis of Affordability." *Middle East Institute*, 5 May. http://www.mei.edu/content/at/egypts-deregulated-property-market-crisis-affordability

Singerman, D. (2007). "Cairo Cosmopolitan, Citizenship, Urban Space, Publics, and Inequality." In Drieskens, B., Mermier, F., and Wimmen, H. (eds), *Cities of the South: Citizenship and Exclusion in the 21st Century*, 80–112. Beirut, Lebanon: Institut Français du Proche-Orient.

Sims, D. (2010). *Understanding Cairo: The Logic of a City out of Control*. Cairo, Egypt: The American University of Cairo Press.

– (2015). *Egypt's Desert Dreams: Development or Disaster*? Oxford, UK: Oxford University Press.

Smierciak, S. (2015). "Egypt Has Already Proved Fertile Ground for Emaar Chairman Alabbar." *The National*, 25 March. http://www.thenational.ae/business/economy/egypt-has-already-proved-fertile-ground-for-emaar-chairman-alabbar
Stewart, D. (1999). "Changing Cairo: The Political Economy of Urban Form." *International Journal of Urban and Regional Research* 23, no. 1, 103–27. https://doi.org/10.1111/1468-2427.00182
Wikipedia Contributors. (n.d.). "Dreamland (Egypt)." *Wikipedia*. Accessed 25 April 2016. https://en.wikipedia.org/wiki/Dreamland_(Egypt)

16 Massive Suburbanization, Heterogeneous Suburbs in China

TIANKE ZHU AND FULONG WU

Introduction

Suburbanization in China is driven by land development and urban expansion (Zhang 2000; Zhou and Ma 2000; Feng, Zhou, and Wu 2008). The current form of suburbanization is created by rapid land-use transformation and spatial reconfiguration following the establishment of a land market. Chinese suburbanization is not only dominated by state-led industrial relocation but also driven by the entrepreneurialism of the state in the land development process (Wu and Phelps 2008; Shen and Wu 2017), in which Chinese suburban new towns are outcomes of land reforms and commodity housing development.

The new wave of suburban development, characterized by new town development, represents a consequence of population movement in China and is also an integral part of the transformation from state-led industrialization to a market-oriented system. Despite the dominant role of government-sponsored residential development in Chinese suburbs, all ranges of sectors and capital have been involved in the transformation of rural land for urban land use in contemporary suburban new towns.

Suburban Development in China

Chinese housing reform provided the most important preconditions for space commodification. In the pre-reform era, property development was almost absent from suburban development. Now, developers are pioneers who actively materialize gains from land through property development. Meanwhile, local government has its own stake in revenue generation from local growth, which allows it to improve the suburban environment, as well as to sustain and consolidate itself in the political arena. However, suburban growth in the post-reform period emerged as less planned and was mixed, with work-unit residential areas, fragmented industrial development zones, and urban villages on the metropolitan fringe (Deng and Huang 2004). The spatial pattern of suburban expansion in the early stage of suburbanization in the 1980s and 1990s was scattered and emerged as monotonous residential use, consisting of mostly match-box-style apartment buildings, often built by state-owned industrial enterprises in the suburbs for their staff.

With rising land development and economic growth, suburban expansion has not only played an important role in serving the need of decentralization but has also played the dominant role in growth in regional and even global economies. Furthermore, suburbs are well-planned as new towns, which become places with good environmental quality for residential and commercial development, combined with improved living facilities and services. Clusters of suburban residential settlements are often well-planned and evolve into self-sustained towns.

Many studies emphasize that the market-driven approach indicates China's embrace of neoliberalism (Harvey 2005), and much has been written about the transferring of use rights of state-owned land to private developers (Zhou and Ma 2000; Deng and Huang 2004; He and Wu 2009; Shen and Wu 2013). In the socialist period before economic reform in 1979, the land market was absent. Suburban land development was carried out by state-owned enterprises as part of the industrial development process. In the late 1980s, the land market was introduced into China, and "housing commodification" has dramatically sped up since 1998, when the allocation of public housing was suspended. In the 1990s suburban development was driven by an emerging land development by private developers. However, the state still plays an enabling role in terms of the development of land use, labour and environmental policy, and judicial and legislative frameworks, while private developments take commercial, residential, and industrial forms defined by political and social exclusion. Market-oriented land reform is a state-engineered, state-controlled, and state-led process meant to facilitate accumulation and legitimize the state's power (Wu 2010; Shen and Wu 2017). Housing reform and suburban development form an integral process that reflects the ideology of governmental entrepreneurialism (He and Wu 2009). On the one hand, market transition and spatial restructuring create opportunities for residential mobility and residential land use restructuring;

on the other hand, stratified affordability gives rise to a new system of residential resettlement in the suburbs.

In fact, administrative and fiscal decentralization also boosted large-scale suburban development. Government has not only gained greater discretion to deal with local revenue, but also fully uses its power to manage land leasing and urban development. In other words, the government is not just a regulator but also a market player (Wu and Zhang 2007). Evidently, local government has morphed into an entrepreneurial government, creating various investment methods and platforms to participate in land development. This is broadly similar to other places, where different social, political, economic, and environmental processes also highlight distinct roles in shaping suburban development and suburban life (Phelps, Wood, and Valler 2010; Ekers, Hamel, and Keil 2012; Hamel and Keil 2015).

In this chapter, we reveal the distinctive features of new town construction in China, which exists on a massive scale but also leads to the creation of heterogeneous spaces in the suburb. Existing theories have revealed the features of suburbanization occurring after industrialization in North America (Bourne 1996; Harris and Lewis 1998). Western theories on suburbanization reveal that the middle class moved to the suburbs to escape urban decline and "urban illness" in inner-city areas (Fishman 1987). The notion of "post-suburbia" further suggests a process in the post-industrial economy that led to the densification of suburbs and an increase of service activities there (Phelps et al. 2010). The development of suburbs is driven not only by residential preference, as shown in earlier suburbanization literature, but also by economic restructuring and changing modes of governance (Ekers et al. 2012; Hamel and Keil 2015) – for example, emerging private governance in suburban gated communities and a more neoliberal approach to development. In contrast, the process of Chinese suburbanization is closely associated with industrialization and urbanization. Suburbs in the context of rapid urbanization provide not only a new living environment developed by the private housing market but also an opportunity for land development and investment (Shen and Wu 2017). The new middle class is moving to suburbs for a better living environment, especially for better housing quality, rather than escaping from urban decline. Their purchase of suburban housing is part of their social class formation processes (Zhang 2010). But they also buy suburban housing as a second property for investment purposes. A significant number of residents in inner urban areas have been relocated to the suburbs because of demolition and renewal. In addition, as Chinese suburban development is an integral process of industrialization, the development of suburban industries has attracted an influx of migrants from other rural areas. As a result, Chinese suburban development is also a process of urbanization of the rural areas. A study of Chinese suburban development will thus enrich our understanding of suburbanization and post-suburbia development, and reveal heterogeneous suburban landscapes during this massive suburbanization process as well as the interwoven forces of state, market, and society, with different motivations, constraints, and preferences. The Chinese suburb developed under massive suburbanization is heterogeneous because suburban residents are from different places, with different motivations, and the process operates through different governance modes and development approaches.

A Case Study of Jiangning, Nanjing

Nanjing, located 330 kilometres west of Shanghai, is the capital city of Jiangsu Province and one of the most developed cities in China. It is also a central city in the Yangtze River Delta. The Jiangning district in Nanjing has experienced significant suburbanization and has the highest amount of per capita housing space in Nanjing. The landscape is a typical suburban one. Within Jiangning district, there are different residential areas, showing different types of suburban developments

A Brief Review of the Process of Suburbanization in Nanjing

From 1992–1999: The Development of Blue-Collar and Rural Enclaves

Suburbanization in Nanjing started with the development of industrial zones and decentralizing industries. The initial spatial pattern was planned for the urban fringe in order to accommodate enterprises relocated from the central area. Residential clusters were planned around industrial parks, which provided a convenient living environment for relocated employees. The residential clusters emerged in distinctive patterns with existing surrounding settlements. The social status of residents in residential clusters showed homogeneity, as represented by blue-collar workers.

From 2000–2003: Middle-Class Relocation

Rapid population growth in the central city of Nanjing resulted in declining housing conditions, and areas of poverty began to emerge.

Municipal government began the process of urban development in the peripheral areas to address the issue of housing congestion in central areas. Residential areas with extremely poor housing conditions in the urban centre were demolished for commercial land use. Meanwhile, the government energetically developed infrastructure and housing projects to boost the local housing market and economic growth. Areas close to the urban centre with natural environmental advantages were considered prime places for accommodating a decentralized population and providing better housing. In that situation, the municipal government adjusted the administrative boundary of its districts to address the problem of insufficient land for developing large-scale infrastructure. Meanwhile, alongside increasing demands for housing by the new middle class, many residential projects were built in the urban fringes with comparatively low housing prices. For instance, small- and medium-sized condominiums and high-end residential communities emerged during this period, with an increasingly middle-class population. However, most homebuyers were speculative investors, as living facilities such as schools, hospitals, and public transport were still inadequate in the Nanjing suburbs. In general, suburban communities in Nanjing were not livable places.

From 2003–Present: Booming of the Suburban New Town

The strategy of rural-urban planning has been applied to Chinese suburbanization since 2003 in order to improve the living environment and conditions in the suburbs. This is in essence an urbanization strategy that transforms the former rural landscape. In addition to continually supplying luxury housing in the urban fringe, the improvement of facilities in the central areas of suburban new towns was expedited. Small- and medium-sized communities were launched at the same time. Consumers wanted a new quality of housing, and real estate developers provided commodity housing to satisfy distinctive housing choices and preferences. Moreover, suburban gated communities have increased dramatically in number and are planned as self-contained clusters to address security and privacy issues. Rural migrants from other cities concentrated in the suburban industrial zone to pursue employment opportunities.

Suburbs in Nanjing were not only regarded as locations where urban and rural areas were connected but also fulfilled an urban function that transformed rural areas to produce economic growth and increase levels of urbanizations. On the one hand, residents living in the suburbs had diverse living demands, especially the middle class, whose pursuit of high-quality environment and facilities stimulated urban fringe areas, transforming them from supporting urban central areas to becoming self-contained developments. On the other hand, the government gradually relocated large industrial enterprises with environmental impacts, and thus improved the quality of the suburban environment. Meanwhile, the suburban residential pattern formed ringed layers in Nanjing due to the diverse demands of the housing market. Town centres and high streets are located close to the main city. High-quality residential communities are distributed in the inner ring of the suburbs. Relocated urban households and high-income migrants are their main residents. In the middle ring, industrial parks, rural enterprises, and dormitories for employees serving nearby enterprises and the rural population are the main occupants. The outer ring has no clear boundaries with the middle ring and has a landscape of rural areas.

The Location of Jiangning New Town

Jiangning new town is located in the southwestern suburbs of Nanjing. It is one of eight new towns which were officially established in the 2006–20 Jiangning Master Plan. The planned land area is 36 square kilometres by 2020 (figure 16.1).

Since 2000, suburban development in Nanjing has evolved into a new form, with vast investments involved in property development. The suburban form was created by the decentralization of manufacturing, administration, retailing, and public services. Today, the suburban pattern increasingly involves mixed land uses.

The new round of suburbanization in Jiangning was started by the Jiangning Economic and Technological Development Zone (ETDZ) as a platform to develop export-oriented industries and to accommodate capital flow. By improving transport infrastructure, such as building highway connections between Jiangning, the airport of Nanjing, and other metropolitan areas, as well as the development of the largest high-speed railway station in China, the government tried to form an environment for foreign investment. In terms of economic growth, the percentages of agriculture output value in Jiangning have changed from 21.63 per cent of GDP in 1995 to 7.42 per cent of GDP in 2005; meanwhile, the percentage of services sectors in GDP has

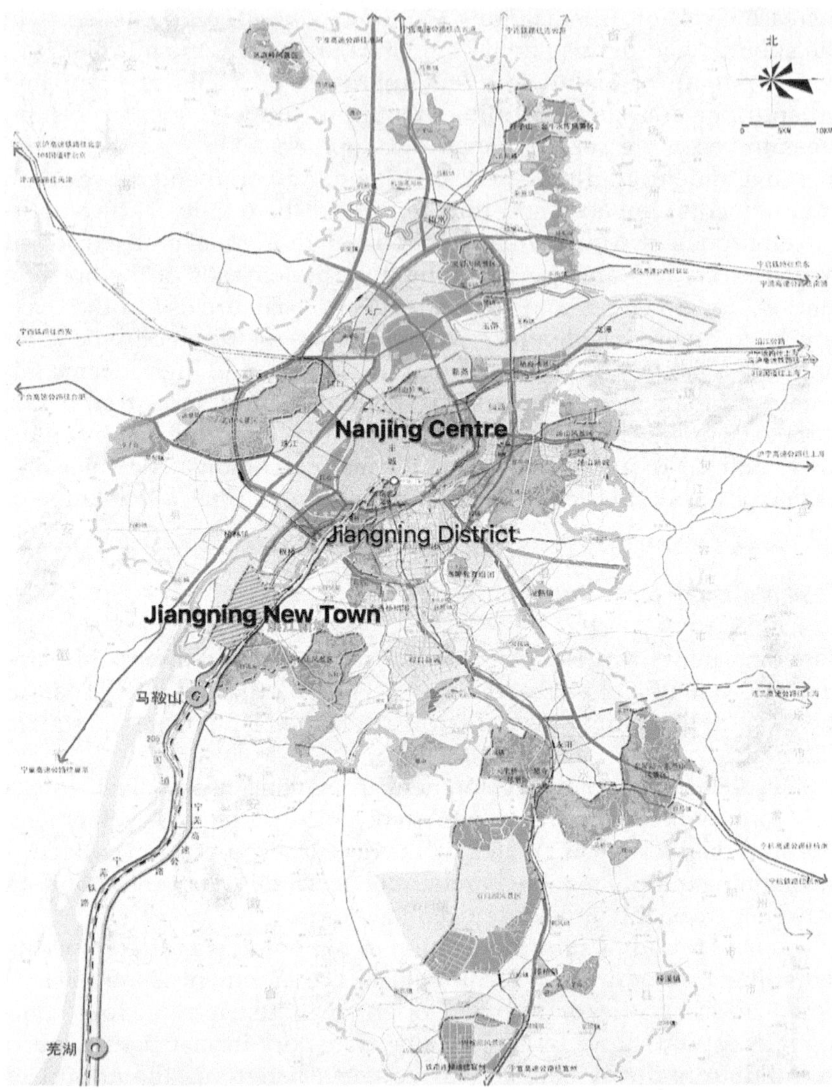

16.1 The location of Jiangning new town in Nanjing. Modified from 2006–20 Jiangning Master Plan.

changed from 17.85 per cent to 31.23 per cent. The level of urbanization – that is, the percentage of urban population to the total population of Jiangning – rose from 22.2 per cent in 2000 to 61.8 per cent in 2005, indicating that Jiangning has become urbanized due to its economic growth and structural adjustment (Jiangning Statistical Bureau 2009).

Massive Suburbanization

Suburbanization as a Development Strategy

The suburbanization process in Jiangning started with Dongshan township. The government of Nanjing set up an ETDZ by relocating the existing rural population and developing commodity housing. Since 1990, Nanjing had experienced massive population growth. Spatially, the city rapidly sprawled into the inner suburbs. At that time, the inner ring just around the central city still consisted of a mostly rural landscape. Surrounded by rural land, the ETDZ was set up to solve the problem of scattered developments. Evidently, the initial development of Jiangning was influenced by a process of industrialization. Furthermore, local enterprises mainly relied on manufacturing and exports. The municipal government aggressively expanded the spatial boundary to allow decentralization.

When the export sector encountered difficulties during the 1998 Asian financial crisis, local enterprises suffered tremendous decline, which resulted in an economic downturn. However, the government shifted its role to boosting domestic demand by stimulating the housing market and investing in infrastructure to sustain economic growth. Evidence shows that in 1999 residential land already accounted for 26.1 per cent of total land use in Jiangning. Meanwhile, residential clusters had formed on the suburban fringe.

In order to further boost the growth of Jiangning, the municipal government adjusted the administrative boundary in 2000, which expanded Jiangning and allowed it to acquire more land for development. In 2000, the municipal government raised the administrative level of Jiangning from a county to a district. In order to attract the commercial and financial sectors, the built environment has been improved by attracting real estate developers in commodity housing. A vast amount of investment in residential projects was injected into Jiangning.

The new round of the 2000–30 Nanjing Master Plan also proposed a strategy of boosting the growth of the land market. The Master Plan presented an intention to develop Nanjing as a metropolitan area and to develop service industries rather than export manufacturing industries. First, the ultimate goal is to build Nanjing as a partner of Shanghai and an international city. The plan aims to build a new polycentric pattern for population and service industrial concentration at a larger metropolitan scale. Second, the policy of rural-urban integration becomes a driving force to promote economic structural change, innovation, and service industries. Third, policies have been proposed by Nanjing to gain a range of investments in property. Importantly, massive land and housing development becomes the crucial force. The average housing price per square metre has reached more than 14,000 Yuan within ten years, from only 3,800 Yuan in 2005; the entire population of Jiangning reached 935,659, including an increase of 23,345 local registered residents. This number indicates that massive land development and the adjustment of economic structure had attracted a vast number of migrants to live in Jiangning, which not only increased the urbanization process but also provided a labour force.

State entrepreneurialism at the district level is also being used to receive profits from land leasing and contribute to fast GDP growth in a competitive environment. Crucially, the suburbs themselves had already undergone a degree of transformation due to economic reform. Additionally, the marketization of land development pushed township enterprises from traditional rural industries into urban agriculture, such as urban farms, and tourism areas in the suburbs.

However, huge profits from land development also led to small-scale developments in the surrounding towns. The central government had to reinforce its regulation of land development due to the pressure on agricultural land. Its policy aims to control scattered development in the suburbs by processing the planned annual quota for the total amount of land allocated to urban construction. In 2004, the central government also initiated a new policy on quota allocation and indicated that if an amount of built-up land in the countryside was reclaimed for agricultural uses, an equal amount could be added to the total quota for construction land in the urban area. In these circumstances, the Nanjing municipal government initiated a new strategy for suburban land development known as "One City, Three Districts." The strategy proposed a new metropolitan structure for Nanjing to implement the strategy of "Three Concentrations" – concentration of industry towards planned industrial parks, population towards cities and towns, and land development towards larger settlements. In this strategy, Jiangning was selected as one of the new metropolitan districts in which to concentrate at least 600,000 urban residents. The new metropolitan layout of Nanjing shows a polycentric structure consisting of different sizes of settlements.

In sum, the development of new towns integrated employment and residential uses, which was a strategy for growth. Property development in the new towns contributed to local revenue and taxes. The municipal government aimed to adjust the industrial structure by promoting a service economy in the new towns. The diverse suburban residential areas reflected the development of a new economic pole for population concentration in the Chinese suburbs.

Diverse Suburban Populations

Housing consumption intensifies social differentiation and creates an identity that signals lifestyle formation. The process of social differentiation has also led to growing disparities within Chinese suburbs. Before economic reform, Chinese suburbs covered a sparse residential area (danwei housing) developed by stated-owned industries. Large numbers of young families and blue-collar workers occupied this type of settlement to serve the nearby factories. Chinese housing reform brought housing commodification. Residents were able to obtain housing through the housing market, with or without subsidies. With the establishment of a housing market, commodity housing became the primary means of new housing supply. Commodity housing involves a range of packaging and branding activities to attract consumers. Higher-income or better-off households moved to the suburbs in search of an aesthetically pleasing suburban living environment. Packaging and branding practices manifested aesthetic features of living conditions that represented a "civilized modernity," which can distinguish better-off suburbia from backward rural and traditional old neighbourhoods (Pow 2009; Wu 2010; Zhang 2010). Suburban commodity housing has produced stratified standards of housing (Huang and Clark 2002). In order to stimulate housing demand, all commercial banks were encouraged to offer low-interest mortgages, which provided the possibility for existing urban households to become homeowners. Suburban homeownership is now consciously promoted as a symbol of well-being by both the government and developers.

With greater purchasing power, the emerging urban rich and new middle classes have become homebuyers in the suburban housing market. Purchasing a new home in the suburbs does not just mean the choice of a better place to live but also means choosing a lifestyle. For these social groups, the imagined suburban lifestyle refers to the establishment of their own social distinction. The aesthetics of suburban gated communities with strong branding and packaging practices materialized their sense of desirability (Wu 2004). The distinctive suburban landscape offers a private location for the new rich to realize their aspirations through spatial exclusion and lifestyle practices (Zhang 2010). With the rise of suburban property values, purchasing a second home in the suburb is an efficient and stable investment method of increasing personal wealth. The suburban housing market has become an object of speculation. Due to the policy of land quotas to protect land resources, luxury and low-density suburban villas became extremely rare. Housing prices are far beyond the affordability of ordinary households; thus, affordable housing was developed in the suburbs. Individual homebuyers are able to borrow from the Housing Provident Fund. As relatively cheap housing, affordable housing is controlled by the state. Residents who used to live in the central area benefited from the process of urban regeneration. Although there are different standards of compensation for relocation, the huge amounts of cash compensation for resettlement and a number of additional houses immediately turned these residents into the new rich. Selling or renting out redundant houses became a way of increasing capital, which gave the new rich a greater role in shaping the suburban real estate market.

Another group of suburban residents is formed by rural migrants. They come to large cities in search of better job opportunities, while often earning the lowest income and having the lowest occupational status, such as serving suburban middle-class families or nearby industries. Since they lack household registration in the area, migrants are unable to acquire subsidized housing, which makes them unable to acquire full property rights, nor can they access local schools or welfare programs. The rental of cheap housing or living in underredeveloped fringe areas became common residential choices (Wu 2008). The living conditions in these neighbourhoods are usually bad, typically characterized by overcrowding and poor facilities. However, the existence of rural housing and redeveloped neighbourhoods in peri-urban villages with extremely low rental prices but with a convenient location for the workplace is the most important factor that attracts rural-urban migrants to the suburbs (Wu 2008). Migrants are thus concentrated in suburban areas, forming informal suburban settlements.

Heterogeneous Suburban Residential Landscapes

The Villas of the Expanding New Rich

Different forms of housing are not only represented as personal preferences but are also closely associated with different lifestyles. The "good life" refers to a large house with a large yard. Thus, developers always advertise their properties with aesthetic architectural styles and imagined "noble" lifestyles. With the intention of satisfying different preferences in the growing affluent class, villa housing with different styles is provided in the housing market. Meanwhile, developers are also aware that spatial forms can transform lifestyles and eventually Chinese society itself (Wu 2010). The development of suburban villas has gone beyond the economic sphere and will reshape Chinese suburbanization by creating different kinds of suburban lifestyles, as suggested by a real estate manager: "The first phase of a project is developing suburban villas from 300 square metres to 800 square metres per household. We invited a design team from France to provide a pure European continental lifestyle for our house buyers. Afterwards, the second phase of housing development will concentrate on developing large-scale apartments from 200 square metres to 400 square metres per household" (Interview with a real estate manager from Vanke Real Estate, 7 November 2015).

The reaction from developers has clearly verified middle-class housing preferences as the suburban housing market expanded. Various high-end housing choices are provided for middle-class buyers who want to upgrade their living standards by changing their houses, as a suburban villa is considered luxurious in the increasingly stratified housing market. In the Chinese context, luxury housing has an atmosphere of aristocracy and is privileged. The density of suburban areas is planned and controlled, which results in a limited supply of villa-type housing. However, a larger living space is still a priority for homebuyers who wish to buy a home in the suburbs. In particular, along with the growing population of younger new rich in Nanjing, modern and luxury serviced apartments with a large floor space have become a popular choice. A real estate agent commented: "Previously, homebuyers came to our showroom and the first question asked was how much is one apartment? Several years later they began to care about the possibilities to buy a villa and criticized our housing form and façade. Now we have found that a villa is not the first choice for many home buyers, because they prefer larger apartments rather than villas" (Interview with a real estate seller from Greenland Holding Group, 7 November 2015).

However, suburban villas with large gardens are still the most common housing type in high-end suburban communities in Jiangning. Take, for example, the suburban villa community Top Regent Park Villa (TRPV), developed by Top Regent Real Estate (TRPE), a company founded by Sun Yat-sen Mausoleum Authority (SYMA) and China State Construction International Holdings Ltd, HK (CSCI). TRPE indicated that "their target house buyers are exclusively foreign businessmen and affluent senior government officers by using notions of 'power,' 'luxury,' and 'exclusivity' to distinguish them from other suburban housing" (Interview with a former member of real estate developer from TRPE, 8 November 2015). TRPV was the first luxury suburban villa community in Nanjing before 1995. The emerging suburban villa community rapidly attracted the attention of the affluent class. Thus, developers announced that their properties had sold out a few months after construction. Nowadays, suburban villa communities account for about 9 per cent of the total number of gated communities in Nanjing.

In recent suburban residential developments, the theme of exclusiveness is still used in the advertisement of luxury communities, but with more selling points. Developers attempt to create higher-quality living environments by using distinctive themes such as "low-density," "greener," "ecological," "luxury," and "livable" (Wu 2010). The luxury suburban villa community developed by private developers was targeted at upper-class housing consumers as well as foreigners; they were developed along the lines of exclusive lifestyle enclaves with condominium facilities such as swimming pools, club houses, restaurants, tennis courts, and other amenities. Some housing projects even contain on-site schools and medical facilities that serve only residents. Overall, exclusive high-class enclaves emerged and have led to urban segregation.

Cultural elements are widely used to help produce fresh waves of newness and diverse niche markets and to dazzle the populace with glamour consumption (Jameson 1991). The uniqueness of Western-themed villas in China brings an image of the "good life" to Chinese households who aim to change their lifestyle by changing their houses. Another example of luxury suburban communities is Master-Land Villas. The developer adopted this outlandish English name to create this suburban community with a sense of exoticism but without any relation to its meaning. The aim is not to refer to a Western living style, but to refer to the social status of the householders. As pursuing a luxury suburban villa in China is an emotional experience through which house buyers attempt to fulfil their aspirations to a Western luxury lifestyle, the names of suburban luxury communities like Master-Land Villas need to be constantly enchanting with symbolic meaning. Buying a luxury suburban villa is also a socially embedded process, which gives the Chinese new rich a way of establishing their identity through a distinctive lifestyle (Zhang 2010).

Since 1995, due to the popularity of Top Regent Park Villa, the developer, Jinling Real Estate, has tried to bring so-called themes of affluent elegant taste such as antique collection, painting, and wine, into their suburban villas, distinct from other previous suburban villa projects, which only use natural landscapes and high privacy. The intention is to satisfy the housing demands of a rising affluent class by letting them use their luxury suburban villa to show that they are not only wealthy but also have taste. Therefore, despite using Italian architectural elements in the housing forms, other cultural symbols and slogans from the imagination of European lifestyle are also put into the brochures. Meanwhile, the developer sets up a series of cultural events exclusively for selected affluent house buyers. However, this strategy is not always successful, as a house buyer remarked: "I believed that Master-Land Villas was the best luxury community in Nanjing right now as the architectural form and interior design are very similar to what I saw when I had a tour to Europe last summer. I have seen many villa communities in Jiangning, but they are just luxury houses and not my taste!" (Interview with an anonymous middle-aged female house buyer, 19 June 2015).

In terms of the architectural form of luxury villa homes, developers in suburban home development always proudly announce that their residential project is designed by an architectural team with high-prestige educational backgrounds and a full understanding of Western tastes. Their projects are often labelled to show cultural sophistication, with symbolic elements of ancient Roman architectural style as well as, for instance, marble bridges, gardens, artificial lakes, etc. A sales staff member commented: "To be honest, our designer is from Guangdong with one year's overseas experience only in France. Our boss has complained several times that the form of the villa is not like what he saw in Italy. We attached a cultural theme to this project in order to distinguish the properties from most other luxurious villa projects you may find in Nanjing, especially in Jiangning. Luckily, our clients quite liked it and bought it!" (Interview with a senior house sales staff member from Master-Land Villas, 19 June 2015).

16.2 The villas of the Master-Land project in Jiangning, Nanjing. Photo by Tianke Zhu.

Like other luxury suburban communities, Master-Land Villas is gated and controlled by a young and professional security team, and the entire community is monitored by 24-hour CCTV. To access the community requires a security pass or fingerprint. In addition, security has the right to refuse visitors, who attempt to enter the neighbourhood without appropriate reasons. As a security staff person commented, "You know, even potential house buyers are required to provide at least 2 million funds certificate in order to have a viewing access. Our homeowners include television stars, celebrities, basketball players and some entrepreneurs. If their privacy is violated and they get angry, we will be fired by our boss as well" (Interview with a security leader from Master-Land Villas, 19 June 2015).

In terms of the economy, district government, developers, and financial institutions are interested in developing high-end residential projects. Driven by political achievement and fiscal income, municipal and district governments enthusiastically keep on selling suburban land. House buyers aim to sustain their capital value by seeking and purchasing high-quality housing in suburbs. Thus, the combination of the above factors leads to the development of suburban communities. Furthermore, high-end suburban residential development brings huge profits from land development, leading to a lack of supply of housing for young families.

Banks and financial institutions lend money to developers to get profits. Credit is a major source of funding for suburban residential development. Mortgages are low risk, which increases housing prices. As explained by a real estate developer:

> In fact, the government encourages us to develop high-end communities. They are more willing to sell the land to a high-price project, which helps boost the price of surrounding land. Sometimes, we have to promise the government the lowest housing price in our residential project; otherwise, they will not lease the land. The aim of government is to raise the land price and get more profit in the further process of land leasing. We cooperate with a selected bank and help house buyers to get a mortgage from this selected bank, and the bank is willing to lend money to us for high-end residential development, as they get more profits from both sides. (Interview with the manager of a real estate company, 30 January 2015).

Commodity Housing Apartments for the Middle Class

Besides luxury residential projects, most properties in the suburbs are for homebuyers who want to have a new home to settle into before marriage. Jiangning attracts a large number of migrants, who come to Nanjing for better jobs. Many universities have set up their campuses in Jiangning in order to expand their space. An industry park was established in 2005 by the district government, which aimed to serve new entrepreneurs by providing low-rent offices and a tax-free strategy to encourage young talents to establish their businesses in Jiangning. Thus, suburban apartments have become the most common residential type, accounting for about 56 per cent of the total number of suburban homes in Nanjing. The housing projects are mainly developed as mid- and high-rise apartments with facilities. The number of residents of a gated community range from hundreds to thousands. The level of social class diversity in these suburban apartments project varies based on housing prices. More importantly, smaller apartments with an affordable housing price became attractive to those who are not married but wish to buy properties, as the timing of marriage in China is becoming delayed in general. For instance, developers used different strategies

16.3 The middle-class gated community of Vanke Paradiso in the new town of Jiangning, Nanjing. Photo by Tianke Zhu.

to create a community with different housing types and facilities in different phases. Their primary aim was to create a "high-profile" and self-contained "livable" suburban community. A real estate salesman commented: "The smaller size, one or two-bedroom apartment, is our best-selling type. The price for each property has already covered the furnishings and decoration. This development project is for young families, so we provide a living room larger than other similar properties in the market. The living room can be divided as part of your working area or you can invite your friends to your house to have a party on the weekend. Meanwhile, a twenty-four-hour steward will be in the luxury lobby for any assistance needed" (Interview with a real estate salesman from Vanke Real Estate, 8 November 2015).

Suburban land remains cheaper than land in the inner city of Nanjing. The comparatively low cost gives young couples a chance to upgrade their housing by moving to the suburbs. Before 1995 in Nanjing, many developers were invited by the district government to develop second homes, which produced an impression that Jiangning was a rural place without sufficient urban facilities. To advertise suburban apartments to young families, developers created a discourse of "community" that gave an image of new homes for young couples and established property management teams to provide all ranges of service to suit young homebuyers' needs, in order to meet their concerns about moving out of the urban centre. A twenty-six-year-old homeowner noted appreciatively: "My husband and I moved here two years ago after marriage, the price was 10,000 Yuan per square metre which is affordable for both of us. We decided to move here because the new home has already been furnished by developers, which saved a lot of money for us!" (Interview with female homeowner, 4 November 2015).

As the Chinese urban population experienced a "consumer revolution" thanks to rising income (Davis 2000), they desired to improve their housing conditions. Thus, developers tried to provide services in commodity housing estates in the suburbs to attract young families to move out of the city. The local government is responsible for infrastructure provision – for example, roads for commuting (Lin 2007). Young to middle-aged couples, sophisticated-looking middle-class professionals, seek to own a stylish home in the city, ensuring life quality and family stability. Young families represent a growing middle class with stable incomes and strong ability to pay a mortgage. They purchase a property to become a member of the homeowner class. Rising house prices attract them to own property as young as possible in order to secure their lives. With the traditional belief that owning a home means owning security, young Chinese couples tend to save to purchase a home. Migrants from other cities want to buy a property to change their hukou (residential registration) in Nanjing to gain access to social welfare benefits and the right to attend local schools.

Housing for Relocated Households

Rapid suburban residential development still leads to insufficient housing supply. On the one hand, developers intend to attract more middle-class residents by providing high-quality suburban apartments in order to maintain the growth of the housing market; on the other hand, government aims to transfer industries into suburbs to increase population concentration. In this process, both migrants and existing suburban residents experience the increasing living costs of suburban life.

However, poverty has also emerged in the Chinese suburbs. In Nanjing, the municipal government began to redevelop existing settlements in the peripheral area by relocating residents to new towns. A vast number of rural settlements were demolished to develop infrastructure and public facilities. The strategy was to replace old buildings with new homes in the area and compensate via equal housing area based on the number of family members in the new residential development. Two methods were used. One was to relocate residents into a community that was built for rehousing. The other was to relocate them to a commodity housing estate to live with ordinary house buyers. The land in the town centre was developed for commercial uses. Furthermore, the local government also changed the rural hukou of local residents to urban hukou to make them have access to social welfare, in order to facilitate the relocation process. Most local residents were compensated with more than one house by relocation. Their vacant houses were rented out to migrants and young graduates. Apartments with two or three bedrooms were often subdivided and leased to several migrant families separately. This type of rental is called qunzun (literally, co-renting), and has become a way for migrants to find low-cost rental housing in the suburbs.

Because large numbers of vacant suburban apartments were rented out to migrants, housing conditions declined. Some suburban rehousing areas became poor neighbourhoods. The district government realized that the previous policy of concentrating relocated residents in the same neighbourhood and a compensation policy based on the number of household members were inappropriate. There is also another, hidden, reason. On the one hand, as land prices increased, it was difficult to build large-scale rehousing enclaves because of the high cost of construction. On the other hand, large-scale resettlement enclaves led to the decline of land prices in the surrounding areas. In the case of Jiangning, most developers were required by the local government to accommodate a certain number of relocated residents in their new commodity housing projects in order to implement land leasing. Therefore, the recent suburban apartment community in Jiangning emerged as a heterogeneous social area composed of both homebuyers and relocated householders. Driven by profit, developers usually provide apartment buildings for relocated residents, with a higher density than other commodity housing projects within the same neighbourhood. It became common to see mixed types of housing in the same community, with high-level condominiums and mid-level apartments. In order to maintain community quality, the façades of relocation apartments and commodity apartments are similar, but the interior furnishings and layout

16.4 The estate of Tiandi new town for relocated households in Jiangning, Nanjing. The image shows relocation housing mixed into a suburban commodity residential community in Jiangning, contrasting the extreme high density with other less dense forms. At the same time, the project contains a large number of co-renting housing units. Photo by Tianke Zhu.

are different. As social diversity is strongly suggested by the master plans, developers increasingly promoted their housing projects by using phrases such as "high-end community," "international community," "bourgeois communities," etc., even though the development concentrated many relocated residents. The structure of the population is heterogeneous. This development strategy is widely used in newly built suburban apartment communities to reduce the negative impacts of a heterogeneous suburban living environment.

Conflicts can arise between commodity housing owners and relocated households in the same neighbourhood. Some developers have to build another gate inside the neighbourhood to separate the sections of commodity housing from the rehousing area, and even apply different property management fees. This method is also used in estates with mid-level apartments and villa-type housing. Developers tend to apply different themes to each residential cluster or use different phases of development to reduce the impact of diverse housing types within the same estate, while in fact the demand for homogeneous living by young property buyers has increased. In addition to a better living environment, young home buyers choose to move to Jiangning because of the expectation of increasing property values in the China's growing housing market. The residents of gated communities have a strong sense of property rights and care about their new settlements: "I strongly feel that my property value will be decreased if relocated households do not care about the environment of community" (Interview with an angry female homeowner, 25 October 2015).

Furthermore, marketing also targets young children, because they spend a lot of money and have an influence over their parents' expenditure. Parents want the best for their children. They prefer a socially homogenous environment in the suburbs and regard this as a necessary condition for their moving to the suburbs. Although developers want to build suburban communities for residents with similar social backgrounds and preferences, the mode of suburban development means that the Chinese suburbs have heterogeneous living styles.

Conclusion

Chinese suburbs are heterogeneous places. This chapter has described the development of large-scale suburban estates in a new suburban district of Nanjing, China. Through a systematic scrutiny of diverse suburban residential types, we have selected representative residential neighbourhoods for an in-depth study. The roles of the state and market were mapped along with the residential typology. For each category, we have attempted to reveal how the residents eventually landed in these estates. The governance of these estates continues to reflect how they were initially built, leading to quite different suburban worlds. We argue that the theoretical generalization of the modalities and mechanisms of suburban governance (Ekers et al. 2012) must be enriched through unpacking the typology of suburban residential spaces at a much finer scale. It has been shown that both suburban patterns and population compositions are more diverse than what is traditionally imagined. However, the diversity of spatial forms of suburban residences varies with market demand, residential preferences, and consumption. It was also found that the diversity of suburban population composition varies with distance from the central city and the income levels of the residential neighbourhoods. The diverse features of suburban residential development also can be found in the motivations of various groups.

ACKNOWLEDGMENTS

In addition to grant support from the Major Collaborative Research Initiative "Global Suburban Governance, Land and Infrastructure in the 21st Century" from the Social Sciences and Humanities Research Council of Canada, I would like to acknowledge the support of the UK ESRC research project "The Financialisation of Urban Development and Associated Financial Risks in China" (ES/P003435/1).

REFERENCES

Bourne, L.S. (1996). "Reinventing the Suburbs: Old Myths and New Realities." *Progress in Planning* 46, no. 3, 163–84. https://doi.org/10.1016/0305-9006(96)88868-4

Davis, D. (2000). *The Consumer Revolution in Urban China.* Berkeley: University of California Press.

Deng, F.F., and Huang, Y. (2004). "Uneven Land Reform and Urban Sprawl in China: The Case of Beijing." *Progress in Planning* 61, no. 3, 211–36. https://doi.org/10.1016/j.progress.2003.10.004

Ekers, M., Hamel, P., and Keil, R. (2012). "Governing Suburbia: Modalities and Mechanisms of Suburban Governance." *Regional Studies* 46, no. 3, 405–22. https://doi.org/10.3138/9781442663565-005

Feng, J., Zhou, Y., and Wu, F. (2008). "New Trends of Suburbanization in Beijing since 1990: From Government-Led to Market-Oriented." *Regional Studies* 42, no. 1, 83–99. https://doi.org/10.1080/00343400701654160

Fishman, R. (1987). *Bourgeois Utopias: The Rise and Fall of Suburbia.* New York, NY: Basic Books.

Hamel, P., and Keil, R. (eds). (2015). *Suburban Governance: A Global View.* Toronto, ON: University of Toronto Press.

Harvey, D. (2005). *A Brief History of Neoliberalism.* Oxford, UK: Oxford University Press.

Harris, R., and Lewis, R. (1998). "Constructing a Fault(y) Zone: Misrepresentations of American Cities and Suburbs, 1900–1950." *Annals of the Association of American Geographers* 88, no. 4: 622–39. https://doi.org/10.1111/0004-5608.00115

He, S., and Wu, F. (2009). "China's Emerging Neoliberal Urbanism: Perspectives from Urban Redevelopment." *Antipode* 41, no. 2, 282–304. https://doi.org/10.1111/j.1467-8330.2009.00673.x

Huang, Y., and Clark, W.A. (2002). "Housing Tenure Choice in Transitional Urban China: A Multilevel Analysis." *Urban Studies* 39, no. 1, 7–32. https://doi.org/10.1080/00420980220099041

Jameson, F. (1991). *Postmodernism, or, the Cultural Logic of Late Capitalism.* Durham, NC: Duke University Press.

Jiangning Statistical Bureau (JSB). (2009). *Jiangning Statistical Yearbook.* Nanjing, China: JSB.

Lin, G.C. (2007). "Chinese Urbanism in Question: State, Society, and the Reproduction of Urban Spaces." *Urban Geography* 28, no. 1, 7–29. https://doi.org/10.2747/0272-3638.28.1.7

Phelps, N.A., Wood, A.M., and Valler, D.C. (2010). "A Postsuburban World? An Outline of a Research Agenda."*Environment and Planning A* 42, no. 2, 366–83. https://doi.org/10.1068/a427

Pow, C.-P. (2009). *Gated Communities in China: Class, Privilege, and the Moral Politics of the Good Life.* Abingdon, UK: Routledge.

Shen, J., and Wu, F. (2013). "Moving to the Suburbs: Demand-Side Driving Forces of Suburban Growth in China." *Environment and Planning A* 45, no. 8, 1823–44. https://doi.org/10.1068/a45565

– (2017). "The Suburb as a Space of Capital Accumulation: The Development of New Towns in Shanghai, China." *Antipode* 49, no. 3, 761–80. https://doi.org/10.1111/anti.12302

Wu, F. (2002). "China's Changing Urban Governance in the Transition Towards a More Market-Oriented Economy."*Urban Studies* 39, no. 7, 1071–93. https://doi.org/10.1080/00420980220135491

– (2004). "Residential Relocation under Market-Oriented Redevelopment: The Process and Outcomes in Urban China." *Geoforum* 35, no. 4, 453–70. https://doi.org/10.1016/j.geoforum.2003.10.001

– (2010). "Gated and Packaged Suburbia: Packaging and Branding Chinese Suburban Residential Development." *Cities* 27, no. 5, 385–96. https://doi.org/10.1016/j.cities.2010.06.003

Wu, F., and Phelps, N.A. (2008). "From Suburbia to Post-suburbia in China? Aspects of the Transformation of the Beijing and Shanghai Global City Regions." *Built Environment* 34, no. 4, 464–81. https://doi.org/10.2148/benv.34.4.464

Wu, F., and Zhang, J. (2007). "Planning the Competitive City-Region the Emergence of Strategic Development Plan in China." *Urban Affairs Review* 42, no. 5, 714–40. https://doi.org/10.1177/1078087406298119

Wu, W. (2008). "Migrant Settlement and Spatial Distribution in Metropolitan Shanghai." *The Professional Geographer* 60, no. 1, 101–20. https://doi.org/10.1080/00330120701724210

Zhang, L. (2010). *In Search of Paradise: Middle-Class Living in a Chinese Metropolis*. Ithaca, NY: Cornell University Press.

Zhang, T. (2000). "Land Market Forces and Government's Role in Sprawl: The Case of China." *Cities* 17, no. 2, 123–35. https://doi.org/10.1016/s0264-2751(00)00007-x

Zhou, Y. and Ma, L.J. (2000). "Economic Restructuring and Suburbanization in China." *Urban Geography* 21, no. 3, 205–36. https://doi.org/10.2747/0272-3638.21.3.205

Conclusion

Massive Suburbia – From the Legacy of the Habitat to the Financialization of Housing in the Planetary Periphery

K. MURAT GÜNEY, ROGER KEIL, AND MURAT ÜÇOĞLU

This book has contributed to a growing literature making the case that the North American view on suburbanism as a distinct way of life does not suffice anymore in understanding the suburbanization of the planet (Keil 2018). Neither does the North American example itself, and this book has helped broaden the perspective beyond the ground-related residential home, long the symbol of the American way of life, as the only morphology of global suburbanization. There are in fact variegated densities and shapes, from informal settlements in some cities of the Global South to state-sponsored high-rise estates such as those in China and Turkey. Even in North America itself, in the edge cities and edgeless cities along the suburban highways in the United States (Blumgart 2018) – but more specifically in Canada, as the chapter by Young in this book aptly demonstrates – high-rise suburbia has been a massive provider of housing (see also Charmes and Keil 2015). The particular form in all cases varies slightly from national case to national case. As the chapters in this book have shown, whether in Eastern Europe (Bernt; Logan) or in Western Europe (Kipfer; Tchoukaleyska), in Canada (Young), Turkey (several chapters in this book), or in China (Zhu and Wu), the twentieth-century legacy of building and rebuilding large-scale, massive housing on the periphery persists in the twenty-first. While the concrete world of these buildings and neighbourhoods is recognizable as a particular morphological feature of what Lefebvre ([1970] 2003) would have called habitat in the "bureaucratic society of controlled consumption," they were also structural units of late industrial urban life. Planned and built in huge industrial clusters, often with the help of prefab parts (Logan in this book), their size differed in relation to their administration and governance. In East Germany, for example, where some high-rise housing estates had populations upward of 30,000, a unit of a large housing estate (Großwohnsiedlung) might have included 2,500 apartments and 4–5,000 inhabitants, supported by appropriate services, while in West Germany, across the iron curtain, 1,000 housing units was classified as "large." After the fall of the Berlin Wall, a generation of soul searching among planners and rebuilding subsidized by the state, but also differentiation and nuance, ensued that recognized more diverse forms of housing (in the peripheries of both East and West) as "large housing estates" (Altrock, Grunze, and Kabisch 2018). Embedded in a discourse of crisis for a generation, the remaking of these massive housing areas has, as Kipfer tells us in his chapter in this book, always been about more than housing alone. As several chapters in this book have demonstrated, the changing demands of the legacies of large-scale housing in East and West have now long been recognized, the socio-demographic structure and governance systems have been diversified, and the massive housing estate, often pronounced dead in specialist and popular discourses, has found new inhabitants and new purpose – and new challenges (Altrock et al. 2018; Budnik et al. 2018; Kohout and Tittl 2018; Szafranska 2018). The legacy of massive suburbia has been shaping the discourse on spatial justice and the right to the city in many contexts. In particular, it has characterized the large renovation schemes that have been launched across cities in Europe and Canada (Berry-Chikhaoui and Medina 2018; Poppe and Young 2015).

While Eastern Europe (and to some degree Western Europe) can be seen as the heartland of massive suburbanization in large housing estates in the twentieth century, the balance has since shifted towards the Global South and the East, as is documented particularly in the chapters on China and Turkey in this book. In Istanbul alone, 4.2 million residences have been constructed in the past generation, most of them in massive suburban housing estates (Devrim 2016). In China and India, massive housing projects have been rolled out at the peripheries of and in between the large mega-urban regions of those countries (Wang, Ratoola, and Xiangming 2010). In Shanghai alone, nine peripheral new towns with up to 1 million inhabitants form the centrepiece of a giant decentralization strategy also involving sixty mid-size towns and six hundred villages in the region (Ranhagen 2014). These are just two outstanding examples; but around the world, as we saw in the chapters of this book, the suburbs are alive and well and full of diversity of form, structure, and population. In fact, we can speak of an era of massive suburbanization which is now upon us that links the monumentality of traditional modernism with the liberalized financial markets of today, supported and played by often autocratic national governments that have become masters of what Ananya Roy (2011) has termed "inter-referencing."

Such inter-referencing, especially among countries and cities outside the West and North, now determines the shape of things to come in the global periphery. As Isikkaya (2016, 316) notes for the case of Turkey, "government's big cities have resembled to [*sic*] each other by presenting a kind of characterless morphology of Dubai aesthetics ... deprived of its multi-cultural and unique urban and architectural identity according to the Islamic neo-liberal urban manipulations/interventions." Yet, if the example of the older, more mature estates in Europe and Canada are any indication, this "characterless" appearance may be in for a change sooner rather than later, and the inscription of a particular ideological significance ("Islamic neo-liberal" in this case) may melt into air, like the communist mode of housing regulation in the East and the social democratic mode of housing regulation in the West did under neoliberal pressures in the past generation, along with their forms of tenure, solidarity, and social coherence (Budnik et al. 2018; Keil 2018).

Moving further afield in terms of the morphology and structure of the suburban planet, we see a wide variety of massive peripheral developments in African and Asian countries, where the conversion of rural into urban land (uses) is rapid and extensive, and the financialization of land is particularly aggressive and widespread among various classes of urban and peri-urban inhabitants (Denis 2018). One of the emerging problems in the peripheries of the world, where building and real estate industries have been soaring, is the capture of increased values to cover costs for municipal services as citizen demands have risen. There is a broad spectrum here from heavy state involvement to complete state retreat (Saunders 2018). The complex and contested city-building processes in these new massive peripheries are examined in this book in the chapters by Belarbi and Rousseau on Morocco, Coker on the Philippines, Haas on Israel/Palestine, Rubin and Charlton on South Africa, Schmid on Egypt, and Zhu and Wu on China. These detailed empirical case studies burst with the life of the new global suburban periphery, but they also pose new conceptual and theoretical challenges surrounding the framing of what constitutes relationships between the centre and the periphery, the formal and the informal, and different forms of tenure, and about the significance of suburban land and property. They ask questions about the construction of and life in "the city yet to come" (Simone 2004), as citizens and migrants, residents and workers, original settlers and newcomers are negotiating urban differentiation in a growing massive suburban environment. Modalities of governance centred on the state and capital hold sway over the legacy suburbs of the twentieth century in Europe and North America and many of the recently built massive suburban new towns in the state-led planning efforts and developmentalist regimes of Israel, Turkey, China, the Philippines, Vietnam, South Africa, Singapore, and others. Even in Australia, which plans to add about 12 million people by mid-century, new town developments are back in the national discourse (Freestone, Taylor, and Bolleter 2018). The massiveness of the new construction is also linked to techno-utopian excess at an unprecedented scale, a virtual "Babel," as one German newspaper recently called this tendency of Babylonian impulses by often authoritarian states such as Saudi Arabia (Matzig 2017). The African examples here – Cairo, Casablanca, and Johannesburg – allow a clearer peek into the space between what Caldeira (2016) calls "peripheral urbanization" and the actions of the state and the market.

Despite significant diversity in terms of forms, modalities, densities, and types of construction of massive suburbia, a common trend in global suburbanization is the rise in financialization of land and housing, and a subsequent uneven urban development together with a dramatic increase in wealth inequality. Particularly in the developing Global South, but also in the developed economies of the Global North, housing and mega-infrastructure projects located in massive suburbia became the key drivers of economic growth as well as a central mechanism of rent generation. As Saskia Sassen (2016, n.p.) emphasizes, "In mid-2014 and 2015 alone, more than one trillion dollars was invested in real estate in just 100 cities across North America, Europe and Asia ... Urban land – not just buildings, but also undeveloped lots – is considered a good investment at a time when financial markets are shaky." As discussed particularly in the case of Istanbul (in the chapters by Çavuşoğlu and Strutz, and Güney, and Üçoğlu), at the cost of the displacement of local populations and the dramatic transformation of new and old neighbourhoods, governments are competing to attract foreign investors who are interested in urban development through suburban expansion. To do this, governments are introducing the largest airport, canal, bridge, high-speed rail, and mass housing projects of the world one after another. While such massive suburbanization creates enormous economic growth and capital accumulation throughout the world, the distribution of this added value is highly unequal.

For instance, the speculative house and land price increase caused by the construction-led economic growth fostered by finance capital has dramatically decreased the possibility of being a homeowner for low-income groups. Despite the massive scope of suburban expansion, the newly built projects were far from being enough to satisfy the affordable housing needs of the lowest-income families, who remained tenants, indebted and financially vulnerable. Since the global mortgage crisis in 2008 homeownership rates for low-income families have not increased, while the debts of the same group have risen gradually. In the same period, real-estate investors who own multiple apartment blocks and large quantities of land have increased the number of their properties and are the major beneficiaries of urban development through surburban expansion (Sayek-Böke 2015).

Another danger of such financialization is an increasing abstraction of housing as a financially exchanged commodity that is alienated from the actual users of housing, namely the residents. As a result, as Sassen (2016) emphasizes, on the one hand, we no longer know who owns our cities, houses, and neighbourhoods. On the other hand, the owners – corporate or individual investors who consider massive suburban housing, land, and infrastructural projects only as a financially profitable value – care less and less about issues related to the use of the housing that are important for residents, such as quality of life, affordability of housing and rents, and availability of access to education, healthcare, and cultural facilities (See Bernt's chapter in this book).

Clearly, the financial crash of 2008 has changed the rules of the game (Brooker 2018). On the eve of the crash, David Harvey (2008, 37) predicted an urban world where the "planet as building site collides with the 'planet of slums.'" Ten years and many repossessions, dispossessions, and re-evaluations later, we understand that much of that new world is suburban. Despite large-scale re-urbanization and re-centring of investment in the built environment of core urban areas, the periphery keeps sprawling outward in most places. While housing and infrastructure are gradually considered more as the stuff of financial investments and less as meeting basic human needs, massive suburbia, which today's financialized housing regime uses worldwide as its fertile terrain, faces the risk of being the site where global inequalities are not reduced but intensified.

The massiveness of the housing structures we see emerging in the world today is linked associatively, if not causally, to regimes of a semi-authoritarian or authoritarian nature. Here, the great promise of the democratic modernity encapsulated in a built environment of equality is turned into a neoliberal and controlled environment of limited opportunity in privatized, individualized, atomized cubicles of housing in places like China, India, Turkey, Singapore, or Hong Kong. These are the states that, with their totaliatarian gesture of omnipotence, paradoxically symbolize the impossibility of nation states in an age of globalization to control their economic, environmental, and social fate effectively (for a recent essay on the subject, see Das Gupta 2018).

This links to the increased problematization of governance in an era of sustained inter-referencing. Hong Kong, for example, is now eyeing Singapore's planning and governance experience as a relevant model for the city-state's burgeoning growth-induced problems (Wong 2018). This development is also linked to the story of the failed architecture of modernity, which used to be a progressive promise but is not anymore. While in today's suburbia the scale of housing construction and infrastructural projects is massive, so is the scale of financial profits made from these projects. Equally impressive, it seems, are the possibility of failure and the emergence of ruins that litter the post-suburban landscape of planetary suburbanization. The successes and failures also reflect back on the legacy of the modernist promise itself that once came with implications of a better world. In an interview on the meaning of modernism in today's world, architectural theorist Reinier de Graaf, with reference to Thomas Piketty's writings about inequality, notes that today's modernist architects may contribute to increasing inequality as they divest their stylistic innovation from social justice concerns:

> Modern architecture was once based on efficient, rational, fast industrial production to give as many people as possible a decent home. It made buildings cheap, so they were available to many people. That same ethos is still very present in modern architecture, but it makes buildings cheap – not to be sold cheap or to be rented cheaply – it makes buildings cheap so that they generate the highest possible return on investment. That same aesthetical ethos of saving, of an economic minimalism, now serves an entirely different purpose. It serves not the happy many, but the happy few. It's the same architecture in the context of a different system. I don't

> think anybody who practices modern architecture is even vaguely aware of this. They were all educated with Le Corbusier and Gropius, and they consider themselves the heirs of those heroes when they actually operate in a system where they are complicit in things completely at odds with the ideals of that movement and those heroes. (cited in Minkjan 2017, n.p.)

We do indeed need a more comprehensive investigation of the unequal effects of massive suburbanization in a period when housing is increasingly financialized, and then to rethink the "right to the suburb" by questioning how massive suburbia can be redesigned as a site that serves those who are in need of decent housing, and not those who exploit land and housing to extract more profit.

We live, indeed, in a world of continuous massive sub/urbanization. There is no escape from it conceptually or materially. You cannot move away from it. In this sense, the sub/urban doesn't have an outside. We began with a reflection on the limited use, in the twenty-first century, of the American example of suburbanization. We can conclude now that due to the pervasive and extensive crisis of the inner suburbs in that country, the landscape of massive suburbia has gained huge notoriety and importance in the (post)-Trumpian world of socio-spatial dissolution, neighbourhood decline, and disillusionment. As Wilson, Boodram, and Smith note in their chapter on the old suburbs of Chicago, these "most stigmatized zones of America's rapidly changing industrial suburbs" have become key waystations of the extended planetary urbanization we experience. The suburban dream has run dry in the country where it was invented. We also know that the peripheries of cities have been branded as places where "urban rage" takes root, from Ferguson, US, to Husby, Sweden (Dikeç 2017). In fact, the massive Swedish suburbia in that most social democratic of countries has also been the arena of a bizarre colonial boomerang (Abourahme 2018), as weapons of far-away wars such as hand grenades have made it to the streets of Stockholm's Varby Gard suburb and to the hands of local immigrant street gangs there (Barry and Anderson 2018). Yet, Kipfer and Dikeç at the outset of this book also remind us that we must be careful not to saddle the urban periphery with the guilt for all political things that go wrong in this difficult moment in human history. This book has sought to better understand the difficulties faced by the people in massive suburbia as we move into a world where their circumstances might become those of the majority of the world's population. Massive suburbia, in this light, must be understood not as an alien outlier but as a constitutive part of the "continuous city" in which we now live (Lerup 2017).

REFERENCES

Abourahme, N. (2018). "Of Monsters and Boomerangs: Colonial Returns in the Late Liberal City." *City* 22, no. 1, 106–15. https://doi.org/10.1080/13604813.2018.1434296

Altrock, U., Grunze, N., and Kabisch, S. (eds). (2018). *Großwohnsiedlungen im Haltbarkeitscheck Differenzierte Perspektiven ostdeutscher Großwohnsiedlungen* [Sustainability Check for Massive Housing Estates: Different Perspectives on East German Massive Housing Estates]. Wiesbaden, Germany: Springer.

Barry, E., and Anderson, C. (2018). "Weapons of War Escalate Violence." *The New York Times International Weekly*, 10–11 March, 1.

Berry-Chikhaoui, I., and Medina, L. (2018). "'Justice pour le Petit Bard': Contester la renovation et imposer la participation" [Justice for le Petit Bard: Contest the Renovation and Impose Participation]. *Métropolitiques*, 9 April. https://www.metropolitiques.eu/Justice-pour-le-Petit-Bard-Contester-la-renovation-et-imposer-la-participation.html

Blumgart, J. (2018) "Return to Edge City." *CityLab*, 10 April. https://www.citylab.com/design/2018/04/return-to-edge-city/552362/

Brooker, N. (2018). "How the Financial Crash Made Our Cities Unaffordable." *Financial Times*, 15 March. https://www.ft.com/content/cc77babe-2213-11e8-add1-0e8958b189ea?segmentid=acee4131-99c2-09d3-a635-873e61754ec6

Budnik, M., Großmann, K., Haase, A., Hedtke, C., and Kullmann, K. (2018). "Soziale Heterogenität und Zusammenhalt in Leipzig-Grünau" [Social Heterogeneity and Solidarity in Leipzig-Grünau]. In Altrock, U., Grunze, N., and Kabisch, S. (eds), *Großwohnsiedlungen im Haltbarkeitscheck Differenzierte Perspektiven ostdeutscher Großwohnsiedlungen*, 213–42. Wiesbaden, Germany: Springer.

Caldeira, T. (2016). "Peripheral Urbanization: Autoconstruction, Transversal Logics, and Politics in Cities of the Global South." *Environment and Planning D: Society and Space* 35, no. 1, 3–20. https://doi.org/10.1177/0263775816658479

Charmes, E., and Keil, R. (2015). "The Politics of Post-suburban Densification in Canada and France." *International Journal of Urban Research and Action* 39, no. 3, 581–602. https://doi.org/10.1111/1468-2427.12194

Das Gupta, R. (2018). "The Demise of the Nation State." *The Guardian*, 5 April. https://www.theguardian.com/news/2018/apr/05/demise-of-the-nation-state-rana-dasgupta

Denis, É. (2018) "Urban Desires and Lust for Land: The Commodification of Rural Spaces in the Global South." *Metropolitics*, 20 April. http://www.metropolitiques.eu/Urban-Desires-and-Lust-for-Land.html

Dikeç, M. (2017). *Urban Rage*. New Haven, CT: Yale University Press.

Freestone, R., Taylor, E., and Bolleter, J. (2018). "New Cities? It's an Idea Worth Thinking about for Australia." *The Conversation*, 14 March. https://theconversation.com/new-cities-its-an-idea-worth-thinking-about-for-australia-92990

Harvey, D. (2008). "The Right to the City." *New Left Review* 53, September–October. https://newleftreview.org/II/53/david-harvey-the-right-to-the-city

Isikkaya, D.A. (2016). "Housing Policies in Turkey: Evolution of TOKİ (Governmental Mass Housing Administration) as an Urban Design Tool." *Journal of Civil Engineering and Architecture* 10, no. 3, 316–26. https://doi.org/10.17265/1934-7359/2016.03.006

Keil, R. (2018). *Suburban Planet*. Cambridge, UK: Polity.

Kohout, M., and Tittl, F. (2018). "A Fragile Balance: The Legacy and Challenges of Czech Housing Estates." In Altrock, U., Grunze, N., and Kabisch, S. (eds), *Großwohnsiedlungen im Haltbarkeitscheck Differenzierte Perspektiven ostdeutscher* Großwohnsiedlungen, 273–96. Wiesbaden, Germany: Springer.

Lefebvre, H. (1970) 2003. *The Urban Revolution*. Minneapolis: University of Minnesota Press.

Lerup, L. (2017). *The Continuous City*. Houston, TX: Rice Architecture.

Matzig, G. (2017). "Babel-Bubble." *Süddeutsche Zeitung*, 26 October, 11.

Minkjan, M. (2017). "Reinier de Graaf: 'Architecture Is in a State of Denial." *Failed Architecture*, 6 November. https://www.failedarchitecture.com/reinier-de-graaf-architecture-is-in-a-state-of-denial/

Poppe, W., and Young, D. (2015). "The Politics of Place: Place-Making versus Densification in Toronto's Tower Neighbourhoods." *International Journal of Urban and Regional Research* 39, no. 3, 613–21. https://doi.org/10.1111/1468-2427.12196

Ranhagen, U. (2014). *Five New Towns in Shanghai: Present Situation and Future Perspectives.* Stockholm, Sweden: Department of Urban Planning at the Swedish Royal Institute of Technology (KTH). https://www.diva-portal.org/smash/get/diva2:1091458/FULLTEXT01.pdf

Roy, A. (2011). "Urbanisms, Worlding Practices and the Theory of Planning." *Planning Theory* 10, no. 1, 6–15. https://doi.org/10.1177/1473095210386065

Sassen, S. (2016). "Investment in Urban Land Is on the Rise – We Need to Know Who Owns Our Cities." *Global Thought, Columbia University*, 6 August. http://cgt.columbia.edu/news/sassen-investment-urban-land-rise-need-know-owns-cities/

Saunders, D. (2018). "The Global Property Boom Could Help Millions. Too Bad It's Going Untaxed." *The Globe and Mail*, 20 April. https://www.theglobeandmail.com/opinion/article-the-global-property-boom-could-help-millions-too-bad-its-going/

Sayek-Böke, S. (2015). "TOKİ Ranta Çalışmış" [TOKİ Worked to Increase the Rent]. *Hürriyet: Ekonomi*, 27 April. http://www.hurriyet.com.tr/selin-sayek-boke-toki-ranta-calismis-28853859

Simone, A. (2004). *For the City Yet to Come*. Durham, NC: Duke University Press.

Szafranska, E. (2018). "Large Polish Housing Estates in Transformation." In Altrock, U., Grunze, N., and Kabisch, S. (eds), *Großwohnsiedlungen im Haltbarkeitscheck Differenzierte Perspektiven ostdeutscher* Großwohnsiedlungen, 297–326. Wiesbaden, Germany: Springer.

Wang, L., Ratoola, K., and Xiangming, C. (2010). "Building for What and Whom? New Town Development as Planned Suburbanization in China and India." In Clapson, M., and Hutchison, R. (eds), *Suburbanization in Global Society*, 319–45. Bingley, UK: Emerald Group.

Wong, O. (2018). "Hong Kong's Urban Planning Problems Must Be Addressed, Experts Say. Is Singapore's Model the Answer?" *South China Morning Post*, 16 April. http://www.scmp.com/news/hong-kong/economy/article/2141808/problems-hong-kongs-urban-planning-must-be-addressed-experts

Contributors

Wafae Belarbi is an architect and a lecturer in urban planning and urban sociology at the École Nationale d'Architecture (Rabat, Morocco). She defended a PhD in geography on the mobilizations of inhabitants and territorial regulation in the southern periphery of Casablanca in 2011. Her research programs deal with social mobilizations in the urban peripheries of Morocco and strategies to integrate people in informal neighbourhoods.

Matthias Bernt is senior researcher at the Leibniz Institute for Research on Society and Space in Erkner, Germany. His research focuses on urban regeneration and governance, shrinking cities, and gentrification.

Basmattee Boodram, PhD, MPH, is an infectious disease epidemiologist and a research assistant professor in the Division of Epidemiology and Biostatistics at the University of Illinois at Chicago School of Public Health. Her research focuses on the intersection of substance abuse and infectious disease among young, predominantly suburban youth.

Sarah Charlton is based in the Centre for Urban and Built Environment Studies (CUBES) in the School of Architecture and Planning at the University of Witwatersrand, Johannesburg. Her research explores the geographies of home and work, housing, and state infrastructure interventions and people's responses to these. She holds a PhD in Town and Regional Planning from the University of Sheffield.

Abidemi Coker received her PhD in Social and Public Policy from the University of Jyväskylä in Finland. Her research interests include urban informality, housing and social protection policy, and the dynamics between actors. She is Coordinator for Human Rights and Democracy at the Embassy of Finland in Lusaka, Zambia.

Erbatur Çavuşoğlu is a faculty member in the Urban and Regional Planning Department at Istanbul Mimar Sinan University. He has contributed to the scholarship about this city especially with regard to urban renewal, the construction sector, and expropriation. As an activist planner, he has initiated participatory planning processes in several Istanbul neighbourhoods, including Gülsuyu-Gülensu and Sulukule, and has published extensively in Turkish media on this topic.

Mustafa Dikeç is a professor of Urban Studies at the École d'Urbanisme de Paris, and visiting professor at Malmö University. His most recent book, *Urban Rage*, was published by Yale University Press in 2017. He is also the author of *Space, Politics and Aesthetics* (Edinburgh University Press, 2015) and *Badlands of the Republic: Space, Politics and Urban Policy* (Blackwell, 2007). He is one of the editors of *International Journal of Urban and Regional Research* (IJURR).

K. Murat Güney is a lecturer in the sociology department at Acıbadem University in Istanbul. Previously he taught in the anthropology department at Boston University and worked as a postdoctoral researcher in the City Institute at York University. His research interests include urban anthropology, anthropology of development, and political economy, with a particular focus on suburbs in the Global South. He received his PhD in anthropology from Columbia University. His dissertation research focuses on the negative side effects of the rapid yet uneven economic development in Turkey through an ethnographic study of chronic work accidents in Istanbul's industrial suburb Tuzla. Güney is the editor of the books *Başka Dünyalar Mümkün* (Other Worlds Are Possible – Varlık, 2007) and *Türkiye'de İktidar'ı Yeniden Düşünmek* (Rethinking Power in Turkey – Varlık, 2009) and the political and cultural criticism journal *Davetsiz Misafir* (The Uninvited Guest).

Oded Haas is a Vanier Scholar and PhD candidate in the Faculty of Environmental Studies at York University, Toronto. He received his BA in Architecture and MA in Interdisciplinary Studies in the Arts from Tel Aviv University, Israel. His research focuses on urbanization processes in settler colonial states, and specifically the role of housing in the political economy of Israel and Palestine. His PhD dissertation is titled "Housing Policy, Design and Struggle: Producing the Colonized Space of Israel/Palestine," and it examines the planning process of, and grassroots resistance to, a project of a new "Arab city" in Israel.

Roger Keil is York Research Chair in Global Sub/Urban Studies in the Faculty of Environmental Studies at York University in Toronto. A former director of York University's City Institute, he researches global suburbanization, urban political ecology, cities and infectious disease, and regional governance. He is the editor of *Suburban Constellations* (Jovis, 2013) and co-editor (with Pierre Hamel) of *Suburban Governance: A Global View* (University of Toronto Press, 2015), and the author of *Suburban Planet* (Polity, 2018).

Stefan Kipfer teaches urbanization, politics, and planning in the Faculty of Environmental Studies at York University. He has published widely on city and country, space and time in Marxist and countercolonial traditions, and on urban politics in cities like Toronto, Paris, and Zurich. He is co-editor of *Space, Difference, Everyday Life: Reading Henri Lefebvre* (with Kanishka Goonewardena, Christian Schmid, and Richard Milgrom – Routledge, 2008), *Gramsci: Space, Nature, Politics* (with Mike Ekers, Gillian Hart, and Alex Loftus – Wiley-Blackwell, 2013), and *Governing Cities Through Regions: Canadian and European Perspectives* (with Roger Keil, Pierre Hamel, and Julie-Anne Boudreau – Wilfrid Laurier University Press, 2017).

Steven Logan teaches at the Institute of Communication, Culture, Information and Technology at the University of Toronto, Mississauga. His forthcoming book, *In the Suburbs of History* (University of Toronto Press), compares Canadian and Czech postwar suburban modernism. Steven was also the co-editor of a special issue of the journal *Public* on "Suburbs: Dwelling in Transition." The issue also included a catalogue of the Leona Drive Project (2009), a site-specific art project in the Toronto suburb of Willowdale on which Steven served as researcher and co-prepared installations based on his archival research. He is currently working on the intersections of cybernetics and urbanism in 1960s Prague.

Max Rousseau is a politist and urban geographer at the CIRAD (France). Currently based at the Institut National d'Aménagement et d'Urbanisme (Rabat, Morocco), he is also a lecturer in several Masters' and PhD programs in France and Morocco. A member of the Montpellier-based research unit in social sciences UMR 5281ART-Dev, his current research programs deal with alternative development policies in shrinking cities and urban-rural relationships in both the Global South and the Global North.

Margot Rubin is a senior researcher in the South African Research Chair in Spatial Analysis and City Planning in the School of Architecture and Planning in the Faculty of Engineering and the Built Environment at the University of the Witwatersrand in Johannesburg. Her research has focused on concerns around urban governance and has investigated the intersections between socio-economic rights, litigation, and urban development, as well as the processes of housing provision and informality. She holds a joint PhD from the Schools of Architecture and Planning and Politics, Wits University.

Karl Schmid teaches in the Department of Anthropology, Faculty of Liberal Arts and Professional Studies at York University. He has conducted research in Egypt on inequality and spatial control, including the development of the city of Luxor by the Egyptian government, World Bank, UNESCO, and the UNDP. His current projects include grasping the diversity of suburban Cairo and the relationships between its highly segregated areas and the potential social and cultural implications of an energy transition within the Greater Toronto Area.

Jasmine Smith is a graduate student in the division of Epidemiology and Biostatistics at the University of Illinois at Chicago School of Public Health.

Julia Strutz is a member of the research group on "Urban Ethics: Conflicts about the good and proper conduct of life in 20th and 21st century cities" at Ludwig-Maximilians University in Munich, with a project on heritage politics in Istanbul. She conducts research in the fields of memory politics, urban renewal, the urban history of Istanbul, and the various ways this history gains importance today in heritage and non-heritage contexts.

Roza Tchoukaleyska is an assistant professor in Environmental Studies at Grenfell Campus, Memorial University of Newfoundland. Her research examines the politicized rhetoric around public space usage in France and Canada and considers the role of urban cultures of consumption in the formation of public space. She has published on urban diversity, public space redevelopment, and municipally led gentrification in the journals *Geoforum, Area, Social & Cultural Geography*, and *Urban Geography*.

Murat Üçoğlu is a PhD candidate and course director in the Faculty of Environmental Studies at York University. He is also a research member of the City Institute at York University. He was a visiting scholar in the Department of Human Geography at Frankfurt Goethe University in the summer of 2014. His research focuses on the financialization of housing markets, the political economy of suburban expansion in Istanbul and Toronto, and urban sustainability.

David Wilson is director of Graduate Studies and professor of Geography, Urban Planning, African American studies, and the Unit for Criticism and Interpretive Theory at the University of Illinois at Urbana-Champaign. He researches the racial economy and political economy of cities and suburbs in the Global West. He is the author of *Chicago's Redevelopment Machine and Blues Clubs* (Palgrave Macmillan, 2018), *Cities and Inequalities in a Global and Neoliberal World* (with Faranak Miraftab and Ken Salo – Routledge, 2016), and *Cities and Race: America's New Black Ghetto* (Routledge, 2006), and the editor of *The Politics of the Urban Sustainability Concept* (Common Ground, 2016). He is completing a book on the redeveloping of Chicago's South Side blues clubs and the politics of subterranean resistance to this.

Fulong Wu is Bartlett Professor of Planning at University College London. His research interests include urban development in China and its social and sustainable challenges. He is the author of *Planning for Growth: Urban and Regional Planning in China* (Routledge, 2015).

Douglas Young is an associate professor in the Urban Studies Programme, Department of Social Science at York University, Toronto. In the past, he has worked as an architect, a local government urban planner, and a developer in the non-profit cooperative housing sector. His current research interests are in two areas: processes of suburban decline and renewal in Toronto, and governing the legacies of twentieth-century socialist and modernist urbanisms in Berlin, Hanoi, and Stockholm. Publications include *In-Between Infrastructure: Urban*

Connectivity in an Age of Vulnerability (Praxis, 2011), co-edited with Patricia Wood and Roger Keil, and *Changing Toronto: Governing Urban Neoliberalism* (University of Toronto Press, 2009), co-authored with Julie-Anne Boudreau and Roger Keil.

Tianke Zhu is a PhD student at the Bartlett School of Planning, University College London. His research focuses on the suburbanization of Nanjing, China.

Index

Aalbers, Manuel, 16
Abouhani, Abdelghani, 227
Acre, 286. *See also* Israel/Palestine, Tantour as planned suburb
activism. *See* resistance
addiction. *See* Chicago, public housing and drug addiction
Africa, 346–7. *See also* Cairo, periphery and cultural assemblages; Casablanca, Lahraouiyine pirate suburbs; Johannesburg, state housing and backyarding
agriculture: in Cairo, 303–6; in China, 325, 328; in Istanbul, 168–9, 202–3; in Morocco, 227, 230
AKP. *See* Turkey, governance (AKP)
Alavi, Hamza, 154–5
Alter, Lloyd, 107
Am Südpark. *See* Halle (Saale), Am Südpark, state restructuring of housing
anthropology, comparative urban, 9–11. *See also* everyday life and suburbanism; research in suburban studies
architects and architecture: about, 349–50; city as product vs. work of art, 94–5, 96f, 99–101, 106–7; crane urbanism in Prague, 95–6, 99–107; housing renewal in Toronto, 113; inter-referencing in Cairo, 303–4, 309–12; inter-war period, 147; Judaization of space, 293; modernist socialist suburbs, 98, 102, 104, 106–7; social justice ideals, 156–7, 349–50; Western themes in Nanjing housing, 331–3. *See also* plans and planners
Asia, 346–7. *See also* China; Manila, informal settlements; Nanjing, Jiangning heterogeneous suburbs
Aurora, Illinois. *See* Chicago, public housing and drug addiction
Australia, new towns, 347

backyarding. *See* Johannesburg, state housing and backyarding
Başakşehir, 165, 187f. *See also* Istanbul, Kayaşehir housing
Belarbi, Wafae: biography, 355; on Morocco's pirate suburbs, 20, 223–40
Benjamin, Walter, 107
Berner, Erhard, 280

Bernt, Matthias: biography, 355; on restructuring estates in former East Germany, 22, 81–93
Bever, Elke, 104
Bloch, Robin, 244
Bogaert, Koenraad, 224
Boodram, Basmattee: biography, 355; on Chicago public housing and drug addiction, 56–78
Borel-Saladin, Jacqueline, 248
Bosporus. *See* Istanbul; Turkey
Bram Fischerville, 242f, 253–4, 254f, 259. *See also* Johannesburg, state housing and backyarding
Brenner, Neil, 57, 60
bribery. *See* corruption
bricolage policies, 20, 243–5, 248–50
Brown, Denise, 106
"Building and Rebuilding the Urban Periphery" MCRI, Istanbul (2015), 20–1, 24n2. *See also* research in suburban studies
Butcher, Siân, 244

Cairo, periphery and cultural assemblages: about, 23, 303–19; agricultural land, 305, 306; Al Rehab City, 303, 313–15; ashwaiyats (slums), 304, 306, 308, 314–16; Bhagat Group, 310–12; class divisions, 314–16; corruption, 308, 312; cultural assemblages, 303–4; demographics, 304, 312–13; Dreamland, 303, 310–12, 316; educational institutions, 311, 316; foreign investment, 309, 311, 313; gated communities, 303, 306, 308–15; historical background, 304–7; housing prices, 313; infrastructure, 308, 314; inter-referencing in themes, 303–4, 309–12; leisure amenities, 310–11, 312, 313; lived experience, 303, 312–15; mall culture, 308–9, 310; mega-projects, 316; middle class, 303, 311, 314–15; neoliberalism, 303, 307–9, 311; New Cairo City, 311, 313; new towns, 304–7, 313; non-occupancy, 313–14; oligarchs, 312, 316; planning, 305; pollution, 310, 314–15; security, 308–10, 311, 314–15; segregation, 308–10, 314–15; simulacra projects, 310–11; social divisions, 315–16; speculation, 308–9, 311, 313; state and military role, 303–4, 307–8, 312, 316; transnational developers, 309–12; transportation, 314
Caldeira, Teresa, 4–5, 19, 241, 261, 261n1, 347
Canada: Indigenous peoples, 149; neocolonialism, 149–50, 154, 156; racism, 149–50; social housing, 122, 156–7; state intervention, 156. *See also* Ontario; Toronto, comparative perspectives on redevelopment; Toronto, housing renewal
capitalism: financialization, 13; global capitalism, 5, 6f, 15; *habitat* and *habiter*, 146, 147–8; housing markets, 205; production of space, 13, 15; real estate capitalism, 11–18; resistance to, 286. *See also* financialization; foreign investment; neoliberalism; speculation
Casablanca, Lahraouiyine pirate suburbs: about, 20, 23, 223–37; administrative divisions, 232–4;

corruption, 228, 231–2, 237; demographics, 224, 230, 233; elections, 227–8, 234; embargos, 231–4; foreign investment, 228, 231; grey areas, 19–20, 226, 237; historical background, 224, 226–9, 236–7; illegal, non-regulatory housing, 224, 237, 238n2; infrastructure, 229, 235–7; Islamic fundamentalism, 228–9; Lahraouiyine, 225f, 229–30, 230f; local associations, 228, 234–6; local authorities, 231, 233; Makhzen (political elites), 227–8, 231, 236–7; map, 225f; national associations, 235; notables (political elites), 227–8, 233; pirate suburbs, 20, 23, 225–9, 230f, 237; poverty, 229–30, 237; punitive regulation, 224–5, 231–4, 237; reintegration of pirate suburbs (INDH), 19–20, 224, 229, 234–7; relocation, 229–30, 237; resistance, 229, 231, 233–4; roll-back neoliberalism, 224–7, 231–4, 236–7; roll-out neoliberalism, 224, 229, 234–7; security, 225, 228, 231–3, 238n4; state social control (pasha and caïd), 226, 233

Çavusoğlu, Erbatur: biography, 355; on Kayaşehir housing in Istanbul, 22–3, 165–78

centres and peripheries: about, 7, 11–12, 37–9, 145–6; centrality/peripherality meanings, 145–6, 148–9, 151; Lefebvre on, 38–9, 42, 146; vs. nested, 7; order vs. disorder, 38, 149; peripheral urbanization, 4–5, 11, 227, 241, 261–2n1, 282, 347; politics as disruptive force, 40–2; power relations, 39, 291; resistance in periphery, 292; right to the city, 42; spatial and socio-political locations, 145–6; terminology, 4–5, 38; territorial hierarchies, 38–9; trends, 11–12

Charlton, Sarah: biography, 355; on Johannesburg's state-led housing, 23, 241–66

Chicago, public housing and drug addiction: about, 21, 56–75; Aurora public housing, 57–9, 58f, 59t, 68–9; business improvement district (BID), 68; Cicero public housing, 57–9, 58f, 59t, 68–9; core, 68–71; demographics, 68; entrepreneurialization, 21, 60–4, 69–71, 73; explosion/implosion, 60, 68, 70, 73; gentrification, 64, 70–1; globalization, 62–4, 68–9, 74; heroin, 57, 59, 65–7, 72–4, 75n1; hopefulness, 67; housing for labourers, 64; housing quality ratings, 59t; immigrants, 64–5, 68–70; lived experience, 59, 65–7, 71–4; map, 58f; neoliberalism, 61, 63–7, 74; planetary urbanization, 56–62, 68–74; poverty, 57, 58, 64, 66, 71–3; race, 56, 58, 69, 74; service economy, 64–5, 68–9, 71, 73; stigma, 56, 64–5, 69–70, 71–3; toxic families, 65–6, 72–3; youth, 57, 72–4, 75n1

China: about, 23, 320; state entrepreneurialism, 320; suburban diversity, 18. *See also* Nanjing, Jiangning heterogeneous suburbs

Cicero, Illinois. *See* Chicago, public housing and drug addiction

class: in Cairo, 314–16; ethnoclasses, 288–92, 294; in Israel/Palestine, 288–90, 294, 296–8, 296–300n3, 299–300n3; in Istanbul, 169–72; in Nanjing new towns, 322–5, 330–1, 335, 337–8; in Paris, 148–51; race and class in France, 36, 43, 46–9; social mixing, 153; spatial dimensions of class relations, 39–40, 148–9; spatial reification and politics, 36, 43, 46, 48–9; suburban profile, 14; Trump supporters, 35–6; vulnerabilities, 43. *See also* luxury housing; middle class; poverty; segregation; wealth inequality; working class
Clerval, Anne, 153
climate change, 9, 205, 208. *See also* environmental issues
Clinton, Hillary, 35
Coker, Abidemi: biography, 356; on informal settlements in Manila, 23, 267–84
Colini, Laura, 92n1
colonialism: Israel as settler-colonial society, 287, 290–1, 293; neocolonialism and state theory, 149–50, 154–6
community activism. *See* resistance
corruption: in Cairo, 308, 312; in Casablanca, 228, 231–2, 237; corruption index, 206–7; in Montpellier, 137–8; in Turkey, 173, 176–7, 186, 206–10, 212–15
cronyism. *See* corruption
Czech Republic. *See* Prague, Jižní Město as socialist suburb

daily life. *See* everyday life and suburbanism
de Graaf, Reinier, 349–50
Denis, Eric, 309
Dikeç, Mustafa: biography, 356; references to, 130, 131; on spatial reification and politics, 21, 35–55
drug addiction. *See* Chicago, public housing and drug addiction
Dubai: aesthetics, 310–11, 346; foreign investment, 313; property developers, 312; simulacra projects, 310–11, 346; social divisions, 315–16

Easterling, Keller, 18, 22, 94, 95, 107
Eastern Europe. *See* Europe
ecology, as term, 202. *See also* environmental issues
economic growth: about, 10; housing-sector-led growth, 10; in Istanbul, 22–3, 188, 190–1, 197, 207–8, 215; living standards, 207–8; in Nanjing new towns, 321, 325–6; neoliberal developmentalist growth, 205, 210, 212, 214–15
Egypt: desert cities, 304–6; educational institutions, 311; new capital, 316; new towns, 313. *See also* Cairo, periphery and cultural assemblages
Ekers, Michael, 118, 243, 244
Ekman, Mattias, 63
elections. *See* politics and political geography
Engels, Friedrich, 142, 147–8
environmental issues: charcoal production in Manila, 269, 270f; climate change, 9, 205, 208; consent for degradation, 202, 208–9, 215–16; danger zones in Manila, 269–71, 269f, 277, 282; earthquakes

in Turkey, 203, 206, 207; ecology, as term, 202; noise pollution, 310; pollution in Cairo, 310, 314–15. *See also* agriculture; Istanbul, environmental issues
Erdoğan, Recep Tayyip, 165, 168, 170, 177, 186, 201. *See also* Turkey, governance (AKP)
essentialism, 43, 47–8
ethnicity. *See* race and ethnicity; religion
Europe: housing estates, 6–7, 97–8; housing in West vs.q East, 345; recent refugees, 90. *See also* France; France, politics and spatial reification; Germany; Halle (Saale), Am Südpark, state restructuring of housing; Montpellier, La Paillade (La Mosson) public space; Paris, comparative perspectives on redevelopment; Prague, Jižní Město as socialist suburb
everyday life and suburbanism: about, 9–10, 14, 59–60; in Cairo, 303, 312–15; in Chicago public housing, 59, 65–7, 71–4; economic growth, 207–8; financialization, 13; *habitat* and *habiter*, 146, 147–8, 150–1; homeownership, 9, 13; inequalities, 42; mortgages, 13, 15, 16–17, 193; race and ethnicity, 14; research gaps, 9–10; social mobility, 9, 170, 189–90; suburban profile, 14; urbanist sensibility, 59–60. *See also* Cairo, periphery and cultural assemblages; Chicago, public housing and drug addiction; class; environmental issues; homeownership; Montpellier, La Paillade (La Mosson) public space; race and ethnicity; religion; space

Fanon, Frantz, 41–2, 144, 154
farms. *See* agriculture
financialization: about, 8–9, 13, 347–50; commodification of housing, 8, 13, 16–17, 205; foreign investment, 15; in Halle, 83–4, 88, 91; interest rates, 193, 329; in Istanbul, 22–3, 186, 189, 191–6, 201, 206, 209–10, 212–13, 215; mortgages, 13, 15, 16–17, 193; municipal housing companies, 83–4; in Nanjing, 329, 335, 337; neoliberalism, 13; privatization, 22; speculation, 13; in Toronto, 119–21; wealth inequality, 13, 347. *See also* foreign investment; Istanbul, TOKİ; neoliberalism; speculation
foreign investment: about, 15; in Cairo, 309, 311, 313; in Casablanca, 228, 231; in Dubai, 313; for infrastructure, 15, 347–8; reciprocity laws, 193–4; in Turkey, 192–4, 209–10, 216n2. *See also* financialization
Förste, Daniel, 92n1
Forty, Adrian, 101
France: *banlieues*, 36, 38, 42–3, 47, 50n1, 143, 145, 150, 153; Daesh attacks, 155; globalization and social categories, 43; lack of data on race, ethnicity, and religion, 130; national redevelopment projects (ANRU), 142–3, 146–8, 152; neocolonialism, 36, 149–50, 154–6; postwar immigrant

housing, 150; state-managed gentrification, 146, 158n3; state racism, 149–50; terrorist attacks, 155; Zone Urbaine Sensible (ZUS), 131. *See also* Montpellier, La Paillade (La Mosson) public space; Paris, comparative perspectives on redevelopment
France, politics and spatial reification: about, 21, 35–50; anti-fascism, 42, 49–50; centre and periphery, 36–7, 43–4, 47–50; Front National (FN), 35, 37, 42, 44–9; Guilluy's competing peripheries, 36–7, 43–9, 50nn1–2; Lefebvre's centre and periphery, 38–9, 42; Lévy's urbane resistance to populism, 36, 44–5, 47, 49; maps and reification, 36, 39, 44–6, 49–50; politics as dynamic force, 37, 40–2, 44, 50; race and class, 36, 43, 46–9; Rancière's social relations, 38–41, 47; social-liberal sensibilities, 47–8; spatial reification, 37–40, 49–50; working class, 36, 43, 46–9
Frank, Thomas, 35–6, 48
Friedman, Benjamin, 207–8
Front National (FN), 37, 44–9

Gans, Herbert, 106
Gardner, David, 254
gated communities: in Cairo, 303, 306, 308–15; Judaization of space, 292; in Nanjing, 322, 324, 330, 332, 336f, 340. *See also* Cairo, periphery and cultural assemblages; Nanjing, Jiangning heterogeneous suburbs; security
Gauteng, 244
gentrification: in Canada, 149; in Chicago, 64, 70–1; in France, 146, 158n3; in Istanbul, 188; racialization of state and space, 149–50; redevelopment as state-managed gentrification, 146, 158n3; state-managed gentrification, 146, 158n3
geography, political. *See* politics and political geography
Germany: housing in West vs. East, 345; recent refugees, 90. *See also* Halle (Saale), Am Südpark, state restructuring of housing
Giedion, Sigfried, 101
global suburbanism: about, 5, 6f; defined, 181; population growth, 16; social categories, 36–7, 43, 48–9. *See also* massive suburbanization
governance. *See* state role
Graham, Stephen, 3
Gramsci, Antonio, 41–2, 144, 154
grey spaces, 19–20, 226, 237
Guilluy, Christophe, 36, 43–9, 50nn1–2
Güney, K. Murat: biography, 356; conclusion by, 344–53; introduction by, 3–34; on speculation and mega-projects in northern Istanbul, 22–3, 181–200
Gürtler, Konrad, 204

Haas, Oded: biography, 356; on ethnocracy in Israel/Palestine, 23, 285–302
habitat and *habiter,* 146, 147–8, 150–1
Halle (Saale), Am Südpark, state restructuring of housing: about, 22, 81–91; Am Südpark, 81–2;

complexity, 91; cooperatives, 84, 85f; demographics, 82, 90; federal programs, 87; financialization, 83–4, 91; historical background, 81–5, 85f; housing in West vs. East Germany, 345; housing prices, 88; infrastructure, 86, 87–8; institutional investors, 83–5, 87, 91; ownership, 84, 85f, 91; poverty, 89–92, 89t; privatization, 82–5, 85f, 88–91; recent immigrants, 90; restructuring plans, 88; shrinkage and rightsizing, 86–91; social downgrading, 89–92, 89t; social housing, 86–7; welfare dependency, 89t; welfare reforms, 85–7
Halles de la Mosson, 129f, 136–7
Hamel, Pierre, 118, 243
Harris, Richard, 15, 243
Hart, Gill, 49
Harvey, David, 116–17, 348
Hatherley, Owen, 106–7
Häussermann, Hartmut, 98
Heeg, Susanne, 17
heroin use. *See* Chicago, public housing and drug addiction
Hirt, Sonia, 6, 18, 101, 103, 182
homeownership: about, 9, 16–17; class, 14; commodification of housing, 8, 13, 16–17, 205; financialization, 13, 16–17; imaginaries, 9, 14, 16–17; investment vs. use, 348–9; in Istanbul, 170–3, 190–1, 195–7, 196f, 208–10, 209t, 212; low-income families, 13, 348; in Nanjing, 324, 329–30, 337; neoliberalism, 16–17; precarization, 16–17; race and ethnicity, 14. *See also* financialization; speculation
housing: *habitat* and *habiter*, 146, 147–8, 150–1; as home vs. real estate, 16. *See also* gated communities; homeownership; imaginaries; informal housing; luxury housing; massive suburbanization
Hrůza, Jiří, 99, 100, 105

illegal settlements. *See* informal housing
Illinois. *See* Chicago, public housing and drug addiction
imaginaries: about, 14, 17–18; competing imaginaries in Israel/Palestine, 285; homeownership, 9, 14, 16–17; housing as imagined centre in Istanbul, 22–3, 177; new symbolic economy, 17–18; suburbs, 14
immigrants. *See* migration; race and ethnicity
informal housing: diversity, 19–21; in Istanbul, 182, 188, 191; in Morocco, 223; in Nanjing, 330; pirate suburbs, 20, 23, 223–4, 226, 230f. *See also* Casablanca, Lahraouiyine pirate suburbs; Johannesburg, state housing and backyarding; Manila, informal settlements
infrastructure: about, 19; in Cairo, 308, 314; in Casablanca, 229, 235–7; foreign investment, 15, 347–8; in Halle (Saale), 86, 87–8; in Johannesburg, 249–50, 253–4, 259; mega-projects, 170–1, 174, 176, 183–8, 192, 203–4, 207, 348; in Montpellier, 128, 135; in Nanjing, 324, 325, 337; in Prague, 101,

104–5, 107; as urban structure itself, 107. *See also* Istanbul, infrastructure; transportation
inter-referencing, 6f, 346
Isikkaya, Devrim A., 346
Israel/Palestine, Tantour as planned suburb: about, 23, 285–99; "Arab city," 285–6, 288–91, 293–4, 296–8; architects and planners, 289–90, 293; centre/periphery relations, 291–2; commuters, 296–7; competing imaginaries, 285; displacement, 289–90; ethnic logic of space, 290–4, 298; ethnoclasses, 291–2, 294, 297, 299–300n3; ethnonational ideology, 291; fear, 296; historical background, 287–8, 291–3; Judaization of space, 290–3, 297–8, 300n4; Judeida-Makr, 286–9, 293, 295; Khirak activist group, 295–7, 299, 300n6; middle-class Palestinians, 288–90, 289, 294, 296–8; neoliberalism, 285, 295, 298; PCI (Palestinian citizens of Israel), 285; plans (National Master Plan), 286–9, 293, 295–6, 299n2; privatization, 288, 294, 298; resistance, 288, 290, 292–9, 300nn6–7; segregation, 287, 293–4, 297; as settler-colonial society, 287, 290–1, 293; Tantour, 285–6, 289–90, 293–9; women, 295
Istanbul: about, 22–3, 165–7, 181–2, 345; AKP-led municipality, 186–7; Ataşehir, 211f; Başakşehir, 165, 187f; beneficiaries of development, 173–7, 190–1, 195–7; corruption, 176–7, 206–10, 212–15; demographics, 181–2, 184, 216n1; discourse of growth, 207–10, 213–16; earthquakes, 203, 206, 207; economic growth, 188, 190–1, 197, 207–8, 215; financialization of housing, 186, 189, 191–6, 201, 209–10, 212–13, 215; foreign investment, 192–4, 209–10, 216n2; Gezi movement, 189, 213–14, 216n3; as global mega-city, 216n1; Halkali, 204f; historical background, 182; homeownership, 170–3, 190–2, 195–7, 196f, 206–10, 209t, 212; housing prices, 181, 191–5, 192f, 212; housing sales (2013 to 2016), 209t, 212; informal housing, 182, 188, 191; luxury housing, 167, 174f, 175–6, 184, 186–7, 196, 204f, 211f, 214f; Master Plan, 168, 186; middle class, 170, 186–91, 193, 196, 206, 210, 213, 216n3; migration, 181–2; neoliberalism, 201, 205, 210, 212–15; poverty, 172, 188–90; relocation, 175, 188, 191, 207; resistance, 176, 189, 213–14, 216n3; social mobility, 170, 189–90; speculation, 22, 173, 185, 188, 191–2, 197, 212; workshop site (2015), 22. *See also* Turkey; Turkey, governance (AKP)
Istanbul, environmental issues: about, 201–16; agriculture, 168, 202–3; air pollution, 203; biodiversity, 202–4; climate change, 205, 208; consent for degradation, 202, 208–9, 215–16; earthquakes, 203, 206, 207; ecology, as term, 202; forests, 183–4, 186, 188–9, 201–2, 203–4, 207; greenbelt, northern, 188–9, 202, 203; historical background, 201–3; protection of north, 183,

186; resistance to degradation, 188–9, 213–14; water, 186, 202–3
Istanbul, infrastructure: about, 183–8; airports, 20, 170–1, 183–6, 185f, 195, 203–4; bridges, 168, 170, 183–4, 186, 195, 203–4; canals, 170–1, 183, 186, 195; commercial space, 171; environmental harm by, 189–90; health and education, 170, 171; highways, 203, 214f; impact on housing prices, 168, 174, 176–7, 187f, 194–5; mega-projects, 170–1, 174, 176, 183–8, 192, 203–4, 207; national pride, 184, 185f; public transport, 170, 171; water, 165, 203–4
Istanbul, Kayaşehir housing: about, 22–3, 165–78, 166f; ambiguity, 166–7, 170–4, 176–7, 185; beneficiaries of development, 173, 190–1; construction companies, 173–5; demographics, 167t, 168–70, 181–2; expropriations and lawsuits, 166–9, 176, 183; historical background, 165–8; homeownership, 170–3, 177–8n2, 195–7, 196f; housing prices, 171–3, 175–6, 181, 194–5; as imagined centre, 22–3, 177; industries, 165, 168, 170, 171; infrastructure, 169, 171; luxury housing, 167, 174f, 175–6, 184, 186–7, 190–2, 196; plans, 167–9, 167t, 175–6; requirements of residents, 170, 177–8n2; social class, 169–72; social housing, 170–3, 172f, 184, 186–7, 189, 192, 196; social media, 167, 177–8n2; speculation, 173; state role, 166–9
Istanbul, TOKİ: about, 165, 173–7, 181–5, 205–6; ambiguity, 166–7, 174, 185–7; authoritarian governance, 185; beneficiaries of development, 173–7, 190–1; construction companies, 173–7, 185, 188, 190; Emlak Konut public real estate company, 174–5, 186–7, 190–1, 196, 201; expropriations and lawsuits, 166–9, 176, 183; high-rise apartments, 172f; historical background, 165, 182–3; Kayaşehir project, 166–8, 186–7; luxury housing, 186–7, 196, 211f, 214f; non-profit housing mandate, 186–7; plans, 167–9, 167t, 175–6, 186; profits, 166; public institution as entrepreneurial company, 166, 185–7, 205–6; speculation, 185, 197, 212. *See also* Istanbul, Kayaşehir housing; Turkey, governance (AKP)

Jacobs, Jane, 127
Jiangning. *See* Nanjing, Jiangning heterogeneous suburbs
Jižní Město. *See* Prague, Jižní Město as socialist suburb
Johannesburg, state housing and backyarding: about, 20, 23, 241–61; ANC role, 252, 254–5, 259; anti-shack sentiments, 255; backyarding, 241, 245–50, 252f, 254f; bricolage policy process, 20, 243–5, 248–50, 261; building materials, 250–1, 255; businesses, 241, 247, 252–4, 258; community self-management, 253–5, 258–9; enforcement of regulations, 241, 248–54, 257; houses (core/starter),

245–6, 245f; houses (post-2001), 246–8, 246f, 247f, 251f, 252f, 254f, 256, 258f; infrastructure, 249–50, 253–4, 259; local policy, 244–5, 250–3, 257, 259–61; map, 242f; national policy, 244–9, 260–1; peripheral urbanization, 241, 261–2n1; provincial policy, 244–5, 248–50, 255–61; public-private partnerships, 250–3, 258–9; relocation, 249, 256–8; resistance, 252–3, 261; right to housing, 260; toilets, 248–9, 256

Johannesburg, state housing and backyarding, suburbs: Alexandra, 242f, 250, 255–6; Bram Fischerville, 242f, 253–4, 254f, 259; Cosmo City, 242f, 250–3, 251f, 252f, 259; Fleurhof, 242f, 250, 253; Golden Gardens, 257; Lehae, 242f, 254–5, 259; Olievenhoutbosch, 242f, 250, 252–3; Orange Farm, 242f, 254–5, 259; Orlando, 242f, 250, 257, 258f; Sandton, 242f, 256; Soweto, 242f, 249–50, 253–5, 257; Zola, 242f, 250, 257

Judeida-Makr, 286–9, 293, 295. *See also* Israel/Palestine, Tantour as planned suburb

Kayaşehir. *See* Istanbul, Kayaşehir housing

Keil, Roger: biography, 356–7; conclusion by, 344–53; introduction by, 3–34; references to, 116–17, 118, 243

Khrushchev, Nikita, 99–101

King, Anthony, 13

Kipfer, Stefan: biography, 357; on comparative perspective on housing in Paris and Toronto, 22, 142–61; on spatial reification and politics in United States and France, 21, 35–55

Klassen, Jerome, 156

Kovachev, Atanas, 6, 18

Lahraouiyine. *See* Casablanca, Lahraouiyine pirate suburbs

La Paillade (La Mosson), Montpellier, 128–9. *See also* Montpellier, La Paillade (La Mosson) public space

Lasovský, Jiří, 100, 103, 104, 105–6

Lefebvre, Henri: capitalist urbanization, 57; centre and periphery, 38–9, 42, 146, 154; explosion/implosion, 11–12, 60, 68, 70, 73; *habitat* and *habiter,* 12, 146, 147–8, 344–5; hegemonic projects, 153; planetary urbanization, 56–7, 60, 63; production materials, 101–2; right to the city, 42, 285–6; spatialization of capitalism, 290; urban revolution, 13–14, 290, 299n1

legacy housing, 21–2. *See also* Chicago, public housing and drug addiction; Halle (Saale), Am Südpark, state restructuring of housing; Montpellier, La Paillade (La Mosson) public space; Paris, comparative perspectives on redevelopment; Prague, Jižní Město as socialist suburb; Toronto, comparative perspectives on redevelopment; Toronto, housing renewal

Lenin, Vladimir, 157

Le Pen, Marine, 35, 45, 49

Levittown, 19, 94, 95–6, 106
Lévy, Jacques, 36, 44–5, 47, 49
lived experience. *See* everyday life and suburbanism
Logan, Steven: biography, 357; on socialist suburbs in Prague, 22, 94–110
luxury housing: in Cairo, 309; in Istanbul, 167, 174f, 175–6, 184, 186–7, 196, 204f, 211f, 214f; in Morocco, 223; in Nanjing, 324, 330–5, 334f

Mabin, Alan, 244
Macron, Emmanuel, 48–9, 147
Madden, David, 16–17, 205
Maghreb, defined, 238n1. *See also* Montpellier, La Paillade (La Mosson) public space
Maier, Karel, 97
Major Collective Research Initiative (MCRI), 20–1, 24n2
Makr (Judeida-Makr), 286–9, 293, 295. *See also* Israel/Palestine, Tantour as planned suburb
Manila, informal settlements: about, 23, 267–83; associations for housing access, 268, 271–8, 275t, 280; businesses, 279; charcoal production, 269, 270f, 282; danger zones, 269–71, 269f, 277, 282; decentralization, 273; demographics, 270; Gawad Kalinga, 271, 273–4, 276; housing rights, 267; local government, 271–7; national policies, 268, 271–3; new housing, 268, 271–2, 281f, 283; NGOs, 267–8, 271–7, 275t, 280–3; participation by squatters, 272–7, 275t, 282; Pasay, 281f; peripheral urbanization, 280–3; poverty, 267–71, 279; professional squatters, 20, 268, 272, 277–80; relocation, 267, 271–3, 275t, 277, 280, 282; safety, 278; social preparation, 274–8, 275t; social relations, 279–80; squatters (informal settler families), 267–71, 279–80; squatters as landlords, 277–9; stages of housing process, 274–7, 275t; sweat equity, 274, 275t; Ulingan, 269, 269f, 270f, 277, 279, 279f, 282; values formation, 274–6, 275t, 278, 282
Marcuse, Peter, 16–17, 205
massive suburbanization: about, 3–11, 6f, 17–21, 344–50; authoritarian governance, 156, 349; centre/periphery, 7; commodification of housing, 8, 13, 16–17, 205; complexity, 91; densities, 18; discourse of progress, 7; diversity, 7, 10, 19–21, 346; everyday life (suburbanism), 14; as explosion, 5, 11–12; financialization, 8–9, 13, 21–2, 118, 347–50; grey spaces, 19–20, 226, 237; housing supply, 14; imaginaries, 14, 17–18; key questions, 8; modalities, 19–20; morphology, 7, 10, 344–6; overview of this book, 20–3; peripheral urbanization, 4–5, 11, 227, 241, 261–2n1, 282, 347; population growth, 16; post-suburbanization, 12–13, 18, 117–19, 322; poverty, 15; race and ethnicity, 14; right to the suburb, 345, 349–50; segregation, 7; sídliště (socialist suburb), 97;

social justice ideals, 156–7, 349–50; suburbanization, 12; symbolic economy, 17; terminology, 7, 12; types, 19–20; urban/suburban distinction, 15; wealth inequality, 349–50. *See also* centres and peripheries; class; everyday life and suburbanism; financialization; political economy; research in suburban studies; resistance; segregation; state role

massive suburbanization, informal housing, 19–21. *See also* Casablanca, Lahraouiyine pirate suburbs; Johannesburg, state housing and backyarding; Manila, informal settlements

massive suburbanization, legacy housing, 21–2. *See also* Chicago, public housing and drug addiction; Halle (Saale), Am Südpark, state restructuring of housing; Montpellier, La Paillade (La Mosson) public space; Paris, comparative perspectives on redevelopment; Prague, Jižní Město as socialist suburb; Toronto, comparative perspectives on redevelopment; Toronto, housing renewal

massive suburbanization, new housing, 23, 347. *See also* Cairo, periphery and cultural assemblages; Israel/Palestine, Tantour as planned suburb; Nanjing, Jiangning heterogeneous suburbs; *and entries beginning with* Istanbul

McGee, Terry, 18

MCRI (Major Collective Research Initiative), 20–1, 24n2

Merrifield, Andy, 60

middle class: in Cairo, 303, 311, 314–15; in China, 322, 324–5, 329–33, 336f, 337; homeownership, 9, 17, 170, 193, 196, 329–33; in Israel/Palestine, 288, 289, 294, 296–8; in Istanbul, 170, 186–91, 193, 196, 206, 210, 213, 216n3. *See also* class

migration: global increase, 65; recruitment of workers, 68–9; refugees in Europe, 90; service sector workers, 64–5; suburban profile, 14. *See also* race and ethnicity

Mitchell, Katharyne, 130

modernism: housing as failed promise, 349; socialist suburbs, 98, 102, 104, 106–7; Toronto, 22, 114–15. *See also* Prague, Jižní Město as socialist suburb; Toronto, comparative perspectives on redevelopment; Toronto, housing renewal

Montpellier, La Paillade (La Mosson) public space: about, 22, 126–39; corruption, 137–8; demographics, 127–9; elevated walkway (Le Grand Mail), 128, 133–5, 134f, 137–8; high-rise social housing, 126–8, 130; historical background, 127–31, 127–33; immigrants, 128–30; infrastructure, 128, 135; La Paillade, 126, 129–31; Montpellier, 127–8; outdoor and indoor markets, 127, 128, 129f, 133, 136–8; private spaces as parallel "public" spaces, 127, 136–8; public space

as social site, 126–8, 132–3, 136–9; race, ethnicity, and religion, 129–31, 137; redevelopment projects, 127, 131–5, 134f, 138–9; security, 127, 132–6; transportation, 128, 135. *See also* France; space
Morocco: informal urbanization, 223; map, 225f; neoliberalism, 224–6; urban safety, 225. *See also* Casablanca, Lahraouiyine pirate suburbs
morphology: diversity of, 7, 10, 344–6. *See also* massive suburbanization
mortgages, 13, 15, 16–17, 193. *See also* financialization
municipalities. *See* state role
Musil, Jiří, 97, 98–9, 100
Muslims, 137, 177, 228–9, 304–5

Nanjing, Jiangning heterogeneous suburbs: about, 23, 320–41; affordable housing, 330; apartments, 226f, 335–7; co-renting, 338, 339f, 341n1; demographics, 327, 328; developers, 320–1; as development strategy, 327–30; development zone (ETDZ), 325, 327; diversity, 23, 322–3, 331–42; economic growth, 321, 325–6; entrepreneurial government, 320–2, 328, 340; environment, 321, 324, 325; finance, 329–30, 335; gated communities, 322, 324, 330, 332, 336f, 340; historical background, 320–5, 336–7; homeownership, 324, 329–30, 337; housing prices, 328, 330, 337; industries, 323, 325, 327–9, 335; informal housing, 330; infrastructure, 324, 325, 337; Jiangning new town, 323, 325–7; leisure amenities, 332; local government, 320–4, 327–8, 338; luxury housing, 324, 330–5, 334f; map, 326f; marketing, 329–33, 337, 339–40; market system, 320; Master Plan, 325–9, 326f; middle class, 322, 324–5, 329–33, 336f, 337; migration, 320, 321, 330, 335, 337; mixed land uses, 325; Nanjing, 323, 326f; neoliberalism, 321–2; new towns, 320–2, 324–7; relocated households, 322–4, 327, 330, 337–40, 339f; ring pattern, 325, 328–9; security, 324, 334; segregation, 332; social class, 322–5, 330–1, 335, 337–8; speculation, 324, 330, 335; villas, 324, 330–5, 334f; Western themes, 331–3
national role. *See* state role
neoliberalism: about, 5, 6f, 116–17; austerity, 123; "common sense" policies, 111, 116–17, 123; as a constructed project, 115; developmentalist growth, 205, 210, 212, 214–15; entrepreneurial subjects, 17, 210; financialization, 13; homeownership, 16–17; impact of economic crisis (2008), 116; individualization of risk, 13; political governance, 116–17; "post" neoliberalism, 116–17, 118–22; roll-back neoliberalism, 116–17, 224–5; roll-out neoliberalism, 116; roll-with-it neoliberalism, 116–17; segregation, 308; small government, 182; speculation, 308–9. *See also* capitalism; financialization; state role

Neustadt. *See* Halle (Saale), Am Südpark, state restructuring of housing
new housing, 23, 347. *See also* Cairo, periphery and cultural assemblages; Israel/Palestine, Tantour as planned suburb; Johannesburg, state housing and backyarding; Nanjing, Jiangning heterogeneous suburbs; *and entries beginning with* Istanbul

Ong, Aihwa, 303
Ontario: downloading of redevelopment to municipalities, 146–7; funding for tower renewals, 121; inclusionary zoning, 121. *See also* Toronto, comparative perspectives on redevelopment; Toronto, housing renewal
ownership, home. *See* homeownership

Palestine. *See* Israel/Palestine, Tantour as planned suburb
Paris, comparative perspectives on redevelopment: about, 22, 142–58; ANRU projects, 142–3, 146–8, 152; capitalism and housing, 147–8; centrality/peripherality meanings, 145–6, 148–9, 151; collective consumption, 146–7; contradictions in strategies, 150–2; demolition/renovation meanings, 147, 148–9, 152; gentrification, 146, 158n3; geographical vs. socio-political locations, 145–6; as housing issue, 146–7; internationalization, 154–6; marketized deconstruction of social housing, 143, 156–7; national role, 142–3, 146–7; neocolonialism, 149–50, 154–6; racism, 149–52; resistance, 151–3; segregation, 141, 145–6, 151–3; social class, 148–51; social mixing, 149–53; socio-political vs. spatial locations, 145–6; state theory, 154–6; territorialized state interventions, 152–3; trends, 143. *See also* France
Peck, Jamie, 115, 116, 224
peripheries. *See* centres and peripheries
Phelps, Nicolas, 118
Philippines. *See* Manila, informal settlements
Piketty, Thomas, 349
pirate suburbs, 20, 23, 223–4, 226, 230f, 237. *See also* Casablanca, Lahraouiyine pirate suburbs
planetary urbanization: about, 5, 6f; binaries as obsolete, 299n1. *See also* massive suburbanization
plans and planners: for Halle (Saale), 88; for Istanbul, 167–9, 167t, 175–6, 186; for Prague, 102–4; social justice ideals, 156–7, 349–50; for Tantour in Israel/Palestine, 286–9, 293, 295–6, 299n2. *See also* architects and architecture; Israel/Palestine, Tantour as planned suburb
police. *See* security
political economy: about, 5, 6f, 8–9, 14; financialization, 8–9, 13; housing development agencies, 9. *See also* capitalism; financialization; massive suburbanization; state role

politics and political geography: about, 35–7; anti-fascism, 42, 49–50; census, 39; housing redevelopment and alliances, 148–9, 152–3; maps and reification, 36, 39, 44–6, 49–50; place-based social mixing, 148–53; politics as dynamic force, 37, 40–2; populism and spatial reification, 21, 35–9; resistance to populism, 36, 44–5, 47, 49; spatial reification, 35–40, 44–5, 49–50; urban uprisings, 41–3, 148–9, 152–3. *See also* France, politics and spatial reification
population growth, 16
populism, 21, 35–9, 44–5, 47, 49. *See also* politics and political geography
"post" as prefix, 116–22
post-suburbanization, 12–13, 18, 117–19, 322. *See also* massive suburbanization
Poulantzas, Nikos, 154, 156
poverty: about, 15; in Casablanca, 229–30, 237; in Chicago, 57, 58, 64, 66, 71–3; in Halle, 89–92, 89t; homeownership, 13, 348; in Istanbul, 172, 188–90; in Manila, 267–71, 279. *See also* class
power, state. *See* state role
Prague, Jižní Město as socialist suburb: about, 22, 94–107; architects and planners, 94–5, 97, 99–101, 103–7; city as product vs. work of art, 94–5, 96f, 99–101, 106–7; crane urbanism, 95–6, 99–107; as garden suburb, 102f, 105f, 106–7; historical background, 96–9, 102–5; infrastructure, 101, 104–5, 107; migration, 98; mix of urban and suburban traits, 98–9; modernism, 98, 102, 104, 106–7; plans, 102–4; public spaces, 96f, 97–9, 101, 102f, 103–4, 106; separation of cars and pedestrians, 103–4, 105f, 106; sídliště (socialist suburb), 95–102, 102f; single-family homes, 104; transportation, 97–9, 104–5
privatization of housing: about, 22; in Halle (Saale), 82–5, 85f, 88–91; in Israel/Palestine, 288, 294, 298
professional squatters, 20, 268, 272, 277–80. *See also* Manila, informal settlements
provincial role. *See* state role
public housing, North America. *See* Chicago, public housing and drug addiction; Toronto, comparative perspectives on redevelopment; Toronto, housing renewal
public space, 126–7, 138–9. *See also* Montpellier, La Paillade (La Mosson) public space; space

race and ethnicity: about, 14; anti-fascism, 42, 49–50; in Canada, 149–52; in Chicago, 56, 58, 69, 74; ethnoclasses, 291, 299–300n3; in France, 36, 43, 46–9, 130, 149–50; in Montpellier, 129–31, 137; in Paris, 149–53; place-based resistance, 152–3; racialization of state and space, 149–50; racism, 35, 41, 130, 149–52, 157; redevelopment and racialization of state and space, 149–50; social mixing, 148–53; spatial dimensions of social relations, 39–40; suburban profile, 14; symbolic violence, 149; Trump supporters, 35–6; uncollected data

on, 130; urban uprisings, 41–3, 152; in the US, 35, 41, 149. *See also* religion; segregation
Rancière, Jacques, 38–41, 47
real estate. *See* financialization; homeownership; infrastructure
redevelopment as gentrification. *See* gentrification
redevelopment of legacy housing, 20–1, 146–7. *See also* Chicago, public housing and drug addiction; Halle (Saale), Am Südpark, state restructuring of housing; Montpellier, La Paillade (La Mosson) public space; Paris, comparative perspectives on redevelopment; Prague, Jižní Město as socialist suburb; Toronto, comparative perspectives on redevelopment; Toronto, housing renewal
refugees, 90. *See also* migration
reification in social research, 37–9. *See also* France, politics and spatial reification
religion: Muslim Brotherhood, 304–5; Muslims in Casablanca, 228–9; Muslims in Montpellier, 137; Muslims in Turkey, 177. *See also* Israel/Palestine, Tantour as planned suburb; race and ethnicity
research in sub/urban studies: about, 5, 6f, 8–11, 20–1; arrival cities, 10; assemblages, 11; centres and peripheries, 11, 37–8; comparative anthropology, 9–11; essentialism, 37; MCRI workshop, Istanbul (2015), 20–1, 24n2; methodological individualism, 37, 44–5, 47; production of space, 15, 37; "repeated instances," 10; research gaps, 9–10, 118, 156; spatial reification, 37–9, 44–5, 47; spatial turn, 37, 39–40; "thinking with elsewhere," 8, 10; trends, 10–11; urban problems, 15; urban theory, 10–12; walking-while-talking method, 132. *See also* centres and peripheries; everyday life and suburbanism; Lefebvre, Henri; massive suburbanization; political economy; state role
resistance: to capitalism, 286, 349–50; in periphery vs. centre, 292; to political populism, 36, 44–5, 47, 49; politics as disruptive force, 40–2; to redevelopment, 151–3; social justice ideals, 156–7, 349–50; to Tantour as planned suburb, 288, 290, 292–9, 300nn6–7; urban uprisings, 41–3, 148–9, 152
right to the city/suburb, 42, 285–6, 345, 349–50
Rivière, Jean, 46
Robinson, Jennifer, 10
roll-out and roll-back neoliberalism, 224–7, 229, 231–7. *See also* neoliberalism
Rousseau, Max: biography, 357; on Morocco's pirate suburbs, 20, 223–40
Roy, Ananya, 17, 19, 143–4, 346
Rubin, Margot: biography, 357–8; on Johannesburg's state-led housing, 23, 241–66

safety. *See* security
Sassen, Saskia, 347, 348
Schmid, Christian, 60

Schmid, Karl: biography, 358; on Cairo's planned periphery, 303–19
security: about, 19; authoritarian statism, 156; in Cairo, 308–10, 311, 314–15; in Casablanca, 225, 228, 231–3, 238n4; in France, 150; in Montpellier, 127, 132–6; in Nanjing, 324, 334; state and capital accumulation, 19. *See also* gated communities
segregation: about, 7, 153; in Cairo, 308–10, 314–15; in Nanjing, 332; neoliberalism, 308; in Paris, 141, 145–6, 150–3; place-based resistance, 152–3; socio-political vs. spatial locations, 145–6, 152–3; in Tantour as planned suburb in Israel/Palestine, 287, 293–4, 297, 299–300n3; in Toronto, 141, 151–3. *See also* class; Israel/Palestine, Tantour as planned suburb; race and ethnicity
Serfati, Claude, 156
Shanghai, new towns, 345–6
sídliště, as term, 97. *See also* Prague, Jižní Město as socialist suburb
Sieverts, Tom, 112
Simone, Abdul Maliq, 59, 223, 226, 347
simulacra projects, 310–11
Smith, Jasmine: biography, 358; on Chicago public housing and drug addiction, 56–78
social activism. *See* resistance
social class. *See* class
social housing. *See* Chicago, public housing and drug addiction; Halle (Saale), Am Südpark, state restructuring of housing; Istanbul, Kayaşehir housing; Montpellier, La Paillade (La Mosson) public space; Toronto, comparative perspectives on redevelopment; Toronto, housing renewal
socialist suburb, 95, 98, 102, 104, 106–7. *See also* Prague, Jižní Město as socialist suburb
social justice, 156–7, 349–50. *See also* resistance
social research. *See* research in suburban studies
Soja, Edward W., 61–2
South Africa. *See* Johannesburg, state housing and backyarding
Soweto, 242f, 249–50, 253–5, 257. *See also* Johannesburg, state housing and backyarding, suburbs
space: about, 14–15; capitalist production of space, 13, 15; ethnic logic of space, 290–4, 298; global processes, 14; grey spaces, 19–20, 226, 237; hierarchies of, 15; Judaization of space, 290–3, 297–8, 300n4; private spaces as parallel public spaces, 127, 138–9; public spaces in Prague, 96f, 97–9, 101, 102f, 103–4, 106; public spaces in Toronto, 113–14; racialization of state and space, 149–50; as site of social interaction, 126–7; socio-political vs. spatial locations, 145–6; spatial reification, 37–9. *See also* France, politics and spatial reification; Montpellier, La Paillade (La Mosson) public space
speculation: about, 13; in Cairo, 308–9, 311, 313; in Istanbul, 22, 173, 185, 188, 191–2, 197, 212; in Nanjing, 324, 330, 335;

neoliberalism, 308–9. *See also* financialization
squatters. *See* informal housing
Stanek, Łukasz, 102
Stanilov, Kiril, 98
state role: about, 5, 6f, 14, 154–7; authoritarianism, 156, 349; in Cairo, 303–4, 307–8, 312, 316; in China, 320–2, 328, 340; decommodifying housing, 156–7; diversity, 19–21; entrepreneuralism in China, 320–2, 328, 340; neocolonialism and internationalization, 154–6; "post" as prefix, 116–22; research gaps, 118, 156; as symbolic policing, 38; theories of, 154–6. *See also* Cairo, periphery and cultural assemblages; Istanbul, TOKİ; Johannesburg, state housing and backyarding; Nanjing, Jiangning heterogeneous suburbs; neoliberalism; Turkey, governance (AKP)
Strutz, Julia: biography, 358; on Kayaşehir housing in Istanbul, 22–3, 165–78
substance abuse. *See* Chicago, public housing and drug addiction
suburbanism, as term, 14. *See also* everyday life and suburbanism
suburbanization, as term, 12. *See also* massive suburbanization
Sweden, 350
Swyngedouw, Erik, 117
Sýkora, Luděk, 98
Szelenyi, Ivan, 89

Tafuri, Manfredo, 101
Tantour. *See* Israel/Palestine, Tantour as planned suburb
Tchoukaleyska, Roza: biography, 358; on redevelopment in Montpellier, 22, 126–41
Teige, Karel, 95, 99
territorialization of social relations, 37–9. *See also* France, politics and spatial reification
Theodore, Nik, 115
Tickell, Adam, 116, 224
TOKİ. *See* Istanbul, TOKİ
Toronto, comparative perspectives on redevelopment: about, 142–58; capitalism and housing, 147–8; centrality/peripherality meanings, 145–6, 148–9; collective consumption, 146–7; contradictions in strategies, 150–2; demolition/renovation meanings, 147, 148–9; geographical vs. socio-political locations, 145–6; marketized deconstruction of social housing, 143, 156–7; municipal role, 146–7; neocolonialism, 149–50, 154; provincial role, 146–7, 148; public housing authority (TCHC), 112, 146–7, 148; racism, 149–52; redevelopment as housing issue, 146–7; resistance, 151–3; segregation, 141, 151–3; social mixing, 148–53; social reproduction, 147–8; state theories, 154–6; territorialized state interventions, 152–3; trends, 143
Toronto, housing renewal: about, 22, 111–24; affordable housing, 111, 115, 117, 121–3; architects, 113; "common sense" policies, 111, 116–17, 120, 123; conceptualization of suburbs,

116–18; demographics, 112, 115; everyday life, 121; financialization, 119–21; funding, 115, 121, 122–3; historical background, 112, 114–15; in-between city, 112, 119, 120; inner suburbs, 112, 114–15; local governance, 119–20; modernism, 22, 114–15; need for new housing, 123; neoliberalism, 111–12, 115–18, 121, 123; non-profit housing (TCHC), 112, 123; ownership, private and non-profit, 112–14, 122–3; plan, 121; post-political governance, 118; "post" suburban and neoliberalism, 116–22; progressive discourse, 22, 111–12, 117, 121–3; public spaces, 113–14; rent control, 115; retrofitting, 118, 119, 120; social exclusion, 22, 120; statistics, 112–13, 115, 119; tower housing, 22, 111–15, 113f, 114f, 122–3; Tower Renewal program, 111–12, 115, 119–23; transportation, 113, 120; urban politics, 116–18; wait lists for affordable housing, 115, 123

transportation: in Cairo, 314; in Montpellier, 128, 135; in Paris, 146; in Prague, 97–9, 104–5; in Toronto, 113, 120. *See also* infrastructure; Istanbul, infrastructure

Trump, Donald, 35–6

Turkey: authoritarian governance, 22, 182–3, 185; Başakşehir, 165; earthquakes, 203, 206, 207; homeownership, 22, 190–1, 195–7, 196f; housing prices, 191–5, 192f; housing sales (2013 to 2016), 209t, 212; scholarship on, 189–90; wealth inequality, 190, 197, 212. *See also entries beginning with* Istanbul

Turkey, governance (AKP): about, 173–7, 182–3, 191–5, 201–2; ambiguity, 185; authoritarian, 22, 182–3, 185; beneficiaries of development, 173–7, 190–1, 195–7; corruption, 173, 176–7, 186, 206–10, 212–15; developmentalist growth, 201–2, 205, 207–8, 210, 212, 214–16; economic policies, 191–3, 197, 201; elections, 165, 170; Erdoğan, 165, 168, 170, 177, 186, 201; foreign investment, 192–4, 209–10, 216n2; KİPTAŞ, 201, 206; Master Plan, 186; neoliberalism, 201, 205, 210, 212–15; resistance to, 176, 188–91, 216n3. *See also* Istanbul, TOKİ

Turok, Ivan, 248

Üçoğlu, Murat: biography, 358; conclusion by, 344–53; on ecology of Istanbul's periphery, 23, 201–19; introduction by, 3–34

United Arab Emirates. *See* Dubai

United Kingdom: spatial reification and politics, 35

United States: election (2016), 35–6; immigrants, 64–5; imperial dynamics, 156; politics and spatial reification, 21, 35–6, 48; racism and politics, 35, 41. *See also* Chicago, public housing and drug addiction

urban/suburban distinction, 15

urban uprisings, 41–3, 148–9, 152. *See also* resistance

Venturi, Robert, 106

Wacquant, Loïc, 149
wealth inequality: financialization, 13, 347; massive suburbanization, 349–50; Turkey, 190, 197, 212
Wilson, David: biography, 358–9; on Chicago public housing and drug addiction, 56–78
Wilson, Japhy, 117
working class: financialization of homeownership, 13, 348; place-based resistance, 152–3; spatial reification and politics, 36, 43, 46, 48–9; in suburbs, 13–14; vulnerabilities, 43. *See also* class; poverty
Wu, Fulong: biography, 359; on heterogeneous suburbs in Nanjing, 23, 320–43; references to, 18, 118

Yiftachel, Oren, 19, 299–300n3
Young, Douglas: biography, 359; on progressive neoliberalism in Toronto, 22, 111–25

Zarecor, Kimberley, 106
Zhu, Tianke: biography, 359; on heterogeneous suburbs in Nanjing, 23, 320–43

GLOBAL SUBURBANISMS

Series Editor: Roger Keil, York University

Published to date:

Suburban Governance: A Global View / Edited by Pierre Hamel and Roger Keil (2015)

What's in a Name? Talking about Urban Peripheries / Edited by Richard Harris and Charlotte Vorms (2017)

Old Europe, New Suburbanization? Governance, Land, and Infrastructure in European Suburbanization / Edited by Nicholas A. Phelps (2017)

The Suburban Land Question: A Global Survey / Edited by Richard Harris and Ute Lehrer (2018)

Critical Perspectives on Suburban Infrastructures: Contemporary International Cases / Edited by Pierre Filion and Nina M. Pulver (2019)

Massive Suburbanization: (Re)Building the Global Periphery / Edited by K. Murat Güney, Roger Keil, and Murat Üçoğlu (2019)

www.ingramcontent.com/pod-product-compliance
Lightning Source LLC
LaVergne TN
LVHW090802070826
844660LV00022B/1057